Structural Integrity

Volume 23

The *Structural Integrity* book series is a high level academic and professional series publishing research on all areas of Structural Integrity. It promotes and expedites the dissemination of new research results and tutorial views in the structural integrity field.

The Series publishes research monographs, professional books, handbooks, edited volumes and textbooks with worldwide distribution to engineers, researchers, educators, professionals and libraries.

Topics of interested include but are not limited to:

- Structural integrity
- Structural durability
- Degradation and conservation of materials and structures
- Dynamic and seismic structural analysis
- Fatigue and fracture of materials and structures
- Risk analysis and safety of materials and structural mechanics
- Fracture Mechanics
- Damage mechanics
- Analytical and numerical simulation of materials and structures
- Computational mechanics
- Structural design methodology
- Experimental methods applied to structural integrity
- Multiaxial fatigue and complex loading effects of materials and structures
- Fatigue corrosion analysis
- Scale effects in the fatigue analysis of materials and structures
- Fatigue structural integrity
- Structural integrity in railway and highway systems
- Sustainable structural design
- Structural loads characterization
- Structural health monitoring
- Adhesives connections integrity
- Rock and soil structural integrity.

**** Indexing: The books of this series are submitted to Web of Science, Scopus, Google Scholar and Springerlink ****

This series is managed by team members of the ESIS/TC12 technical committee.

Springer and the Series Editors welcome book ideas from authors. Potential authors who wish to submit a book proposal should contact Dr. Mayra Castro, Senior Editor, Springer (Heidelberg), e-mail: mayra.castro@springer.com

More information about this series at https://link.springer.com/bookseries/15775

Shahrum Abdullah · Salvinder Singh Karam Singh ·
Noorsuhada Md Nor
Editors

Structural Integrity Cases in Mechanical and Civil Engineering

Springer

Editors
Shahrum Abdullah
Faculty of Engineering and Built Environment
Universiti Kebangsaan Malaysia
UKM Bangi, Selangor, Malaysia

Salvinder Singh Karam Singh
Faculty of Engineering and Built Environment
Universiti Kebangsaan Malaysia
UKM Bangi, Selangor, Malaysia

Noorsuhada Md Nor
Civil Engineering Studies
Universiti Teknologi MARA
Cawangan Pulau Pinang
Permatang Pauh, Malaysia

ISSN 2522-560X ISSN 2522-5618 (electronic)
Structural Integrity
ISBN 978-3-030-85645-8 ISBN 978-3-030-85646-5 (eBook)
https://doi.org/10.1007/978-3-030-85646-5

This Springer imprint is published by the registered company Springer Nature Switzerland AG
The registered company address is: Gewerbestrasse 11, 6330 Cham, Switzerland

Preface

Structural integrity is a scientific area that studies the integrity of a support structure taking into account the structural loads for which the structure was designed, but also taking into account the loads operated during the life of the structure. Besides, this discipline studies past structural failures to be taken into account in structural assessments of existing engineering structures as well as in future designs.

Structural integrity and failure analysis include a set of topics for structural assessments to be developed such as: structural integrity, failure analysis, structural durability, degradation and conservation of materials and structures, dynamic and seismic structural analysis, fatigue and fracture of materials and structures, risk analysis and safety of materials and structural mechanics, fracture mechanics, damage mechanics, analytical and numerical simulation of materials and structures, computational mechanics, structural design methodology, experimental methods applied to structural integrity, multiaxial fatigue and complex loading effects of materials and structures, fatigue corrosion analysis, scale effects in the fatigue analysis of materials and structures, fatigue structural integrity, structural integrity in railway and highway systems, sustainable structural design, structural loads characterization, structural health monitoring, adhesives connections integrity, rock and soil structural integrity, etc.

The 5th Symposium on Damage Mechanism in Materials and Structures (SDMMS 2021) took place in Universiti Teknologi MARA, Cawangan Pulau Pinang, Malaysia on 8–9 March 2021. This symposium provided a venue for researchers and engineers in damage mechanisms in the materials and structures field from academia, industry, and government to meet in a forum where the latest research results are presented and prospects for future developments are discussed. The symposium was organized by the Universiti Teknologi MARA (UiTM) Cawangan Pulau Pinang, Faculty of Civil Engineering and Universiti Kebangsaan Malaysia (UKM), Computational and Experimental Mechanics (CEM) Research Group. This event was the fifth in a series of conferences that started in August 2016, Malaysia. The 1st Symposium on Damage Mechanisms in Materials and Structures (SDMMS), 2016, was organized by the Universiti Kebangsaan Malaysia (UKM), Centre for Automotive Research (CAR),

Faculty of Engineering and Built Environment, Malaysian Association of Computational Mechanics (MACM), and Universiti Malaysia Perlis (UniMAP). The 2nd Symposium on Damage Mechanisms in Materials and Structures (SDMMS), 2017, took place in Hotel Bangi-Putrajaya, Bandar Baru Bangi, Selangor on October 24th, 2017. The 3rd SDMMS were held in Hotel Bangi-Putrajaya, Bandar Baru Bangi, Selangor, in 2018. The 4th SDMMS which were held in Hotel Bangi-Putrajaya, Bandar Baru Bangi, Selangor on October 20–22th, 2019.

This volume of the Structural Integrity series with 24 chapters covers the subject related to the damage mechanism and structural integrity in the scope of mechanical engineering and civil engineering. Thus, the failure pattern of various materials and structures in both engineering filed was agreed upon as the main subject matter for the discussion. The topics approached are in the scope of fatigue damage, fatigue crack initiation and propagation, life prediction techniques, computational fracture mechanics, dynamic fracture, damage mechanics and assessment, Non-Destructive Test (NDT), concrete failure assessment, failure on soil structures, structural durability and reliability, structural health monitoring, construction damage recovery, and any relevant topics related to failure analysis. The book is suitable for those who involve in this field, such as academicians, engineers, practitioners, students, and researchers. They are the main players who want to obtain an up to date view of the recent advances in the area of damage mechanisms in both the engineering field of mechanical and civil.

The keynote lectures of the 5th SDMMS 2021 were given by: Prof. Ir. Dr. Ahmad Kamal Ariffin Mohd. Ihsan of the Universiti Kebangsaan Malaysia, Malaysia; Associate Prof. Dr. Wonsiri Punurai of the Mahidol University, Thailand; Ir. Hambali Chik of the Petroliam Nasional BerhaD (PETRONAS); Dato' Ir. Dr. Goh Teik Cheong of the M.E.I. Project Engineers Sdn Bhd; and Dr. Musa Bashir of the Liverpool John Moores University, England.

The chairman of the 5th SDMMS, Prof. Shahrum Abdullah (Universiti Kebangsaan Malaysia), acknowledges all authors who have contributed to the success of the event and their contributions to this volume dedicated to the structural integrity cases in civil and mechanical engineering, as well as the organizers, sponsors, scientific committee for their support. Also, Springer is fully acknowledged for its support to this volume.

UKM Bangi, Malaysia — Shahrum Abdullah
Porto, Portugal — José A. F. O. Correia
UKM Bangi, Malaysia — Salvinder Singh Karam Singh
Porto, Portugal — Abílio M. P. De Jesus
May 2021

Contents

1 A Review of Shear Wall Location Response in High-Rise RCC Structures as a Result of Earthquake Effect 1
Shamilah Anuda, Nasr Abdullah Abdulrahman, and Nik Zainab Nik Azizan

2 LCO Flutter Instability on Oscillating Supersonic Wing by Means of Linearized Aerodynamic Small Disturbance Theory 13
Nur Azam Abdullah, Erwin Sulaeman, and Meor Iqram Meor Ahmad

3 Peridynamic Method for Behaviour of Polycarbonate Specimen in Impact Test 29
M. A. Azizi, A. A. Fahad, and S. A. Rahim

4 Enhancement of Mechanical Properties for AZ31B Quenching in Nano Fluid 45
M. M. Mubasyir, M. F. Abdullah, K. Z. Ku Ahmad, R. N. I. R. Othman, and A. H. Isahak

5 Comparison of Lattice Structure Configurations for Suitability in Turbine Blades Using Modal and Harmonic Response Analysis 57
Sajjad Hussain, Wan Aizon W. Ghopa, Salvinder Singh, Abdul Hadi Azman, Shahrum Abdullah, and Hafizan Kamaruddin

6 Systematic Review: Overview on Trends and Future Opportunities of Additive Manufactured Lattice Structures 75
Asliah Seharing, Abdul Hadi Azman, and Shahrum Abdullah

7 Review on Coastal Liquefaction at Sabah Bays 91
Noor Sheena Herayani Harith, Isabella Boy Setanis, Ejria Saleh, and Tze Liang Lau

8 Ground Motion Prediction Equations for Sabah Region 101
Noor Sheena Herayani Harith, Nur Afifah Kassim, Samnursidah Samir, and Azlan Adnan

9 Fatigue Detection on Glass Fibre Reinforced Polymer Material Using Fiber Bragg Grating Sensor 115
Miminorazeansuhaila Loman and Mohd Hafizi Zohari

10 Biomechanical Overloading Factors Influencing the Failure of Dental Implants: A Review 127
Muhammad Ikman Ishak, Ruslizam Daud, Ishak Ibrahim, Fauziah Mat, and Nurul Najwa Mansor

11 Sensitivity Creep Simulation for the Equipment's Component Experiencing Extreme Elevated Temperature 143
Hashim Othman, Nurul Hana Kamaruzaman, and Ir. Hambali Chik

12 Characterization of Fatigue and Crack Growth on AZ31B Magnesium Alloy 155
A. H. Isahak, M. F. Abdullah, M. K. Faidzi, Aidy Ali, and M. M. Mubasyir

13 The Effects of Mean Strain on the Fatigue Life of the SAE 1541 Carbon Steel Based on the Strain-Life Approach 171
Teuku Edisah Putra, Husaini, and Rauzatul Akmal

14 Mechanical Characteristics of Fly Ash-Based Geopolymer Bricks 185
Mahfodzah Md Padzi, Nor Sarah Zahir, and Mohd Ridzuan Mohd Ali

15 Effect of Bio Lubricants on the Dynamic Performance of Rotor Bearing System: A Mathematical Model 195
Muhammad Imran Sadiq, Wan Aizon Wan Ghopa, Mohd Zaki Nuawi, Mohammad Rasidi Mohammad Rasani, and Tajammal Imran

16 Experimental Investigation of Acoustic Emission Characteristics at Different Type of Sensors for Granite 209
Noorsuhada Md Nor, Khairul Afinawati Hashim, Shahrum Abdullah, and Nur Fazlinieza Khairosam

17 Seismic Excitations of Prefabricated Reinforced Concrete Building Performance: Case Study of Kolej Delima Universiti Teknologi MARA, Cawangan Pulau Pinang 223
Mohd Samsudin Abdul Hamid, Norul Mas Diyana Ahmad, Kay Dora Abd Ghani, Siti Hafizan Hassan, and Nurulzatushima Abdul Karim

18 The Effects of Peat Modified Asphalt Binder Concentrations on Viscoelastic Properties 239
Ahmad Suliman B. Ali, Mohd Idrus Mohd Masirin, Abdalrhman Milad, Allam Musbah Al Allam, and Nur Izzi Md Yusoff

19 Experimental Assessment of Steel Fibre Reinforced Concrete Beam Strengthened with Carbon Fibre Reinforced Polymer 253
Noorsuhada Md Nor, Abdul Hakeem Zulkifli, Soffian Noor Mat Saliah, Noor Syafeekha Mohamad Sakdun, Nor Z. Amin, Nor N. A. Anisah, and Azmi Ibrahim

20 Simulation of Structural Analysis on Fatigue Crack (Isotropic) in FRP Laminate 267
Roslin Ramli, Mohd Hisbany Mohd Hashim, Suhailah Mohamed Noor, Anizahyati Alisibramulisi, and Muhammad Azrie Husainy Mohd Jasri

21 Managing Damages Recovery: Adopting Green Road Operation and Maintenance Index Criteria 277
Jeffryl Azniel Adzar, Rozana Zakaria, Eeydzah Aminudin, Dayalan Rainoo Raj, Rozelawati Ishak, Mohamad Faizal Sahadan, Ridzuan Mohd Radzi, Mohamad Hafizudin Syafiq Abd Rashid, Vikneswaran Munikan, Siti Mazzuana Shamsudin, and Sherliza Zaini Sooria

22 Building Defects Assessment Framework for Malaysian Construction Projects 297
Siti Hafizan Hassan, Mohd Samsudin Abdul Hamid, Mohd Faisal Zaini, Syahrul Fithry Senin, Zulfairul Zakariah, and Mohamad Zain Hashim

23 CO_2 Emission and Cost Estimates of Wastepaper Sludge Ash in Controlled Low-Strength Material Towards Sustainability 313
Mohd Azrizal Fauzi, Mohd Fadzil Arshad, Noorsuhada Md Nor, and Noor Idayani Noordin

24 Acoustic Emission Characterisation of Reinforced Concrete Beam at an Initiation Stage of Fatigue Damage 327
N. M. Nor, S. N. Mat Saliah, M. S. Tahir, and N. A. Yahya

List of Figures

Fig. 1.1	Different shapes or geometries of shear walls	4
Fig. 2.1	Present double wedge airfoil	15
Fig. 2.2	Present wing internal ribs division	16
Fig. 2.3	External stores underwing placement	17
Fig. 2.4	Full supersonic wing model	21
Fig. 2.5	V-g plot for determination of flutter for the baseline case (aluminum skin) at sea level	22
Fig. 2.6	Mode shapes 1–6 of presented Case 3	24
Fig. 2.7	V-g plot for determination of LCO flutter for presented Case 3 at sea level	25
Fig. 2.8	V-f plot for determination of LCO flutter for presented Case 3 at sea level	25
Fig. 2.9	V-g plot for determination of LCO flutter for presented Case 3 at various altitudes	26
Fig. 3.1	Nodes (blue) that interact with corresponding node (red)	30
Fig. 3.2	Charpy testing machine and specimen arrangement	31
Fig. 3.3	Dimension of Charpy impact test specimen according to ISO 179–1 standard	33
Fig. 3.4	Definition of ξ and η in PD	34
Fig. 3.5	Flow chart of PD numerical method of Charpy impact test	36
Fig. 3.6	Top view of specimen and boundary conditions in PD Charpy impact test	37
Fig. 3.7	The Ramberg–Osgood model [19]	38
Fig. 3.8	Typical load-time curve shape obtained in PC Charpy impact test [10]	39
Fig. 3.9	PD simulation load-time curves of (**a**) V-notched specimen, (**b**) U-notched specimen, and (**c**) No-notch specimen	40
Fig. 3.10	Shape of specimens after impact test finished	41
Fig. 4.1	Flow diagram of the methodology	47

Fig. 4.2 Geometrical properties of the specimen used for tensile test [17] ... 48
Fig. 4.3 Furnace heat treatment for specimen ... 49
Fig. 4.4 Schematic diagram of AZ31B quenching in nanofluid ... 50
Fig. 4.5 Schematic diagram of optical analysis ... 50
Fig. 4.6 Stress–Strain for all 12 samples ... 52
Fig. 4.7 Percentage differences of specimen compared to controlled sample ... 53
Fig. 5.1 Different configurations of lattice structures ... 58
Fig. 5.2 Flow of the steps involved in this study ... 61
Fig. 5.3 Bounding box used for constructing lattice structure ... 62
Fig. 5.4 Configurations used for analysis ... 63
Fig. 5.5 Loading condition for harmonic response ... 64
Fig. 5.6 Modal analysis results of solid model ... 66
Fig. 5.7 Mode shapes of solid model ... 66
Fig. 5.8 Modal analysis results of octet truss lattice model ... 67
Fig. 5.9 Mode shapes of octet truss lattice model ... 67
Fig. 5.10 Modal analysis results of cubic lattice model ... 68
Fig. 5.11 Mode shape results of cubic lattice model ... 68
Fig. 5.12 Deformation in Y direction for solid model ... 69
Fig. 5.13 Deformation in Z direction for octet truss model ... 70
Fig. 5.14 Deformation in Z direction for cubic lattice model ... 70
Fig. 5.15 Von Mises stress results for first three modes of solid model .. 71
Fig. 5.16 Von Mises stress results for first three modes of octet truss model ... 72
Fig. 5.17 Von Mises stress results for first three modes of cubic lattice model ... 72
Fig. 6.1 Cubic lattice structure ... 76
Fig. 6.2 Flowchart of the systematic literature review conducted ... 78
Fig. 6.3 Number of articles published from 1965 to 2020 ... 79
Fig. 6.4 Number of citations from 1965 to 2020 ... 80
Fig. 6.5 Journal article publication on lattice structure according to countries ... 80
Fig. 6.6 Percentage of publications according to research fields ... 81
Fig. 6.7 Top 10 most frequent keywords used from 2011 to 2020 ... 84
Fig. 6.8 The co-occurrence of index keywords visualization published from 1965 to 2020 ... 84
Fig. 6.9 Number of articles and citations from the year 2007 to 2020 .. 85
Fig. 6.10 Top 20 subject area of fatigue study in lattice structure ... 85
Fig. 6.11 International collaboration connections from 2007 to 2020 ... 86
Fig. 6.12 Lattice structure parameters that influence fatigue behaviour ... 87
Fig. 7.1 The local seismic source and seismicity of Sabah ... 92
Fig. 7.2 Flow chart of methodology ... 95

Fig. 7.3 Example of fault identification at 50 km site radius around at East Sabah (three red circle shows the study area) [18] 96
Fig. 7.4 Sample of soil profile 1 of current study 97
Fig. 7.5 Sample of soil profile 2 of current study 97
Fig. 7.6 Factor of Safety (FS) at BM2 location (red line shows the limit of FS at 1.0) 98
Fig. 7.7 Factor of Safety (FS) at KGU location (red line shows the limit of FS at 1.0) 99
Fig. 8.1 Malaysia location within the ring of fire [2] 102
Fig. 8.2 Recorded time histories in East Malaysia due to the local-fault event occurring on May 30, 2005 (M_W = 5.3, Rhypo = 90.8 km) 105
Fig. 8.3 Distribution of data used in terms of Peak Ground Acceleration (PGA) and distance 106
Fig. 8.4 Distribution of data used in terms of magnitude and distance 106
Fig. 8.5 Comparison of GMPEs attenuation curves and recorded PGA for magnitude between 2.9 and 3.4 108
Fig. 8.6 Comparison of GMPEs attenuation curves and recorded PGA for magnitude between 3.5 and 3.9 109
Fig. 8.7 Comparison of GMPEs attenuation curves and recorded PGA for magnitude between 4.0 and 4.4 109
Fig. 8.8 Comparison of GMPEs attenuation curves and recorded PGA for magnitude between 4.5 and 4.9 110
Fig. 8.9 Comparison of GMPEs attenuation curves and recorded PGA for magnitude between 5.0 and 5.4 110
Fig. 8.10 Comparison of GMPEs attenuation curves and recorded PGA for magnitude between 5.5 and 5.9 111
Fig. 8.11 Comparison of GMPEs attenuation curves and recorded PGA for magnitude between 6.0 and 6.4 111
Fig. 9.1 Fabricated GFRP panel 118
Fig. 9.2 Dimensions of workpiece for tensile and fatigue tests and position of FBG sensor during fatigue test 118
Fig. 9.3 Schematic of fatigue test and FBG data acquisition rig set up 120
Fig. 9.4 Original FBG wavelength before fatigue test 120
Fig. 9.5 FBG wavelength shifts of workpiece at 40% ultimate stress during a fatigue test 121
Fig. 9.6 Fractured workpiece after fatigue test with FBG sensor still intact 122
Fig. 9.7 Intensity of FBG wavelength for workpiece that undergone 40% of ultimate strength load during fatigue test 123
Fig. 11.1 The internal castable refractory spalled-off resulted in the component material had exposed to the process gas 144

Fig. 11.2 Skin metal temperature survey from IR Thermography during the incident ... 146
Fig. 11.3 Skin temperature survey records for the affected component ... 146
Fig. 11.4 Cumulative creep damage simulation results ... 150
Fig. 11.5 In situ surface replica results at the refractory spalling area of the internal surface at 200x magnification ... 151
Fig. 11.6 Close-up view of in situ surface replica result at the affected area. The red arrow indicated the presence of carbide at the grain boundaries ... 151
Fig. 11.7 In situ surface replication at baseline area at 200x magnification ... 152
Fig. 11.8 In situ surface replication at baseline area at 500x magnification ... 152
Fig. 12.1 Process flow for the conducted research study ... 157
Fig. 12.2 **a** Tensile and fatigue AZ31B magnesium alloy specimens, and **b** the dimensions (mm) ... 158
Fig. 12.3 Tensile test machine ... 158
Fig. 12.4 The 100 kN servo-hydraulic fatigue test machine ... 159
Fig. 12.5 **a** CT of AZ31B magnesium alloy specimen, and **b** the dimension (mm) ... 160
Fig. 12.6 American standard test materials (ASTM E647, 11) ... 162
Fig. 12.7 Sample tensile test of stress–strain curves of AZ31B magnesium alloy ... 163
Fig. 12.8 Trend of S–N curve for AZ31B magnesium alloy ... 164
Fig. 12.9 Three samples of AZ31B CT with respective crack lengths ... 165
Fig. 12.10 Trend of a-N curve for AZ31B magnesium alloy at 10 Hz ... 166
Fig. 12.11 Experiment CG rate curves da/dN versus ΔK for CT using 10 Hz ... 166
Fig. 13.1 Details of the strain data developed ... 174
Fig. 13.2 Examples of the developed negative mean value strain data: **a** CAL with tensile-compress mode, **b** CAL with compress-tensile mode, **c** VAL with tensile-compress mode and low–high cycle, **d** VAL with tensile-compress mode and high-low cycle, **e** VAL with compress-tensile mode and low–high cycle, **f** VAL with compress-tensile mode and high-low cycle ... 174
Fig. 13.3 Examples of the developed zero mean value strain data: **a** CAL with tensile-compress mode, **b** CAL with compress-tensile mode, **c** VAL with tensile-compress mode and low–high cycle, **d** VAL with tensile-compress mode and high-low cycle, **e** VAL with compress-tensile mode and low–high cycle, **f** VAL with compress-tensile mode and high-low cycle ... 175

Fig. 13.4 Examples of the developed positive mean value strain data: **a** CAL with tensile-compress mode, **b** CAL with compress-tensile mode, **c** VAL with tensile-compress mode and low–high cycle, **d** VAL with tensile-compress mode and high-low cycle, **e** VAL with compress-tensile mode and low–high cycle, **f** VAL with compress-tensile mode and high-low cycle 175
Fig. 13.5 Strain-life curves based on the Coffin–Manson model: **a** CALs with tensile-compress mode, **b** CALs with compress-tensile mode, **c** VALs with tensile-compress mode and low–high cycle, **d** VALs with tensile-compress mode and high-low cycle, **e** VALs with compress-tensile mode and low–high cycle, **f** VALs with compress-tensile mode and high-low cycle 177
Fig. 13.6 Strain-life curves based on the Morrow model: **a** CALs with tensile-compress mode, **b** CALs with compress-tensile mode, **c** VALs with tensile-compress mode and low–high cycle, **d** VALs with tensile-compress mode and high-low cycle, **e** VALs with compress-tensile mode and low–high cycle, **f** VALs with compress-tensile mode and high-low cycle 178
Fig. 13.7 Strain-life curves based on the SWT model: **a** CALs with tensile-compress mode, **b** CALs with compress-tensile mode, **c** VALs with tensile-compress mode and low–high cycle, **d** VALs with tensile-compress mode and high-low cycle, **e** VALs with compress-tensile mode and low–high cycle, **f** VALs with compress-tensile mode and high-low cycle 179
Fig. 14.1 Research flowchart 187
Fig. 14.2 Sample dimension (**a**) cube and (**b**) cylinder 188
Fig. 14.3 Curing pool 189
Fig. 14.4 Compressive test equipment, (**a**) auto compressive machine and (**b**) bending test machine 189
Fig. 14.5 Splitting Test, (**a**) IPC UTM-1000 Universal Testing Machine and (**b**) Computer to monitor and record the splitting tensile testing 190
Fig. 14.6 Comparison percentage between composition of fly ash and OPC 191
Fig. 14.7 Fractography surface of specimen: (**a**) Cube and (**b**) Cylinder 192
Fig. 14.8 Comparison of compressive strength between this research and previous research 192
Fig. 14.9 Comparison of splitting tensile strength between this research and previous research 193

Fig. 15.1 Jeffcott Rotor, a simple rotor system mounted on bearings and support [10] 197
Fig. 15.2 Equilibrium position of journal bearing. Adapted from [26] 198
Fig. 15.3 Schematic diagram of journal bearing test bench 200
Fig. 15.4 Pressure distribution as a function of angle θ for Soybean oil with different eccentricity ratios 202
Fig. 15.5 Oil film thickness as a function of angle θ for Soybean oil with different eccentricity ratios 203
Fig. 15.6 Pressure distribution as a function of angle θ for three bio lubricants with = ε0.6 203
Fig. 15.7 Sommerfeld number S as a function of eccentricity ratio ε for three bio lubricants 204
Fig. 15.8 Attitude angle ϕ as a function of Eccentricity ratio ε for three bio lubricants 204
Fig. 15.9 Coefficient of friction as a function of Eccentricity ratio ε for three bio lubricants 205
Fig. 15.10 Dimensionless minimum film thickness h/c as a function of Sommerfeld number for three bio lubricants 205
Fig. 16.1 Process flow for determining the acoustic emission characteristics at different sensors 211
Fig. 16.2 Sample of the rock—granite 212
Fig. 16.3 Schematic diagram of the test setup 213
Fig. 16.4 Handling method of the pencil lead on surface of the granite 214
Fig. 16.5 Test setup for the pencil lead fracture in conjunction with acoustic emission test 214
Fig. 16.6 Sensors for the pencil lead fracture **a** VS75-V and **b** VS 150-M 214
Fig. 16.7 Percentage difference of the energy between sensors VS75–V and VS150–M 216
Fig. 16.8 Energy distribution with respect to acoustic emission source location for each sensor 216
Fig. 16.9 Correlation between energy and the acoustic emission source location 217
Fig. 16.10 Frequency trend with respect to acoustic emission source location for each sensor 218
Fig. 16.11 Percentage difference of the peak frequency for sensors VS75–V and VS150–M 219
Fig. 16.12 The correlation between the peak frequency and the acoustic emission source location 220
Fig. 17.1 **a** Hysteresis energy demand in largest yield excursion for far-fault motion. **b** Hysteresis energy demand in largest yield excursion for near-fault motion [15] 227
Fig. 17.2 Research methodology overall flow chart 228

Fig. 17.3 Front view of Kolej Delima UiTMPP which constructed using IBS ... 229
Fig. 17.4 3D view model prototype of Kolej Delima UiTMPP using ESTEEM 8 ... 229
Fig. 17.5 Hysteresis behaviour of Kolej Delima UiTMPP under different seismic excitations. **a** Kunak Earthquake; **b** Bukit Tinggi Earthquake; **c** El-Centro Earthquake; **d** Pacoima Dam Earthquake ... 233
Fig. 17.6 Mode shape under earthquake excitations **a** Kunak earthquake; **b** Bukit Tinggi earthquake; **c** El-Centro earthquake; **d** Pacoima Dam earthquake ... 235
Fig. 18.1 Isochronal plots of the complex modulus G* at 1 rad ... 244
Fig. 18.2 Isochronal plots of the complex modulus G* at 10 rad ... 244
Fig. 18.3 Black Diagrams for original and PNpt modified asphalt binder ... 246
Fig. 18.4 Surface energy of asphalt binder ... 246
Fig. 18.5 Electron images of PNpt modified asphalt binder ... 247
Fig. 18.6 Failure temperatures for the original and PNpt modified asphalt binder ... 248
Fig. 19.1 Process flow for determining the acoustic emission trend of the samples' behaviour ... 255
Fig. 19.2 A setup of the instruments on the beam specimen ... 257
Fig. 19.3 Crack pattern on the beam specimen (a) B0SF0C, (b) B1SF0C, and (c) B1SF1C ... 259
Fig. 19.4 Signal strength, load with respect to time for beam specimen B0SF0C ... 260
Fig. 19.5 Signal strength, load with respect to time for beam specimen B0SF1C ... 262
Fig. 19.6 Signal strength, load with respect to time for beam specimen B1SF1C ... 263
Fig. 19.7 Signal strength with respect to damage modes for all beam specimens ... 264
Fig. 20.1 Example of isotropic laminate ... 269
Fig. 20.2 Nine layers of fiberglass laminate ... 269
Fig. 20.3 Meshing ... 270
Fig. 20.4 +45 and -45 equivalent stress ... 271
Fig. 20.5 +45 maximum principal elastic strain ... 272
Fig. 20.6 −45 maximum principal elastic strain ... 272
Fig. 20.7 +45° Crack initiation length ... 273
Fig. 20.8 -45° Crack initiation length ... 274
Fig. 20.9 −45° crack initiation width ... 274
Fig. 20.10 −45° crack initiation width ... 274
Fig. 21.1 Overall road network in Peninsular Malaysia ... 283
Fig. 21.2 Comparison between road network ... 284
Fig. 21.3 pHJKR scoring by total criteria and points ... 286

Fig. 21.4 pHJKR project rating ... 287
Fig. 21.5 Σ criteria post-FGD analysis ... 290
Fig. 21.6 Total carbon emission per year per kilometer for Federal and State road ... 291
Fig. 22.1 Flowchart methodology ... 299
Fig. 22.2 Total number of respondents and their position ... 302
Fig. 22.3 Building defects framework ... 310
Fig. 23.1 Life Cycle Assessment (LCA) ... 315
Fig. 23.2 The life cycle of building product [14] ... 316
Fig. 23.3 The material life cycle process ... 316
Fig. 23.4 The percentage of CO_2 emission for material ... 320
Fig. 23.5 The percentage of CO_2 costing RM/m^3/T for each sample ... 322
Fig. 24.1 Setup of the third point bending test for the RC beam ... 330
Fig. 24.2 Relationship between signal strength and normalised cycle ... 332
Fig. 24.3 Relationship between signal strength and normalised cycle at an early stage ... 332
Fig. 24.4 Cumulative signal strength at an early stage ... 333
Fig. 24.5 Intensity analysis of the RC beam under fatigue ... 334
Fig. 24.6 Relationship between historical index and normalised cycle ... 335
Fig. 24.7 Relationship between severity index and normalised cycle for the whole testing ... 336
Fig. 24.8 High historical index at an early stage ... 336
Fig. 24.9 Severity index at an early stage ... 337

List of Tables

Table 1.1 The common methods of earthquakes resistant 2
Table 2.1 Supersonic wing planform geometry [9] 15
Table 2.2 High modulus graphite/ epoxy composite material properties [10] 16
Table 2.3 External stores (missiles) [11, 12] 17
Table 2.4 Composite wing skin optimized geometry 22
Table 2.5 Results on weight reduction 23
Table 2.6 Percentage of wing weight reduction 23
Table 2.7 Mode descriptions of Case 3 24
Table 3.1 Experimental result of impact test on PC 39
Table 4.1 Chemical composition of Mg AZ31B alloy [3] 48
Table 4.2 Categorization of specimen 48
Table 4.3 Characterization of Specimen 52
Table 4.4 Metallurgical analysis of the sample for each medium 53
Table 5.1 Boundary conditions for modal analysis 63
Table 5.2 Boundary conditions for harmonic response 65
Table 5.3 Summary of modal analysis results 68
Table 5.4 Summary of harmonic response results 72
Table 6.1 Evolution of research articles in lattice structure from 1965 to 2020 82
Table 6.2 Top 20 productive journal article sources 83
Table 7.1 Magnitude scaling factor [11] 94
Table 7.2 Liquefaction factor of safety (FS) results at 15 locations along Sabah bays 98
Table 8.1 GMPE Model as described in Douglas [13] 107
Table 10.1 Comparison of mechanical properties of titanium, alumina and zirconia [44] 132
Table 11.1 Additional technical information of the component 145
Table 11.2 Microstructure rating chart [5] 149
Table 11.3 Projected remaining life from creep life simulation 150
Table 11.4 Overall hardness evaluation results [6] 153

Table 12.1 Tensile test results for the AZ31B magnesium alloy 163
Table 12.2 Fatigue test results of the AZ31B magnesium alloy 164
Table 12.3 Results of 3 CG length samples using a frequency of 10 Hz 165
Table 12.4 Paris law constant, C and m, for the AZ31B magnesium alloy using a frequency of 10 Hz 167
Table 13.1 Mechanical properties of the SAE 1541 carbon steel [14] ... 176
Table 14.1 Xrf analysis of fly ash 191
Table 14.2 Compressive strength at different fly ash 192
Table 14.3 Splitting tensile strength results 193
Table 15.1 Specifications of the journal bearing test bench 201
Table 15.2 Specifications of Analyzer 201
Table 15.3 Physical properties of selected bio-lubricants referred from [32] 202
Table 15.4 The operating conditions selected for calculations 202
Table 16.1 Type of sensors for the pencil lead fracture 215
Table 16.2 Energy for each source location for both type of sensors 218
Table 16.3 Peak frequency for each source location for both type of sensors 218
Table 17.1 Structural elements dimensions 228
Table 17.2 Past earthquake record adopted 230
Table 17.3 Summary of the result analysis 231
Table 18.1 The properties of the original asphalt binder 242
Table 18.2 Physical properties of peat 242
Table 19.1 The physical properties of the CFRP 256
Table 20.1 The angle of orientation and lamination sequences 269
Table 20.2 Material data for specimen +45° and -45° 269
Table 20.3 Result of the +45° 270
Table 20.4 Result of the -45° 271
Table 20.5 Fatigue experimental results 273
Table 21.1 Green certification by project type and cost 282
Table 21.2 Certification phase by green rating tools 283
Table 21.3 pHJKR (Road) comparison to generic categories average % 288
Table 21.4 Concurrence of pHJKR with the basic conditions and theories of a comprehensive transportation rating system development 289
Table 21.5 Experts attended to FGD workshop 289
Table 21.6 Criteria for expert validation in FGD 290
Table 21.7 Emission source from routine maintenance discovered from data availability survey 291
Table 21.8 Operation and maintenance scorecard derived from expert validations 293
Table 22.1 Likert scale [5] 300
Table 22.2 Years of working experience of the respondents 302

Table 22.3 List of interviewee's details 302
Table 22.4 Descriptive statistics for types of defects 303
Table 22.5 The rank of common types of building defects based on the relative importance index 304
Table 22.6 Descriptive statistics of the contributing factors to building defects 304
Table 22.7 The rank of common contributing factors to building defects based on relative importance index 305
Table 22.8 Highlighted comments from interviewees on the result from Tables 22.5 and 22.7 306
Table 22.9 Answers for consequences of the building defects to the parties involved 307
Table 22.10 Answers for effects of building defects on the buildings 307
Table 22.11 Answers for if no action is taken against the defects 307
Table 22.12 Answers for precautions to reduce the defects 308
Table 22.13 Highlighted comments from interviewees on the result from Tables 22.5 and 22.7 309
Table 23.1 IPCC guideline (2016), emission factor 317
Table 23.2 Mixture proportions and 28-day UCS results for WSA_RFA_CLSM 318
Table 23.3 Value of CO_2 emission for the material 319
Table 23.4 Total CO_2 emission by each sample ($1m^3$) 321
Table 23.5 Total CO_2 emission by each sample (1 m^3/T) 322
Table 23.6 The total cost of CO_2 emission by sample (1 m^3/T) 323
Table 24.1 The cycles for each part 334

Chapter 1
A Review of Shear Wall Location Response in High-Rise RCC Structures as a Result of Earthquake Effect

Shamilah Anuda, Nasr Abdullah Abdulrahman, and Nik Zainab Nik Azizan

Abstract The reinforced concrete shear walls are structural systems used in contemporary high-rise buildings to sustain lateral load structures such as wind, seismic loads, and carrying gravity loads. To make the structural system more efficient in resisting horizontal and gravity loads, ground motions and thus causing less damage to the structure during the earthquake. The high lateral load due to the earthquake is one of the adverse effects for high buildings. Shear walls have more strength, stiffness, and resist loads applied along with their height in the plane. Buildings with correctly constructed and recorded shear walls have performed admirably in previous earthquakes. Several studies on shear wall architecture and seismic force effectiveness have been reported. The main focus of this paper is to conduct comprehensive literature surveys and to evaluate the scope of work relevant to some investigation that is studied by adjusting different locations of the shear wall to identify the best position of the shear wall to produce minimal deflection by deciding parameters such as storey drift, storey shear and displacement.

Keywords Shear Wall · Shear Wall Location · High-Rise RC Building · Earthquake

1.1 Introduction

In recent times, the world has suffered dramatically from the natural disasters that occur from time to time, such as torpedo, earthquake, flood, hurricane and tsunami. Most of these natural disasters are unpredictable. However, the enormous hazards that may affect the existence of a building is an earthquake. Typically, areas are at risk of earthquakes, destroying buildings, loss of life, destruction of public services and infrastructure [1]. Earthquakes are among the natural hazards that have the greatest potential for causing substantial damage to engineered structures. Because earthquake forces are random and unpredictable in nature, engineers methods for studying structures under these forces influence must be sharpened. More than half of the

S. Anuda · N. A. Abdulrahman (✉) · N. Z. N. Azizan
Universiti Malaysia Perlis, 01000 Perlis, Malaysia

S. Abdullah et al. (eds.), *Structural Integrity Cases in Mechanical and Civil Engineering*, Structural Integrity 23, https://doi.org/10.1007/978-3-030-85646-5_1

Table 1.1 The common methods of earthquakes resistant

S. no	Element structure	Descriptive	Author
1	Shear walls	Strategically located stiffened walls using shear walls and is capable of transferring lateral forces from floors and roofs to the foundation	[3]
2	Braced Frames	Vertical frames that transfer lateral loads from floors and roofs to the foundations. Like shear walls, Braced Frames are designed to cater lateral loads but commonly used where shear walls are impractical	
3	Base Isolation	This seismic design strategy involves separating the building from the foundation and acts to absorb shock. As the ground moves, the building moves at a slower pace because the isolators dissipate a large part of the shock	
4	Diaphragm wall	Floors and roofs can be used as rigid horizontal planes, or diaphragms, to transfer lateral forces to vertical resisting elements such as walls or frames	

country is thought to be vulnerable to damaging earthquakes. The country's north-eastern area and the entire Himalayan belt are vulnerable to large earthquakes with a magnitude greater than 8.0. Due to these reasons, researchers and engineers have investigated how to increase the earthquake-resistant structures using several ways such as seismic isolator, high-tensile carbon fiber twine reinforcement buckling-restrained braces, fluid-filled shock absorbers, bracing frame and the shear wall. The studies report that incorporating lateral load resisting systems into the building configuration dramatically improves the structure's earthquake performance [2]. The major criteria for designing RCC structures in seismic zones control lateral displacement resulting from lateral forces.

1.1.1 The Common Methods of Earthquakes Resistant

There are known and practiced measures to protect against seismic threats. According to Abhijeet and Kanchan [3], some common resistant method is used nowadays by engineers worldwide to minimize damage to structures due to the earthquake's events (Table 1.1).

1.1.2 Shear Wall

Shear wall is one of the excellent means of providing earthquake resistance to a multistoried reinforced concrete building. During earthquake motion, the structure's behaviour depends on the distribution of weight, stiffness, and strength in both horizontal and vertical planes. This structural element will transfer the lateral forces created due to the wind and seismic forces to the diaphragm located below or to the foundation. The structural design of buildings for seismic loading is primarily concerned with structural safety during major earthquakes; in tall buildings, it is essential to ensure lateral stiffness to resist lateral load [4]. The continuous concrete vertical wall serves both architecturally as partitions and structurally to carry gravity and lateral loads. Their very high in-plane stiffness and strength make them ideal for tall buildings. In a shear wall structure, such walls are entirely responsible for the building's lateral load resistance [5]. They act as vertical cantilevers in separate planar walls and as non-planar assemblies of connected walls around elevator, stair and service shafts. Because they are much stiffer horizontally than rigid frames, shear wall structures can be economical to about 35 stories. The shear wall thickness for buildings may vary from 150 to 400 mm [6].

1.1.3 Shear Wall Configurations

Shear wall is a structural member in a reinforced concrete framed structure to resist the shear produced due to lateral forces. When a high-rise building subjected to lateral wind and seismic forces, the shear wall typically used to reduce lateral displacement. Shear walls are rectangle in cross-section, i.e. one dimension is much larger than the other. In contrast, the rectangular cross-section is frequently used, such as L- and U-shaped sections. Thin-walled hollow RC shafts around the structure's elevator core also act as shear walls and should be taken advantage of to resist the earthquake forces. The Shear Wall sections are classified into six types, as shown in Fig. 1.1 [7].

1.1.4 Types of Shear Wall

1. RC Shear Wall.
2. Plywood Shear Wall.
3. Mid ply Shear Wall.
4. RC Hollow Concrete Block Masonry Wall.
5. Steel Plate Shear Wall [8].

1.1.5 Advantages of Shear Walls in RC Buildings

1. Shear wall resistance to horizontal lateral force and exposure to the earthquake.
2. It has a very high level of rigidity that resists lateral load.
3. Shear walls are useful for deflection control.
4. RCC shear walls are easy to build detailed reinforcement.

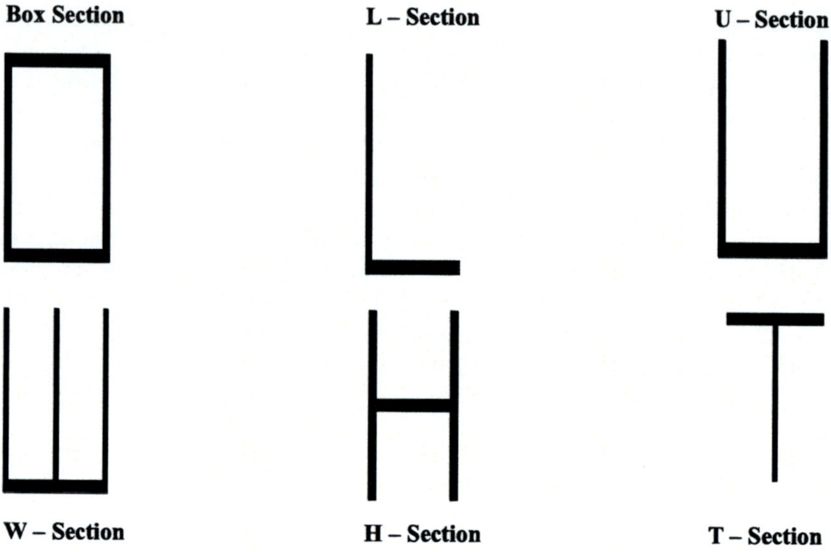

Fig. 1.1 Different shapes or geometries of shear walls

5. It minimizes damage to structural and non-structural damage caused by the earthquake.
6. Well-designed shear walls provide adequate safety and provide a high level of protection against expensive non-structural damage during moderate seismic damage [9].

1.1.6 Strengthening of RCC Building with Shear Wall

In addition to slabs, beams, and columns, reinforced concrete (RC) structures sometimes have vertical plate-like RC walls known as shear walls. These walls typically begin at the foundation level and run the length of the building. In high-rise towers, it can be as thin as 200 mm or as thick as 400 mm [10]. Shear walls, similar to vertically-oriented wide beams that carry earthquake loads downwards to the foundation, are commonly supported along both the buildings' length and width. Buildings with shear walls that have been properly constructed and detailed have performed admirably in previous earthquakes [11]. In seismically active areas, shear walls must be meticulously detailed. However, even buildings with a sufficient amount of not specially detailed walls for seismic performance (but had enough well-distributed reinforcement) were rescued from collapse in previous earthquakes [12]. Lateral loads are the most important in tall buildings, and they increase dramatically as the building rises in height [13]. Strength, rigidity, and stability are all taken into consideration in the construction. The structural system intended to support vertical loads may not be capable of withstanding lateral loads [14]. The lateral load design would increase the structural cost significantly with an increase in the storey even though it has. Special structures to resist lateral loads can be used in tall buildings to achieve cost savings. Below is a list of some of the systems.

1.1.6.1 Moment Resistant Frames (MRF)

MRF consists of linear, horizontal members (beams) in the plane with linear vertical members (columns) by rigid or semi-rigid joints.

1.1.6.2 Braced Frames

Pure rigid frame structures are inefficient for buildings greater than about twenty stories because the deflection caused by the bending of columns and girders is too high, and the buildings drift. A braced frame tries to enhance the effectiveness of pure rigid frames by adding truss members between floor systems, such as diagonals so that the shear is absorbed primarily by the diagonals [15, 16].

1.1.6.3 Shear Wall Structures

A shear wall is a structural system that provides stability against wind, earthquake and blast from inherent structural forms. The shear wall may be either planar, open or closed portions around elevators and stairwells. These structures can be either built-in steel or concrete or solid or perforated, and shear walls act as deep, slender cantilevers [17]. These can be structurally divided into coupled shear walls, shear wall frames, shear panel and staggered wall into two walls coupled with beams on each floor.

1.1.6.4 Tube Structures

A three-dimensional space frame composed of three or more frames braced frames or shear walls joined at or near their edges to form a vertical tube-like structural system capable of resisting lateral forces in any direction by cantilevering from the foundation [18]. There are different types of shear walls which are as given below:

1. Cantilever shear walls
2. Flanged cantilever shear walls
3. Coupled shear walls
4. Shear wall with openings
5. Box system.

1.1.6.5 Parameters Influencing the Response of Shear Walls

1. Height to Width Ratio
2. Type of Loading
3. Flexural Reinforcement
4. Shear reinforcement
5. Diagonal Reinforcement
6. Special Transverse reinforcement
7. Concrete Strength
8. Construction Joint
9. Axial Compressive stress
10. Moment to Shear Ratio.

Thus the design philosophy can be summarized with the following requirements. The structure must be able to withstand low-intensity earthquakes without sustaining any damage. As a result, during minor and regular earthquakes, the entire structure should remain elastomeric. The structure should be able to withstand a moderately strong earthquake with only minor structural damage that is easily repaired. Without collapsing, the structure should be able to withstand a high-intensity earthquake with a return time much longer than its design life. At this time, the shear wall is an essential part of an earthquake-resistant structure. As a result, supplying a shear wall in various positions and determining which position produces the best results is done with the help of software.

1.2 Literature Review

In reinforced concrete buildings, shear walls, which are vertical plates similar to RC walls, are prevalent. Shear walls, which are similar to vertically-oriented wide beams that contain and transfer earthquake loads to the foundation, are used to contain and transfer earthquake loads. These walls start at the foundation and go all the way up to the roof. Their width in a high-rise building will be between 150 and 400 mm [19]. Shear walls are typically supported by columns and run the length and breadth of the structure. These columns' main purpose is to support gravity loads. Existing buildings that are not designed to withstand seismic and lateral loads suffer considerable damage as a result of lateral loads. This type of structure is extremely complicated and impossible to design [20]. Height considerations, section properties considerations, and the number of shear walls to be connected are all examples of different aspects of shear walls classified. Monolithic shear walls are classified as short, squat, or cantilever, depending on their height to depth ratio.

Tall towers and multi-story structures have intrigued mankind from the beginning of civilization, with their construction served first as a defense mechanism and subsequently for ecclesiastical purposes [21]. Because of their height, tall buildings are subjected to lateral forces caused by wind or earthquakes, which can cause the structure to snap in shear or bend over. Rigidity (i.e. resistance to lateral deflection) and stability (i.e. resistance to overturning moments) standards have become increasingly critical in recent years. Shear walls (structural walls) play an important role in lateral stiffness, strength, ductility, and energy dissipation capacity [22]. Due to numerous functional criteria, a regular openings pattern must be supplied in many structural walls to accommodate doors, windows, and service ducts. Depending on the opening's shape, this type of opening decreases the shear wall's stiffness to some extent. This parametric study aims to look into the impacts of different shapes of openings in shear walls on multi-storey building's reactions and behaviours. The software ETABS9 was used to investigate the 8, 10, 12, and 15 storey prototype buildings with different openings and without openings in the shear wall. To lessen the negative impacts of twist in buildings, the opening area percentage in a shear wall should be less than or equal to 30% of the total area of that particular shear wall in the RCC building. Shear walls can form an effective lateral force resistance system by minimizing lateral displacements during earthquakes where they are positioned with the building's proper opening [23].

1.2.1 RC Shear Wall

In addition to slabs, beams, and columns, reinforced concrete (RC) structures have vertical plate-like RC partitions called shear walls. Those walls usually begin at the foundation and continue to the top of the structure [24]. In high-rise buildings, they can be as thin as 150 mm or as thick as 400 mm. Buildings with RC shear walls installed inside their orientation path gain considerable strength and stiffness, reducing lateral sway and minimizing damage to the structure and its components. The overturning results are important because shear walls include large horizontal earthquake forces [25]. In their orientation, RC shear walls provide substantial strength and stiffness to structures, reducing lateral sway and minimizing damage to the structure and its contents. The overturning effects on shear walls are significant because they carry large horizontal earthquake forces. Shear walls are more effective when placed along the building's exterior perimeter; this layout improves the structure's resistance to twisting.

1.2.2 Function of Shear Wall

Shear walls provide the necessary lateral energy for horizontal earthquake forces to be resisted. While shear walls are strong enough, the horizontal forces will be transferred to the next detail in the load path underneath them. Different shear walls, floors, foundation walls, slabs, or footings may be used in some analyses of the RC shear wall's strength at an outstanding position in various ways. Shear walls also provide lateral stiffness, preventing excessive sideways movement of the roof or floor above [26]. Sufficiently rigid shear walls will keep floor and roof framing members from moving out of their support. Furthermore, sufficiently strong buildings are more likely to sustain less non-structural damage [27].

1.2.3 Past Earthquake Effect on High Rise Building

The history of earthquakes shows that if buildings are not properly designed and built to the necessary standard, they will cause significant damage. As a result of this reality, buildings have been made safe from earthquake forces. Subsequently, there is a need to determine a high-rise building's seismic responses to design earthquake-resistant structures by carrying seismic analysis of the structure. A considerable increase in high-rise buildings and modern development headed for more complex structures in recent days. The space constraints, especially in urban areas, face structural engineer with sufficient strength and stability of these tall buildings against the lateral load. Thus, the effect of lateral loads, such as wind loads and earthquake loads, are attaining extreme importance for high-rise buildings [28]. Well-designed shear walls provide adequate safety and provide a great measure of protection against costly non-structural damage during moderate seismic damages. Shear walls provide significant strength and stiffness to buildings in their orientation, significantly reducing the building's lateral sway and reducing damage to the structure and its contents [29].

1.2.4 Performance of Location of Shear Wall for Multi-Storey

Suwalka et al. [30] proposed the shear wall's prime location and its effectiveness of best shear wall in the bare frame system. The study of the 16-storey building presented some investigations analyzed in both the structural system, i.e. shear wall frame structure and without shear wall structure. The building located in Zone-IV, according to IS 1893:2002. The construction model 3D was analyzed using the linear static method, response spectrum and surface messing were done to a model shear wall. Suwalka's research shows that models were compared to various parameters such as lateral displacement in the X and Y directions, tale drift, and axial force in columns using Etabs software. Results assessed the efficiency of varying interface and found that the shear wall located in a building with a shear wall at all corners is the effective approach to cater for the lateral load. The provision of a shear wall with an appropriate location is advantageous. The structure performs better if the optimum configuration and its footprint identified before the entire structure's design was conducted [30].

A study has been carried out by Titiksh and Bhatt [29] to determine the effects of additions of shear walls and also the optimum structural configuration of multi-storey buildings by transposition the locations of the shear wall. Four various shear wall positions for 11-storey buildings have analyzed by computer software ETABS. The framed structure subjected to lateral and gravity loads according to Indian Standards, and the results determine the best location for the shear walls. The results indicate when shear walls placed at the center of the geometry in the form of a box or the corners, the structures behave in a more stable manner. In conclusion, among all other studied possibilities, building with a box-type shear wall at the core of the structure is the optimum framing technique for middle and high-rise buildings. In order to increase the performance of the structure, earthquake resisting techniques such as Seismic Dampers and Base Isolation can be used [10].

Snell (2004) studied a shear wall provision represents a structurally efficient way to stiffen a building structural system because a shear wall's primary function is to increase the rigidity for lateral load resistance. In modern-day tall buildings, shear walls are typically used as a vertical structural element for resisting the lateral masses that can be prompted by the effect of wind and earthquakes, which cause the failure of structure as proven in parent shear walls of varying sections, i.e. square shapes to greater irregular cores consisting of the channel, T, L, Barbell shape, Box and so forth can be used. Provision of walls allows to divide an enclosed area, whereas cores include and bring services together with the elevator. Wall openings are necessarily required for building windows in outside partitions and for doors or corridors in internal walls or carry cores.

Rajkolhe and Khan [31] researched the shear wall's optimum location in high-rise R.C building under lateral loading. In this research, the analysis was done following the earthquake code IS 1893[PART-I] 2002 and wind code IS 875 [PART-I]-1987 [31]. The study was carried away by taking the irregular plan of building 21-storeys on medium [ZONE-II] and severe [ZONE-V] and soil [TYPE-II]. Furthermore, the seismic analysis of the building's irregular plan for static earthquake and static wind analysis. The comparison by providing shear wall at four edges with and without the shear wall in the irregular plan investigated to determine the shear wall's ideal position.

In conclusion, the plan without a shear wall gives more displacement and more drift than a plan with a shear wall lengthwise four edges. Hence, by providing shear wall lengthwise, four edges can diminish storey displacement, storey drift, storey shear and increase the structure's strength and stiffness. These results demonstrated that giving a shear wall along four edges is ideal for a shear wall [2].

Patil et al. [32] studied the optimal location of shear wall in high-rise building subjected to seismic loading. In this project, a high-level building with 15 storeys in zone IV considered all the regular and irregular building models. Two different types of models are considered in every shape of the building by changing the shear wall's location in the structure plan, located at the corners of the building and the building's central core section [32].

In this seismic analysis, storey drift and displacement are essential parameters to be considered and analyzed using standard package Etab-2013—producing a 3D building model for both Equivalent Static Method and Response Spectrum Methods. The storey displacement reported in the models where shear walls provided at corners of the building provide maximum displacement compared with the models with a shear wall at the building's central core location. Hence maximum lateral displacement always occurs at the topmost level for both types of locations of shear walls for all models. In the meantime, storey drift ratios for all stories in the models having a shear wall at the building's central core location are less than those in the models where shear walls provided at corners of the building. Generally, the storey drift ratio is meagre in bottom stories but very high at the middle levels and finally decreases towards the upper levels. This study indicates that the storey drifts were reduced due to the shear wall's primer at the building's core sections, enabling the structure to behave as almost ideally stiff. In this way, the damage risk of the structure & non-structural elements is minimized.

Haque et al. [33] focused on analyzing shear wall location due to earthquake effect in high rise RCC structures. The most common and usual trend of shear wall location and shape have also been discussed here. The methodology is mainly concerned with defining the different location cases for shear wall and their model generation. For this study, five different location cases for a shear wall keep all the other structural elements unchanged.

The cases include:

Case 1: the original plan,
Case 2: moving the shear wall 35 inches inward,
Case 3: moving the shear wall 70 inches inward,
Case 4: moving the shear wall 105 inches inward,
Case 5: moving the shear wall 140 inches inward.

Software ETABS9 has developed the model generation process of these 5 cases considering the loads such as Dead, Live, Wind, Earthquake. The comparative results have shown that shear walls are more effective at the outer periphery and near columns. The structure has analysed without basements. So, their responses to loads are not included in the paper [33].

1.3 Conclusion

From the above literature review, various researchers have studied different types of earthquake-related issues and addressed that shear wall is more prominent in resisting lateral force due to earthquakes. Software review such as Etabs combined with manual work, models created, and shear walls are positioned in various building positions to find the structure's most minor displacement due to shear walls. Researchers have found out that changes in shear wall positions affect the attraction of forces. Shear wall position in any building significantly reduces displacements and reduces the structural effect. Besides dead & live loads, the other loads also effect the lateral displacement of the building. In a frame structure, all the frames are subjected to additional moments and deflections due to earthquake loads. Shear walls can reduce the deflection of a building; shear walls placed at the outer edge portion, periphery and near the column provide more stiffness to the structure resulting in less deflection. Buildings without a shear wall are a source of concern and must be retrofitted in seismic/earthquake and wind-prone areas. Various type of shear walls such as L shaped, O shaped, and I shaped shear walls should be tested for their performances. The reduction of storey drifts due to the shear wall's introduction at the building's core sections enables the structure to behave as almost ideally stiff. In this way, the damage risk of the structure & non-structural elements minimized.

Acknowledgements Universiti Malaysia Perlis (UMP) provided funding for this study (UniMAP); the authors wish to thank the respondents who have spent their precious time and patience participating in this project. I want to express my sincere gratitude to my research supervisor, Dr. Shamilah Anudai @ Anuar, to allow me to conduct this paper and provide invaluable guidance through this paper. His dynamism, vision, sincerity and motivation have deeply inspired me. My appreciation to Assist. Dr. Nik Zainab Nik Azizan for his valuable guidance in the practical aspects related to the project.

References

1. Ahmed MM, Abdel Raheem SE, Ahmed MM, Abdel Shafy AG (2016) Irregularity effects on the seismic performance of L-shaped multi-storey buildings. JES. J Eng Sci 44(5):513–536
2. Abbas T, Othman FM, Ali HB (2017) Effect of infill parameter on compression property in FDM process. Dimensions 12(12.7):25–4
3. Baikerikar A, Kanagali K (2015) Study of lateral load resisting systems of variable heights in all soil types of high seismic zone. Int J Res Eng Technol 03 (10):109–119
4. LovaRaju K, Balaji DK (2015) Effective location of shear wall on performance of building frame subjected to earthquake load. Int Adv Res J Sci Eng Technol 2(1)
5. Baikerikar A, Kanagali K (2014) Study of lateral load resisting systems of variable heights in all soil types of high seismic zone. Int J Res Eng Technol 3(10)
6. Jaswanth B, Surendra YL, Kumar MR (2018) Seismic analysis of multi-storied building with shear walls using ETABS. Int J Curr Eng Technol 8(3):1030–1040. https://doi.org/10.14741/ijcet/v.8.3.24
7. Reddy NJ, Peera DG, Reddy TAK (2015) Seismic analysis of multistoried building with shear walls using ETABS-2013. Int J Sci Res (IJSR) 4(11):1030–1040
8. Dodiya J, Devani M, Dobariya A, Bhuva M, Padhiar K (2018) Analysis of multistorey building with shear wall using etabs software, vol 5, pp 1543–1546
9. Mondal MAA, Bhaskar MGB, Telang MD (2017) Comparing the effect of earthquake on shear wall building and non-shear wall building–a review, pp 1–4
10. Agrawal A, Charkha SD (2012) Study of optimizing configuration of multi-storey building subjected to lateral loads by changing shear wall location. In: Proceedings of international conference on advances in architecture and civil engineering (AARCV 2012), p 1
11. Kevadkar MD, Kodag PB (2013) Lateral load analysis of RCC building. Int J Mod Eng Res (IJMER) (IJMER) 3(3):1428–1434
12. Malik RS, Madan SK, Sehgal VK (2012) Effect of height on seismic response of reinforced cement concrete framed buildings with curtailed shear wall. J Eng Technol 1:52–62
13. Walvekar A, Jadhav HS (2015) Parametric study of flat slab building with and without shear wall to seismic performance. Int J Res Eng Technol 4(4):601–607
14. Cho WS, Lee SH, Chung L, Kim HJ, Kim SJ, Yu EJ (2012) Seismic performance evaluation of reinforced concrete frame with unreinforced masonry infill. J Arch Inst Korea Struct Constr 28(3):31–41

15. Zareian F, Krawinkler H, Lignos DG, Ibarra LF (2008) Predicting collapse of frame and wall structures. In: The 14th world conference on earthquake engineering. Beijing, China
16. Wang Q, Wang L, Liu Q (2001) Effect of Shear Wall Height on Earthquake Response, National Huaqiao University, Quanzhou 362011, Fujian, China. Eng Struct 23:376–384
17. Chandiwala A (2012) Earthquake analysis of building configuration with different position of shear wall. Int J Emerg Technol Adv Eng 2(12):347–353
18. Misal DJ, Bagade MA (2016) Study of seismic behaviour of multi-storied RCC buildings resting on sloping ground and considering bracing system. Int J Eng Res 5(3):690–697
19. Choi HS, Ho G, Joseph L, Mathias N (2017) Outrigger design for high-rise buildings. Routledge
20. Scawthorn C, Chen WF (eds) (2002) Earthquake engineering handbook. CRC press
21. Sharma R, Amin JA (2015) Effects of opening in shear walls of 30-storey building. J Mater Eng Struct 2(1):44–55
22. Sankar PS, Rao PKR (2017) Static and dynamic analysis of a multi-storied building with shear walls at different locations
23. Chandurkar PP, Pajgade DP (2013) Seismic analysis of RCC building with and without shear wall. Int J Mod Eng Res 3(3):1805–1810
24. Agrawal MTSAK, Kushwaha N (2020) A review on effect of positioning of RCC shear walls of different shapes on seismic performance of building resting on sloping ground using STAAD-pro
25. Elnashai AS, Di Sarno L (2015) Fundamentals of earthquake engineering: from source to fragility. John Wiley & Sons
26. Soni P, Tamrakar PL, Kumhar V (2016) Structural analysis of multistory building of different shear walls location and heights. Int J Eng Trends Technol (IJETT), 32(1)
27. Arya AS, Boen T, Ishiyama Y (2014) Guidelines for earthquake resistant non-engineered construction. UNESCO
28. Halkude SA, Konapure CG, Birajdar SM (2015) Effect of location of shear walls on seismic performance of buildings. Int J Curr Eng Technol 5(2):826–833
29. Titiksh A, Bhatt G (2017) Optimum positioning of shear walls for minimizing the effects of lateral forces in multistorey-buildings. Arch Civ Eng 63(1):151–162
30. Suwalka V, Laata MN, Nagar B (2018) Comparative study and modeling of framed structure with shear wall & without shear wall by using etabs, pp 1241–1248
31. Rajkolhe R, Khan JG (2014) Defects, causes and their remedies in casting process: a review. Int J Res Advent Technol 2(3):375–383
32. Patil R, Deshpande AS, Sambanni S (2016) Optimal location of shear wall in high rise building subjected to seismic loading, vol 10, no 10, pp 2678–2682
33. Maksudul Haque MD, Hasibulhasan Rahat MD, Rifat-Al-Saif, Reza Chowdhury S (2018) Analysis of shear wall location due to earthquake effect in high rise RCC structures. In: Proceedings of 105th IASTEM international conference, Putrajaya, Malaysia, pp. 1–5
34. Journal I, Technological F, Patil R, Deshpande AS, Sambanni S (2016) Optimal location of shear wall in high rise building. Raehyrjtyjty 3(10):2678–2682

Chapter 2
LCO Flutter Instability on Oscillating Supersonic Wing by Means of Linearized Aerodynamic Small Disturbance Theory

Nur Azam Abdullah, Erwin Sulaeman, and Meor Iqram Meor Ahmad

Abstract This article offers an optimization procedure in designing much lighter supersonic wing by employing a composite structure by constraining the structural persistence due to flutter speed, a type of aeroelasticity failure. The application of composite material in aeroelasticity contributes to the changes in the expected flutter speed. The composite material such that graphite/epoxy gives high modulus compared to the metallic material such as aluminium where the structural flexibility could be improved. In contrast, the structural weight could be reduced through the optimization process. Technically, an optimization procedure that is utilizing the aeroelastic parameter as a constraint is called aeroelastic tailoring. In this paper, the objective of this tailoring process is to optimize the wing weight while maintaining the flutter boundaries, where the wing design adopted in this research has been analyzed at sea level. As the optimization process is on-going, the flutter speed and the plate manufacturing thickness become the restriction in the wing weight reduction. The investigation shows good agreement to the objective where the removal of weight for the High Modulus (HM) graphite/epoxy wing skin for the skin weight, clean wing and total wing with missile launcher external stores are 75.82%, 61.96% and 22.09%, respectively compared to the baseline aluminium wing model. For the tailoring process, it is found that the flutter Mach number increases more than 81% using as the Graphite/epoxy composite replaced the aluminium as the skin.

Keywords Structural Durability · Limit Cycle Oscillation · Aeroelastic Failure · Supersonic Composite Wing · Optimization

---

N. A. Abdullah (✉) · E. Sulaeman
Structural Mechanics and Dynamics Research Group, Department of Mechanical and Aerospace Engineering, International Islamic University Malaysia, Kuala Lumpur, Malaysia
e-mail: azam@iium.edu.my

M. I. M. Ahmad
Department of Mechanical and Materials Engineering, Faculty of Engineering and Build Environments, Universiti Kebangsaan Malaysia, Selangor, Malaysia

S. Abdullah et al. (eds.), *Structural Integrity Cases in Mechanical and Civil Engineering*, Structural Integrity 23, https://doi.org/10.1007/978-3-030-85646-5_2

2.1 Introduction

2.1.1 Aeroelasticity

Aeroelasticity is referred to as the field of study related to the interaction between aerodynamic forces and elastic structure deformation [1]. For instance, the combination of bending-torsion modes under linear supersonic aerodynamics is the primary problem for fixed-wing aeroelasticity. In that case, this type of aeroelastic failure situation is classified as a linear problem. For this reason, the analysis of eigenvalues and mode shapes are needed to extend the aeroelastic analysis especially in the frequency domain. In another hand, the mode shapes plot could illustrate the type of vibrational modes, i.e. bending, torsion. It is related to the eigenvalues since the range of frequency is calculated through this parameter and responsible for differentiating every change of mode that triggers excitation until fracture at a certain speed (flutter).

At the same time, several approaches have been introduced to improve the aeroelastic failure, such as the aeroelastic tailoring. For example, the forward-swept wing that endured such static aeroelastic failure called divergence due to the forward wing tip compared to the wing root which is attached to the fuselage once the structure body interacts with the aerodynamic loads. In that case, the implementation of composite material in aeroelastic tailoring as in [2] shows the composite material wing only bends without twisting for the forward-swept wing due to the improvement of the wing stiffness. Consequently, this kind of aeroelastic problem could be solved by employing this technique.

Composite materials can be used to produce highly-tailored aircraft structures that meet stringent performance requirements, but these properties also present unique challenges for analysis and design [3]. More comprehensive study on the assignment of composite structures, especially in terms of the composite ply arrangement should be optimized and hence the flight performance could be improved, especially in terms of aeroelastic failure. For graphite/epoxy composite laminate, the loading rate and temperature are sensitive due to being made of high modulus fibers and low modulus polymeric matrix [4].

In the present work, the aeroelastic analysis performed in this study follows the guideline of flutter analysis from the military specification as in [5] since the analyzed wing was constructed based on a jet fighter wing planform. The optimization of aeroelastic tailoring of this study is the continuation of the study performed by [6]. As the analysis performed completed using MSC Nastran, it is guided as in [7]. The study also showed the analysis at negative boundary altitude to validate the flutter analysis in several tropical countries [8].

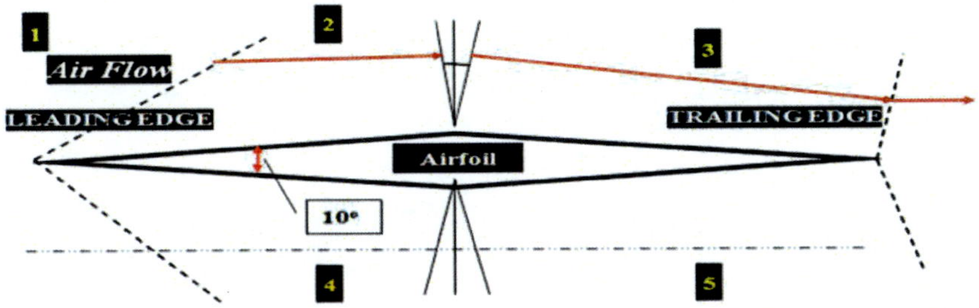

Fig. 2.1 Present double wedge airfoil

2.2 Structural Wing Model

2.2.1 Double Wedge Supersonic Airfoil

Since the analysis in this study involves the operation in a supersonic region, the airfoil of this supersonic wing is a double wedge shape, as shown in Fig. 2.1. The wedge angle of the airfoil is 10°. Along the wingspan, the airfoil is divided into three segments which are the main wing box and two control surfaces at the leading edge and trailing edge. The divisions of the leading edge and trailing edge have been designated as 15% and 20% of the chord length, respectively. A fully design approach of this supersonic wing can be found in [9].

2.2.2 Wing Design

Table 2.1 presents the geometry of the present wing planform. There are 12 ribs inside the wing structure included at the wing root and the wing tip as shown in Fig. 2.2. The wing ribs are constructed based on aluminum material, where the initial design throughout the stress analysis made in [9]. The thickness of each rib is assigned accordingly with the thickness of the skin at each region.

The composite material properties for the High Modulus Graphite/epoxy that have been applied to the formation of the wing presented in Table 2.2 are taken from [10].

Table 2.1 Supersonic wing planform geometry [9]

Wing swept angle, Λ (degree, °)	Taper ratio, λ	Aspect ratio, AR	Average chord length, c (m)	Wing root, C_R (m)	Half span, b/2 (m)
30	0.5	5	1.5	1	3.75

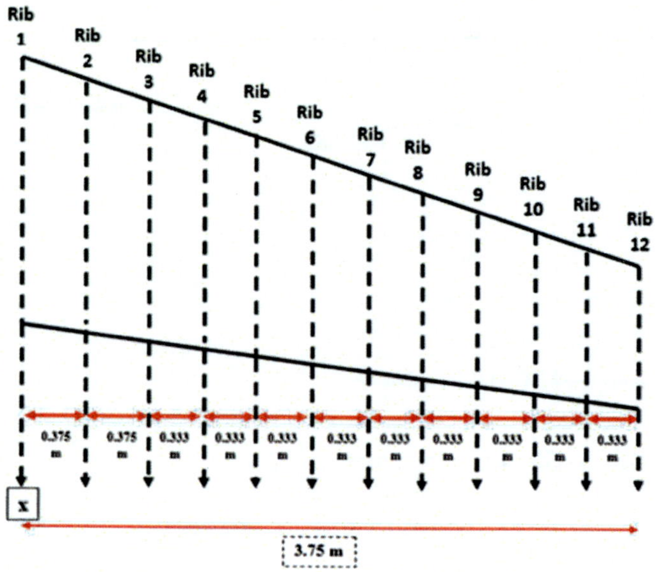

Fig. 2.2 Present wing internal ribs division

Table 2.2 High modulus graphite/ epoxy composite material properties [10]

Elastic properties	E1 (GPa)	230
	E2 (Gpa)	6.6
	G12 (Gpa)	4.8
	ν12	0.25
Physical properties	Density, (kg/m^3)	1630

2.2.3 External Stores

An external store can be specified as an object that is attached to the aircraft component. The external store of fighter aircraft can be in either loaded missile or external fuel tank. For this research, there are three missiles have been attached at the outer part of the supersonic wing as can be shown in Fig. 2.3. The properties of each missile at these three stations are presented in Table 2.3.

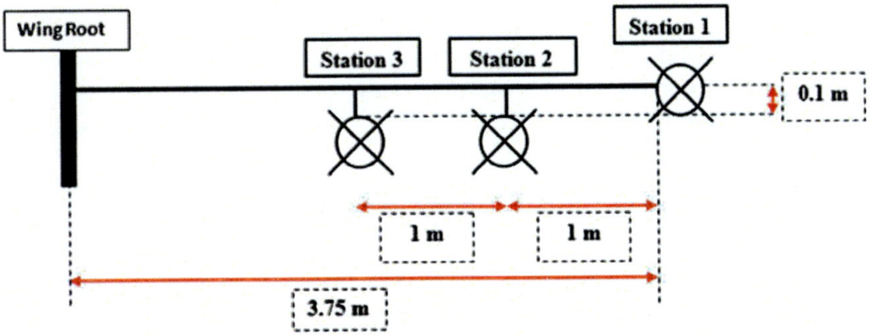

Fig. 2.3 External stores underwing placement

Table 2.3 External stores (missiles) [11, 12]

Station	Missile type	Length (m)	Diameter (m)	Weight (kg)
1	AIM-9 M	2.85	0.128	86.0
2	AIM-120 A	3.66	0.178	157.89
3	AIM-120 A	3.66	0.178	157.89

2.3 Lifting Surface Theory

2.3.1 Integral Equations of the Linearized Small Distribution Equation

Zona 51 is derived from Zona 7, which only incorporates the lifting surface by dismissing the aerodynamics of the aircraft body such as fuselage. Both methods solve the linearized small disturbance equation which is as in Eq. (2.1) which is summarized from [13]:

$$\left(1 - M_{\infty}^{2}\right)\Phi_{xx}+\Phi_{yy}+\Phi_{zz}-2\frac{M_{\infty}}{a_{\infty}}\Phi_{xt} - \frac{1}{a_{\infty}^{2}}\Phi_{xt}= 0 \tag{2.1}$$

Since this work involve in supersonic region, the freestream Mach number, M_{∞} is analyzed at supersonic Mach number. The total potential velocity, ϕ consists of two parts which is denoted in Eq. (2.2):

$$\Phi=\emptyset_0+\emptyset_1 \tag{2.2}$$

In Eq. (2.2), the steady potential $\emptyset_0$ is the wing thickness order while $\emptyset_1$ is the unsteady potential which is responsible in calculating the oscillatory amplitude due to structural modes order.For stability analysis the amplitude of the structural oscillation is assumed to be smaller than wing thickness at all times. Equation (2.3) shows the relation of this assumption.

$$\emptyset_1 \ll \emptyset_0 \tag{2.3}$$

Since the Eq. (2.1) is linear, Eqs. (2.4) and (2.5) are been derived. Equation (2.4) is for the steady linearized small disturbance equation while Eq. (2.5) is for unsteady linearized small disturbance equation.

$$\left(1 - M^2\right)\emptyset_{0xx} + \emptyset_{0yy} + \emptyset_{0zz} = 0 \tag{2.4}$$

$$\left(1 - M_\infty^2\right)\emptyset_{1xx}+\emptyset_{1yy}+\emptyset_{1zz}-\frac{1}{a_\infty^2}\emptyset_{1tt}-2\frac{M_\infty}{a_\infty}\emptyset_{1xt} = 0 \tag{2.5}$$

The generated unsteady potential, $\emptyset_1$ can be written in the reduced form in Eq. (2.6) when an aircraft is performing a simple harmonic motion where ω is the oscillation frequency while ω is the potential of reduced freqeuncy domain.

$$\emptyset_1=\emptyset\ e^{i\omega t} \tag{2.6}$$

To solve modified potential $\overline{\varphi}$ as in Eqs. (2.7), (2.8) is being substituted with L is the reference length for each parameter x', y' and z'. The compressible reduced frequency equation, λ is explained in Eq. (2.9). The Prandtl-Glauert factor, β which provide the measure to λ is denoted as in Eq. (2.10) and reduce frequency, k in Eq. (2.11):

$$\emptyset=\overline{\emptyset}e^{i\lambda M_\infty x'} \tag{2.7}$$

$$x' = L\beta x,\ y' = Ly,\ z' = Lz \tag{2.8}$$

$$\lambda = \frac{kM_\infty}{\beta} \tag{2.9}$$

$$\beta = \sqrt{\left|1 - M_\infty^2\right|} \tag{2.10}$$

$$k = \frac{\omega L}{V} \tag{2.11}$$

The unsteady linearized small disturbance equation as in Eq. (2.5) can be reduced as in Eq. (2.12) using the given equations.

$$\overline{\emptyset}_{x'x'} + \mu\overline{\emptyset}_{y'y'} + \lambda^2\overline{\emptyset} = 0 \tag{2.12}$$

As the analysis will be performed in supersonic, the condition is given that μ is equal to -1 for $M_\infty > 1$. The unsteady source and doublet singularity σ and $\Delta\emptyset$ can be derived over the surface S of the configuration of interest by applying the Green theorem to the Eq. (2.12). The integral solution can be expressed in terms of Ø like in Eq. (2.13) where the field points are denoted as (x_0, y_0, z_0) while the sending points from the sources and doublets as (x, y, z). The E factor is set to be equal to 2 for $M_\infty > 1$, σ is the unsteady source singularity distribution, $\Delta\emptyset$ is the unsteady doublet singularity distribution, $\frac{\partial}{\partial n}$ is the out-normal vector and K is the Kernel function.

$$\emptyset(X_0, Y_0, Z_0) = -\frac{1}{E\pi}\iint_S \sigma(x, y, z)e^{i\lambda M_{\infty}\xi}K\,dS + \frac{1}{E\pi}\iint_{S+W} \Delta\emptyset(x, y, z)e^{i\lambda M_{\infty}\xi}\frac{\partial}{\partial n}K\,ds \tag{2.13}$$

All this gather information is useful to calculate the Supersonic Kernel Function as in Eq. (2.14):

$$K = \frac{cos(\lambda R)}{R} \tag{2.14}$$

where R is expressed in Eq. (2.15) from the generalized coordinates, ξ, η and ζ as in Eq. (2.16):

$$R = \sqrt{\xi^2 + \mu\eta^2 + \mu\zeta^2} \tag{2.15}$$

$$\xi = \left\{\frac{x - x_0}{\beta L}\right\}, \eta = \left\{\frac{y - y_0}{L}\right\}, \zeta = \left\{\frac{z - z_0}{L}\right\} \tag{2.16}$$

The surface of configuration which is called as S can be said as in Eq. (2.17) which will be used as the integral which associated with the unsteady source singularity.

$$S(x, y, z, t) = 0 \tag{2.17}$$

2.4 Aeroelastic Analysis and Steepest Descent Optimization

2.4.1 Analysis Procedure

This study embraces three types of analysis which are used in MSC Nastran as the solution solver. The first is a structural analysis which is comprising the modal analysis using the Lanczos method. Lanczos method will be used to solve the mechanical vibration equation as in Eq. (2.18) where **M** is the mass matrices, and **K** is the stiffness matrices. The final analysis is the flutter analysis which is using the PK method solution as indicated in Eq. (2.19), where the mechanical vibration presented in Eq. (2.18) is now coupled with aerodynamics parameters; **Q** is the aerodynamics influence matrices, and **B** is the damping matrices. The aerodynamics matrices **Q** are separated into two parts, real (R) and imaginary (I).

$$\mathbf{M}\{\ddot{\mathrm{x}}\} + \mathbf{K}\{\mathrm{x}\} = 0 \tag{2.18}$$

$$\left[\mathrm{M_{hh}}\,\mathrm{p}^2 + \left(\mathrm{B_{hh}} - \frac{\frac{1}{4}\rho\bar{c}V\,Q^{\mathrm{I}}_{\mathrm{hh}}}{\mathrm{k}}\right)\mathrm{p} + \left(\mathrm{K_{hh}} - \frac{1}{2}\rho V^2 \mathrm{Q}^{\mathrm{R}}_{\mathrm{hh}}\right)\right]\{\mathrm{u_h}\} = 0 \tag{2.19}$$

The matrix terms in Eq. (2.19) are all real. $\mathbf{Q}^{\mathrm{R}}_{\mathrm{hh}}$ and $\mathbf{Q}^{\mathrm{I}}_{\mathrm{hh}}$ are, respectively, the real and imaginary parts of $\mathbf{Q}^{(\mathrm{m,k})}_{\mathrm{hh}}$.Note that the circular frequency and the reduced frequency are not independent since $k = \omega/2V$, and furthermore, that reduced as displayed in Eq. (2.20).

$$k = \left(\frac{c}{2V}\right) Im(p) \tag{2.20}$$

For the PK method of solution, Eq. (2.21) is rewritten in the state-space form with twice the order. $\{U_h\}$ now includes both modal displacements and velocities. The eigenvalues of the real matrix **A** are either real or complex conjugate pairs. Real roots indicate a convergence or divergence as in the cases of the roll subsidence (rigid body) mode or a structural (torsional) divergence mode.

$$[\boldsymbol{A} - p\mathbf{I}]\{U_h\} = 0 \tag{2.21}$$

Meanwhile, **A** can be represented as the real matrix shown in Eq. (2.22).

$$\mathbf{A} = \begin{bmatrix} 0 & \mathrm{I} \\ -\mathbf{M}^{-1}_{\mathrm{hh}}\left[\mathbf{K}_{\mathrm{hh}} - \frac{1}{2}\rho V \mathbf{Q}^{\mathrm{R}}_{\mathrm{hh}}\right] & -\mathbf{M}^{-1}_{\mathrm{hh}}\left[\mathbf{B}_{\mathrm{hh}} - \frac{\frac{1}{4}\rho c V \mathbf{Q}^{\mathrm{I}}_{\mathrm{hh}}}{\mathrm{k}}\right] \end{bmatrix} \tag{2.22}$$

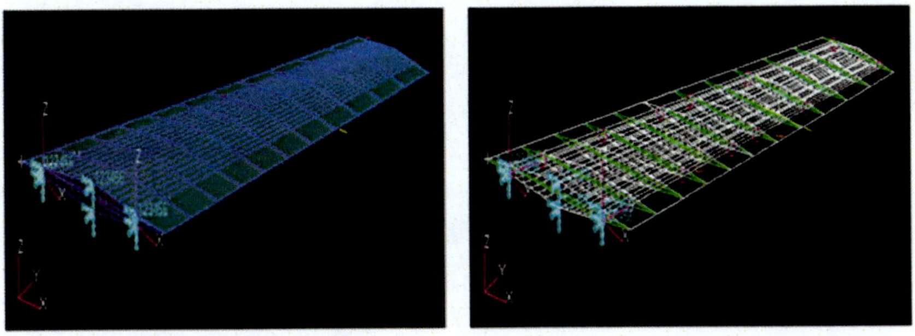

(a) External wing skin model

(b) Internal structure wing model

Fig. 2.4 Full supersonic wing model

For the real roots, the damping, g as shown in Eq. (2.23) is expressed as the decay rate coefficient, which is the distance travelled (measured in chord lengths) to half (or double) amplitude.

$$g = 2\gamma \tag{2.23}$$

2.5 Finite Element Wing Model

2.5.1 Wing Structural Configuration

Figure 2.4a and b show the generated supersonic wing planform based on their presented axis respectively in a finite element interface. As it can be seen in Fig. 2.5a and b, the digit 123,456 is referring to the attachment of spars to the wing-body as the fixed boundary point. In that case, there will be no translational and rotational movements occur at that fixed point when the simulation is conducted. The green diamond shapes referring to the ribs at different location of y along the spanwise direction. The direction of airflow is set along the x-axis parallel to the chordwise direction.

2.5.2 Benchmark: Flutter Speed of Fully Aluminum Wing

This section embarks the flutter speed results that obtained for the baseline, where the supersonic wing skin is constructed by using aluminum plate. V-g plot of this case is plotted as shown in Fig. 2.5.

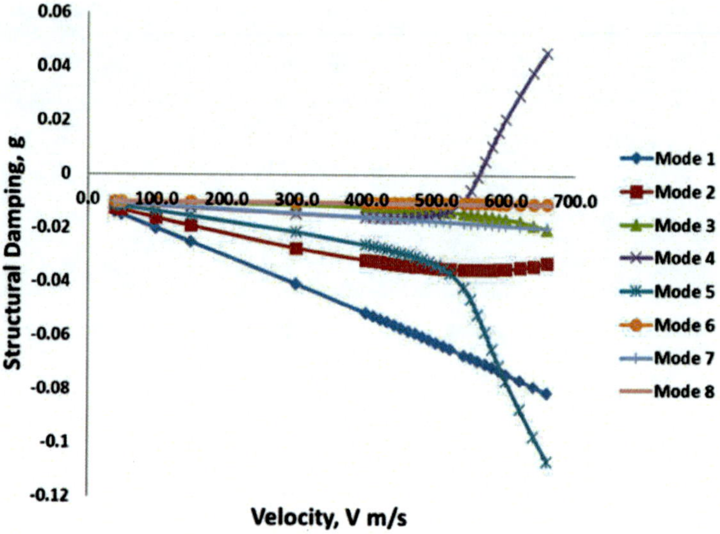

Fig. 2.5 V-g plot for determination of flutter for the baseline case (aluminum skin) at sea level

Based on this plot, the flutter speed at the operational sea level obtained for this case is found to be 581.5 m/s [9]. This result is the benchmark speed that is applied as the constraint speed for the optimization process in the next section.

2.5.3 Optimization Via Steepest Descent Method

The objective of the optimization for this study is to reduce the wing weight using composite material. The fiber orientation of the present composite has been optimized as in [6]. The optimization variable for this study is the plate thickness while the constraint is the flutter speed and manufacturing factor. In this study, the obtained flutter speed is considered the failure speed under the coupling of elastic forces, inertia forces, and aerodynamics load, acting as the limit speed for the optimization. The optimization process is ceased until the possible manufacturing thickness has been achieved. Table 2.4 presents the final optimized geometry of Case 1, Case 2 and Case 3.

Table 2.4 Composite wing skin optimized geometry

Case	Fiber orientation (degree, °)		$t_{composite}/t_{aluminum}$	
	Wing skin	Control surface	Wing skin	Control surface
1	45/20/0/s	90/60/-45/s	3/4	3/8
2	45/20/0/s	90/60/-45/s	1/2	3/8
3	45/20/0/s	90/60/-45/s	3/8	3/8

Table 2.5 Results on weight reduction

Case	Mass (kg)				
	Wing skin	Spars & ribs	Payloads	Launchers	Total
Benchmark: Aluminum Skin	236.3	52.83	401.78	120	810.91
1	85.97	52.83	401.78	120	660.58
2	66.76	52.83	401.78	120	641.37
3	57.14	52.83	401.78	120	631.75

Table 2.6 Percentage of wing weight reduction

Case	Total skin mass reduction (%)	Clean wing mass reduction (%)	Total mass reduction (%)
Benchmark: Aluminum Skin	0.000	0.000	0.000
1	63.618	51.994	18.538
2	71.748	58.638	20.907
3	75.819	61.965	22.094

To monitor the largest reduction part, the weight reduction process divided into several parts such as skin, spars and ribs, and total mass summation as presented in Table 2.5. The payloads and launcher weight are described as external store loads applied in this study.

Table 2.6 shows the final results of the mass reduction for the High Modulus (HM) Graphite/epoxy composite skin wing case. The reason is that the wing mass cannot be reduced further since the flutter speed at lower thickness applied after thickness reduction in Case 3 is very low compared to the baseline. Therefore, the mass optimization process has been ceased until this state.

Based on the data, it is shown that the clean wing weight has been reduced up to 62% from previous aluminum skin through the optimization. Consequently, the structural stiffness of the structure has increased by this application, and hence the flutter speed is technically increased.

2.5.4 *Vibration Modes*

Table 2.7 describes the mode behavior of each vibration mode shape demonstrated in Fig. 2.6. The plotted mode shape influenced by the composite arrangement shown here is essential for computing the aeroelastic flutter speed. For instance, the eigenvector of each mode will be assessed for the structural damping later.

Table 2.7 Mode descriptions of Case 3

Mode	Frequency, f (*Hz*)	Mode description
1	2.592	Wing bending
2	4.652	Wing torsion
3	5.327	Wing bending and torsion
4	6.558	Wing torsion and station 1 store
5	10.254	Wing bending

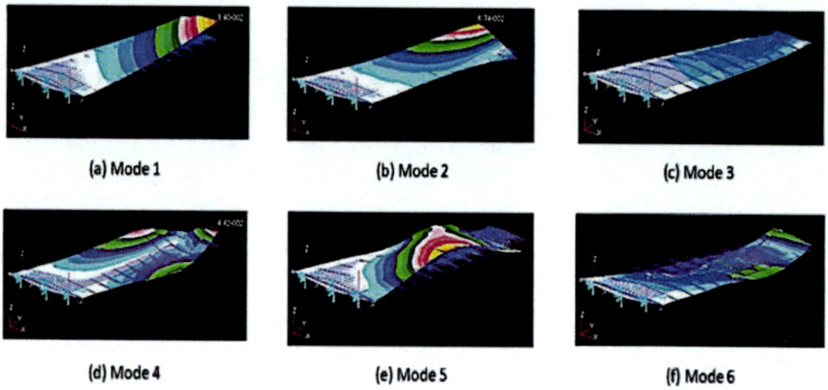

Fig. 2.6 Mode shapes 1–6 of presented Case 3

2.5.5 *Aeroelastic Flutter Assessment*

Subsequent to that vibrational analysis, the oscillating supersonic wing with stores is assessed for the aeroelastic failure. In this case, the flutter speed is predicted by implementing the pk-method of MSC Nastran, where the shell element in the finite element method is coupled with the ZONA 51 of MSC Nastran for supersonic aerodynamic modelling. All mode shapes presented in Fig. 2.6 are analyzed with the existence of aerodynamic loads, and the structural damping of each mode shape is assessed.

Based on the plotted graph in Fig. 2.7, Mode 3 plot has crossed the x-axis where the structural damping is zero (where the assumption of structural failure happens) at speed 585 m/s. However, the graph begins to move into the negative y region (stable state), while the plot is not even reached 3% of the structural damping limit. Hence, the analyzed oscillating supersonic wing with the attached stores is expected to allow an LCO flutter phenomenon occurs at that speed.

The velocity, V versus frequency, f graph is plotted to verify the results as displayed in Fig. 2.8, where the Mode 3 plot is moving near to Mode 2 as the speed increases, starting at 585 m/s. In that case, this phase demonstrates an instability has occurred due to the changes of vibration at that particular speed.

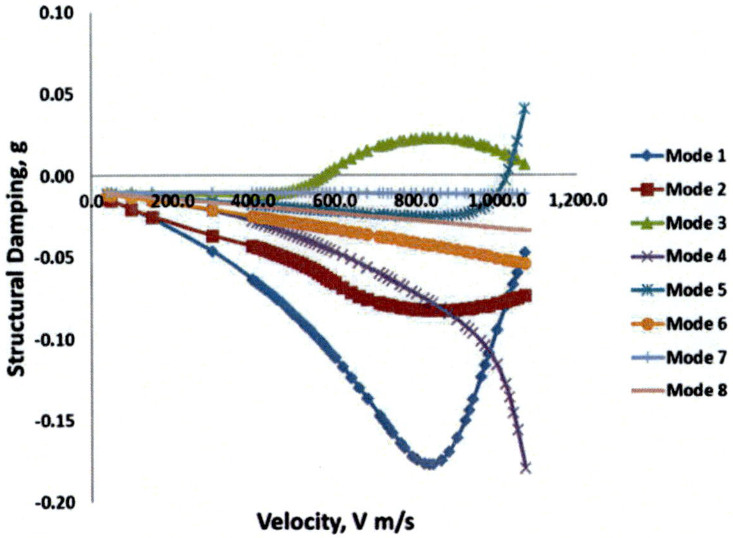

Fig. 2.7 V-g plot for determination of LCO flutter for presented Case 3 at sea level

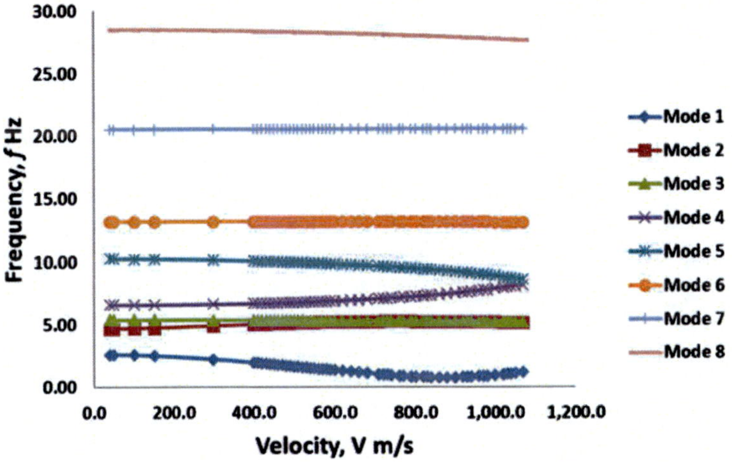

Fig. 2.8 V-*f* plot for determination of LCO flutter for presented Case 3 at sea level

The obtained LCO flutter is extended to observed the trends of flight instability for several altitude variations, presented in Fig. 2.9. From here it is found that changes in term of LCO flutter at different altitudes is due to the influence of altitude density [8].

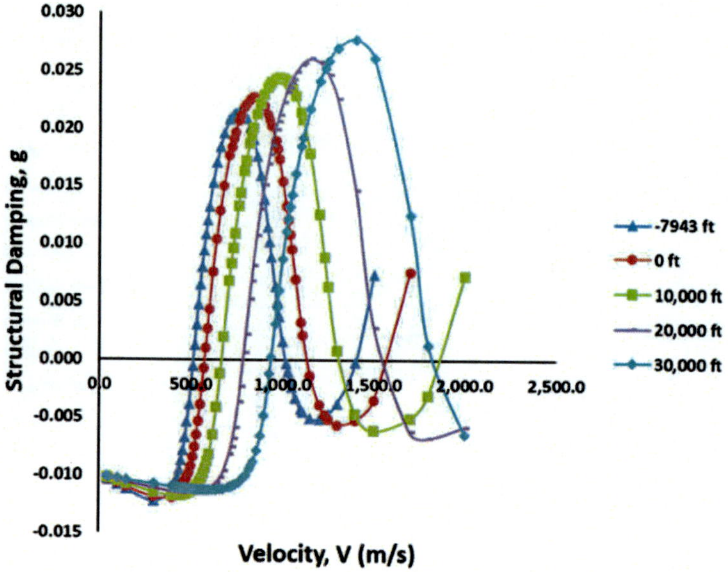

Fig. 2.9 V-g plot for determination of LCO flutter for presented Case 3 at various altitudes

2.6 Conclusion

This paper presents an aeroelastic tailoring approach applied to a supersonic aluminum wing by implementing composite material (HM graphite/epoxy). The aerodynamics modelling of the present supersonic wing is evaluated via the linearized aerodynamic small disturbance theory. The study started with the computational of flutter speed and the assessment of the wing weight. On that sense, the wing structure baseline case has been optimized by constraining the flutter speed at 581.5 m/s and to reduce the weight by the reduction of ply thickness. Based on the finding, Case 3 shows the most substantial reduction of wing mass and highest flutter speed at several altitudes by setting the baseline aluminum skin wing as a constraint. The optimum clean wing weight of the HM Graphite/Epoxy wing is 127.01 kg. The reduction of weight for the HM graphite/epoxy wing skin for the skin weight, clean wing and total wing with missile launcher external stores are 75.82%, 61.96% and 22.09% respectively compared to the baseline aluminum wing. In another hand, the reduction of ply thickness has triggered another type of flutter instability called limit cyclic oscillation (LCO) occurs at 585 m/s due to the changes of vibrational mode that has consequently changed the structural stiffness distribution on the wing. Since the LCO flutter speed of Case 3 is more than the constraint flutter speed (baseline case), hence the optimization process is achieved.

Acknowledgements The authors would like to express their gratitude and thanks to Ministry of Education Malaysia (MOE), Ministry of Higher Education Malaysia (MOHE) and International Islamic University Malaysia for funding this research under the Fundamental Research Grant Scheme for Research Acculturation of Early Career (RACER/1/2019/TK09/UIAM/1).

References

1. Wright J, Cooper JE (2015) Introduction to aircraft aeroelasticity and loads, 2nd ed. Wiley
2. Jones L (1999) Mechanics of composite materials, 2nd edn. Taylor & Francis, Philadelphia
3. Kennedy G, Martins JRRA, Hansen JS (2008) Aerostructural optimization of aircraft structures using asymmetric subspace optimization. In: 12th AIAA/ISSMO multidisciplinary analysis and optimization conference. Victoria, British Columbia, Canada
4. Wu X, Dzenis YA (2001) Rate effects on mode-I delamination toughness of a graphite/eopxy laminated composite. Int J Fract Mech 112:9–12
5. MIL-A-8870C(AS) (1993) Military specification airplane strength and rigidity vibration, flutter, am divergence. Lakehurst, NJ
6. Abdullah NA, Sulaeman E (2014) Aeroelastic tailoring of oscillating supersonic wing with external stores. Appl Mech Mater 464:110–115
7. Rodden WP (1994) MSC/NASTRAN aeroelastic analysis: user's guide, version 68. MSC.Software Corporation
8. Sulaeman E (2012) On the evaluation of negative altitude requirement for flutter speed boundary of transport aircraft and UAV. Appl Mech Mater 225:397–402
9. Azam N, Sulaeman E (2014) Aeroelastic flutter analysis of supersonic wing with multiple external stores. IIUM Eng J
10. Roylance D (2000) Laminated composite plates. Int J Mod Eng Res 15–17
11. Approved Navy Trainning System (2002) Plan for the AIM -9 sidewinder missile system
12. Approved Navy Trainning System (1998) Plan for the AIM-120 advanced medium range air to air missile. Nevada
13. Technology Z (2018) ZAERO versus 9.2 theoretical manual engineers' toolkit for aeroelastic solutions, 3rd edn. ZONA Technology Inc., 9489 E. Ironwood Square Drive, Scottsdale

Chapter 3
Peridynamic Method for Behaviour of Polycarbonate Specimen in Impact Test

M. A. Azizi, A. A. Fahad, and S. A. Rahim

Abstract This paper presents the bond-based peridynamic (PD) method application on predicting polycarbonate (PC) behaviour in Charpy impact test. The purpsose of this application is to apply the capability of the numerical method, i.e. accurate fracture simulation in the prediction of polycarbonate impact fracture behavior. The PD method has integral basis that allows simulation of discontinuities such as crack and dislocation. The material elongation is defined by Ramberg–Osgood formula which includes elastic and plastic deformation properties. The numerical method is validated by comparison with experimental data. There are three types of specimens used in the impact test namely U-notched, V-notched, and No-notch specimens. Two results of the numerical simulation namely force–time curve and crack shape were qualitatively compared with experimental data for validation. Both numerical and experimental data have approximate similarity in force–time curve and the crack shape in specimens. It is shown that the developed PD model for impact test of PC specimens have sufficient similarity with the experimental result and all factors that contributed to the errors were discussed.

Keywords Peridynamic · Impact Test · Fracture Mechanic · Polycarbonate

3.1 Introduction

3.1.1 Background of Peridynamic (PD)

In several fields of mechanical engineering science, fracture mechanics recently is an important feature. Examples of fields requiring high expertise in fracture mechanics are automobile, aerospace, military engineering, and construction structure [1]. A robust method in investigating fracture property of component or material should be acquired.

M. A. Azizi (✉) · A. A. Fahad · S. A. Rahim
Department of Mechanical and Manufacturing Engineering, Faculty of Engineering, Universiti Putra Malaysia (UPM), Serdang, 43400 Seri Kembangan, Selangor, Malaysia
e-mail: muhdazim@upm.edu.my

S. Abdullah et al. (eds.), *Structural Integrity Cases in Mechanical and Civil Engineering*, Structural Integrity 23, https://doi.org/10.1007/978-3-030-85646-5_3

Peridynamic approach has excellent ability to model fracture behaviours, since it has an integral equation basis. PD theory was developed in 2000 by Stewart Silling and it keeps evolving up to now. The key benefit of this approach is to address the drawback of the traditional Finite Element Method (FEM), which unable to model discontinuity such as crack and dislocation because it is based on differential equation. Spatial discontinuity is still valid in method with integral equation basis like PD. Extended Finite Element Method (XFEM) has been implemented to address discontinuity issue, but it still requires complex formulation to define crack propagation behaviour [2].

In PD method, a particle, x only interacts with other particles in the body within the horizon of x as shown in Fig. 3.1. The green circle with radius R is the horizon of particle x. The only particles that interact with particle x are the blue ones, since they are inside the horizon. The horizon has spherical form in 3-dimensional model. Horizon can be defined as a length-scale parameter which gives peridynamic a non-local character. Bond is created to connect two nodes that interact to each other. Each bond consists of material properties such as elasticity, plasticity, and fracture properties.

Fracture in PD happens when some of the bonds in high stress region break. When a bond breaks, stress is concentrated around the broken bond and hence bonds near the broken bond have a tendency to break as well, forming a crack. When broken bonds continue to propagate, total fracture or failure of component may occur. Therefore, crack nucleation and growth in PD is a result of the equations of motion of the material points in the body and not an external crack growth law like in conventional FEM. This PD character makes this new method an ideal candidate for multi-scale material analysis.

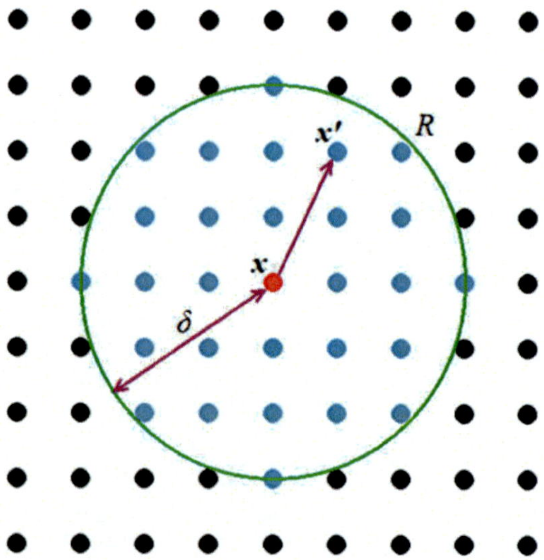

Fig. 3.1 Nodes (blue) that interact with corresponding node (red)

PD method has been applied on various kinds of mechanical tests. Most of the applications were made to investigate the fracture behaviour in each type of the mechanical tests. One of the common tests applied with PD method is impact test since it is closely related to fracture of the specimen. From previous research [3], PD has been applied on various types of impact tests but only few of them are Charpy impact test. From the few application of PD, only one polymer namely polymethyl methacrylate (PMMA) is used as specimen. Therefore, in this work, PD is applied to investigate fracture behavior of other polymer, i.e. polycarbonate (PC) in Charpy impact test.

3.1.2 Charpy Impact Test

One of the standard tests which provide fracture criteria is Charpy impact test. This test is a standardized high strain-rate test that measures the amount of energy absorbed by a material during fracture. Figure 3.2 describes the mechanism of the Charpy impact test machine. Specimen is loosely placed at the holder at the lowest point of the hammer movement. The specimen may have U-notch, V-notch or unnotched. When the hammer is released and hits the specimen, the first maximum height of the hammer is recorded to calculate energy absorbed by the specimen during impact.

This test illustrates whether a material can be categorized as either brittle or ductile. When tested for impact, a brittle material can absorb a small amount of energy, while a ductile material can absorb a large amount of energy. The presence of a fracture

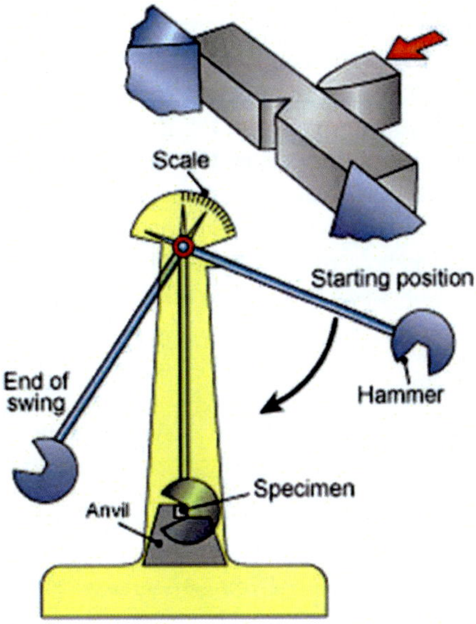

Fig. 3.2 Charpy testing machine and specimen arrangement

surface also provides details about the type of fracture that has occurred; a brittle fracture is bright and crystalline, and a ductile fracture is slender and fibrous. It should be emphasized that Charpy tests are qualitative; the results can only be compared with each other or with a requirement in a specification.

Previously, simulation of Charpy impact test has been performed by using finite element method (FEM). The materials used are mostly metals [4–6]. The outcomes of the simulations are the stress distribution diagram, the force-extension curve, the energy absorbed by the specimen, and the final shape of the specimen. Most finite element simulations in the research did not produce the final shape of the specimen and focused more on the energy absorption criteria. In the papers that include fractured shape of specimen, the crack of the specimen is predefined by refining the meshing at the area in specimen where crack has high possibility to occur [6]. The predefined crack behavior requires complex formulation. By using PD method, the predefined crack behavior is unnecessary, and the crack grows intrinsically depending on the energy distribution [7].

3.1.3 *Polycarbonate Specimen*

In recent years, the demand for PC has become increasingly commercially important to a major engineering plastic. PC is commonly used in blends with other material due to its good compatibility with a range of polymers. PC is amorphous and clear polymers that have 3 main characteristic properties which are toughness, heat resistance and transparency. All these combination properties with good electrical resistance and superior dimensional stability, have led to larger application in electrical parts, optical parts, and automobile parts.

The examination of PC properties is required to increase its reliability in various applications. One of the critical properties that need consideration is fracture property [8]. In this work, PC fracture property is investigated through Charpy impact test, experimentally and numerically. Previously, many investigations on PC behavior undergoing Charpy impact test were carried out [9–11]. Finite element method has been used to simulate the behavior of PC in Charpy impact test but the crack propagation behavior is predefined and did not happen naturally. Therefore, in this work, PD method is applied to achieve better fracture simulation of PC in Charpy impact test.

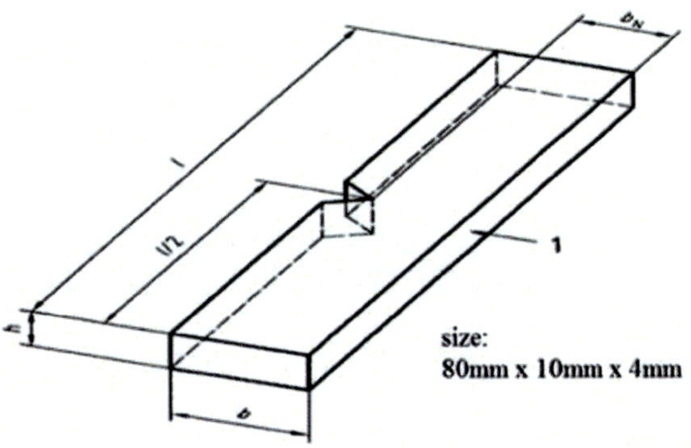

Fig. 3.3 Dimension of Charpy impact test specimen according to ISO 179–1 standard

3.2 Methodology

3.2.1 Charpy Impact Test of PC

Before the PD simulation is performed, the experimental test is executed first. The result of the experimental Charpy impact test is then used as reference and basis of the PD numerical simulation. In this study, three types of specimens are used namely V-notched, U-notched and No-notch specimens. The dimension of the specimen is according to ISO 179–1:2010 standard [12] as shown in Fig. 3.3. For U-notched and V-notched specimen, the notch is 2 mm deep and the notch angle is 45°. There is no notch radius for V-shaped notch while 1 mm notch radius for U-shaped notch.

Following are the procedure of the Charpy impact test:

1. A specimen is placed across parallel jaws in the impact-testing machine.
2. The pointer is set up to 0 J.
3. The hammer is released from the initial height downward towards the sample.
4. Observations and the energy absorbed are recorded and tabulated.

3.2.2 PD Formulation

Silling first proposed the PD theory to overcome the discontinuous problems involving the propagation of fracture and crack [13]. Instead of differentiation, integration is employed in the motion equations to overcome discontinuity problem that appears in FEM due to derivatives of displacement. The differential component in FEM is replaced with integral component as in Eq. (3.1) to form PD formulation. The term $\boldsymbol{f}$ is the bond-force function which describes the deformation properties of the material, $\boldsymbol{u}$ is the displacement of material points, $\boldsymbol{\rho}$ is mass density, $\boldsymbol{H_x}$ is circle of interaction of $\boldsymbol{x}$ with horizon radius of $\boldsymbol{\delta}$, and $\boldsymbol{b}$ is applied body force density field.

$$\rho\frac{\partial^2 u}{\partial t^2} = \int\limits_{H_x} f\big(u(x', t) - u(x, t), x' - x\big)dV_{x'} + b(x, t) \tag{3.1}$$

Bonds are used to describe non-local interactions between material points, and damage in structures appears naturally with the development of broken bonds; therefore, no additional fracture criteria are needed as in FEM when applied to crack growth problems. The PD model can generally be classified as 'bond-based' PD [13] and 'state-based' PD [14], where the former was first proposed while the latter was the refined version. In this work, bond-based PD is applied for the simplicity of the method.

PD concept can be briefly described by a body, *B* as shown in Fig. 3.4. Through a PD bond, only material points, ***x'*** within a prescribed distance from a material point, ***x*** known as the horizon, $\boldsymbol{H_x}$ can interact physically with each other. The horizon is the continuum's characteristic length scale, representing the underlying structure of the material. The radius of the horizon, $\boldsymbol{\delta}$ is unique to each material. In PD method, cracks are modelled as the progressive bond failure between two surfaces.

The material points in a PD model interact with each other directly through the prescribed response or bond-force function, ***f*** which contains all constitutive information associated with the material. The scalar form of the response function, *f* can be defined as Eq. (3.2):

$$f(\eta, \xi) = \frac{\xi + \eta}{|\xi + \eta|}\mu e^{-(|\xi|/l)^2}(c_1 s + c_2\theta) \tag{3.2}$$

The response function given is appropriate for a material behavior at the macroscale. The terms ξ and η are the initial length and elongation of bond respectively as

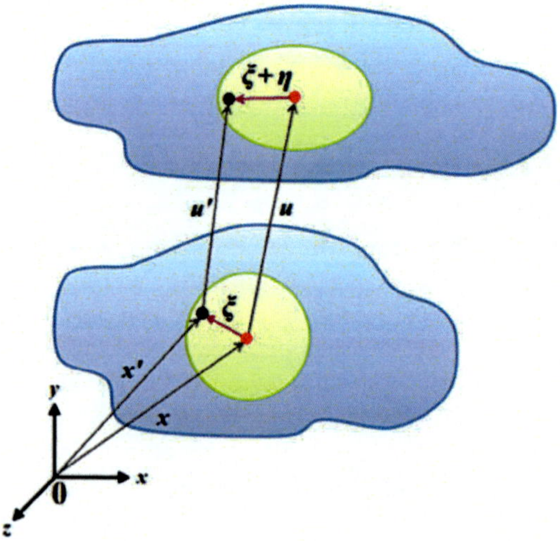

Fig. 3.4 Definition of ξ and η in PD

shown in Fig. 3.4. The response function is linearly dependent on the stretch, s which can be interpreted as the strain in the classical continuum theory. The parameter μ is a step function which depends on the stretch, s and the rupture stretch, s_0 as shown in Eq. (3.3). When s of a bond exceeds s_0, the value of μ change from 1 to 0 and hence the bond is considered as ruptured and eliminated.

$$\mu(\xi, t) = \begin{cases} 1 & \text{if } s(t', \xi) < s_0 \text{ for all } 0 < t' < t \\ 0 & \text{Otherwise} \end{cases} \tag{3.3}$$

The material constant c_1 is the coefficient of elasticity. The value of c_1 corresponding to the classical continuum mechanics bulk modulus k is found by equating the strain energy under isotropic extension from continuum mechanics to the energy density within the horizon in the peridynamic theory for the same deformation [15, 16]. The value of c_1 can be calculated by using Eq. (3.4):

$$c_1 = \frac{18k}{\pi\delta^4} \tag{3.4}$$

In a similar way [15, 16], the rupture stretch, s_0, is related to the energy release rate, G_0. The work to break all the bonds per unit area is considered to be equal to G_0. Thus, the value of s_0 is:

$$s_0 = \sqrt{\frac{10G_0}{\pi c_1 \delta^5}} = \sqrt{\frac{5G_0}{9k\delta}} \tag{3.5}$$

The parameter θ in Eq. (3.2) is the temperature of the bond, while c_2 are the thermal coefficient of material. The temperature is assumed to be constant. Therefore, the temperature component in Eq. (3.2) is neglected and simplified equation is obtained as following:

$$f(\eta, \xi) = \frac{\xi + \eta}{|\xi + \eta|} e^{-(\frac{|\xi|}{l)^2}} c_1 s \mu \tag{3.6}$$

Furthermore, the forces acting among material points decrease with increasing distances; examples of such forces are atomic interactions, gravity, and Lorentz force. Therefore, the response function includes a normal distribution, i.e. the exponential component in Eq. (3.6), which allows the forces to decrease with increasing distance between a pair of material points.

3.2.3 PD Modelling for Charpy Impact Test

The PD programming structure can be briefly described by a flowchart in Fig. 3.5a. The numerical programming of PD method in Matlab software consists of two

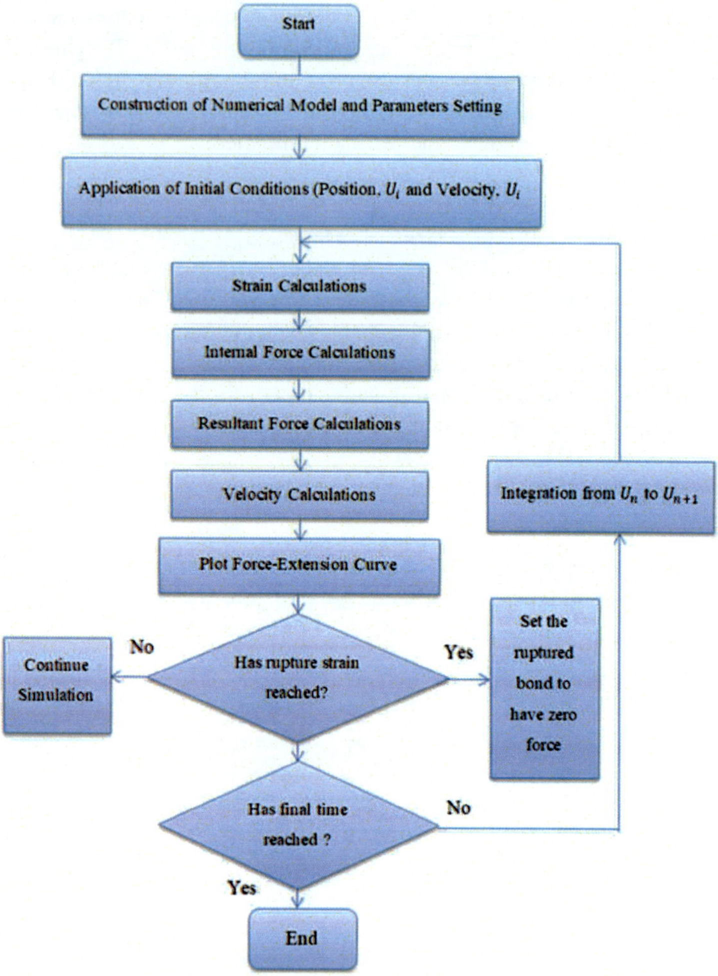

Fig. 3.5 Flow chart of PD numerical method of Charpy impact test

main compartments namely numerical construction and numerical calculation. In the construction file, the dimensions of the numerical specimen are customized according to the real specimen. The size of Charpy impact test specimen is as shown in Fig. 3.3. The parts of the numerical specimen which applied with boundary conditions were identified. In creating notch within the specimen, the coordinates of the nodes inside the notch area were identified for deactivation of the bonds related to the nodes.

Then the boundary conditions were added, i.e. the holders and the hammer as shown in Fig. 3.6. The holders and hammer are considered as undeformable blocks. The positions of the holders are fixed in all directions while the hammer is moving towards the specimen direction, i.e. downward direction in Fig. 3.6 in constant speed.

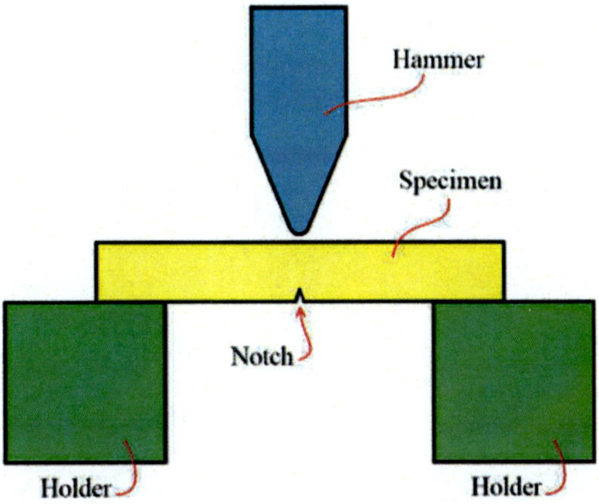

Fig. 3.6 Top view of specimen and boundary conditions in PD Charpy impact test

The speed of the hammer is approximated from the experiment. The simulation is completed once the hammer has fully penetrated the specimen which means the specimen has totally split into two pieces.

In the simulation file, the applied force and the material constants are used to calculate force and strain of each bond, displacement, and velocity of each node. From the calculated strain combined with the external load during a numerical cycle, the force in bond for the following cycle can be obtained. This cycle continues until the final time reached. If the bond strain reaches rupture strain value, the force in the bond is set to be zero denoting the rupture of the bond. After the sets of data are obtained, the force-extension curve can be plotted, and then it can be compared with the experimental curve.

3.2.4 Material Constitutive Equation

PC has brittle and ductile behaviours depending on several factors such as surrounding temperature, crystallization, compounding, and processing features [17, 18]. Ductile elongation constitutive equations are considered since it can also represent brittle behaviour by using suitable values for the parameters. For ductile polymers, the yield point is usually indefinite. The suitable constitutive equation for this type of behaviour is Ramberg–Osgood model as following:

$$\varepsilon = \frac{1}{\eta}\left[1 + \frac{\alpha}{\sigma_R}\left|\frac{\sigma}{\sigma_R}\right|^{m-1}\right] \tag{3.7}$$

where η, α, σ_R, and m are parameters of material [19]. The stress–strain curve of the Ramberg–Osgood model is presented in Fig. 3.6. Since bond-force, f and stretch, s

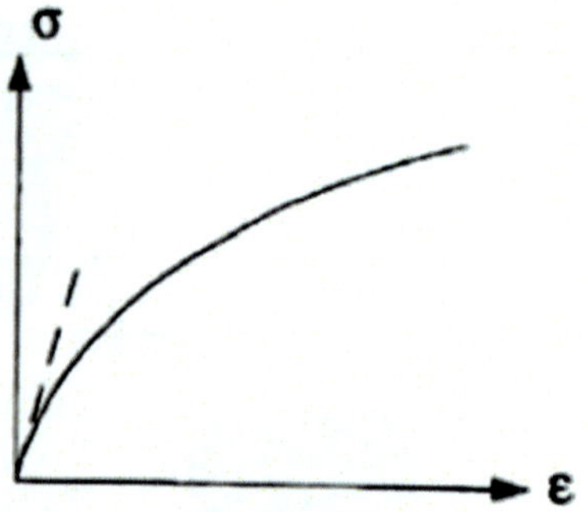

Fig. 3.7 The Ramberg–Osgood model [19]

are used in PD instead of stress, σ and strain, ε, hence σ is replaced with, f/A where A is cross-sectional area of bond and ε is replaced with s. For simplicity, assume $A = 1$ and Eq. (3.7) is converted into:

$$s = \frac{1}{\eta}\left[1 + \frac{\alpha}{\sigma_R}\left|\frac{f}{\sigma_R}\right|^{m-1}\right] \tag{3.8}$$

when Eq. (3.8) is compared with Eq. (3.6), ignoring the exponential component and μ, the function of c_1 with ductile elongation properties is obtained as follows:

$$c_1 = \eta\left[1 + \frac{\alpha}{\sigma_R}\left|\frac{f}{\sigma_R}\right|^{m-1}\right]^{-1} \tag{3.9}$$

It is clear that the value of c_1 is changing with the increment of f to obtain nonlinear plastic elongation as shown in Ramberg–Osgood model in Fig. 3.7.

3.3 Result and Discussion

Result consists of comparison between experimental and numerical PD data. Three types of specimens are involved in this comparison. The overall experimental result is displayed in Table 3.1. For each type of specimen, the impact energies measured during test for 10 specimens are averaged. The conditions of the specimen were also observed and are shown in Table 3.1. The highest energy absorption occurs in No-notch specimen, followed by U-notched and V-notched specimens.

Table 3.1 Experimental result of impact test on PC

Specimen type	Average absorbed energy (J)	Specimen condition
No-notch	12.08	Bent
U-notched	6.82	Broken
V-notched	3.50	Broken

3.3.1 Load Curve Comparison

The energy absorption curve can be obtained in an impact test by planting a sensor at the striker of the impact test machine. From a previous research of PC material [10], a load-time curve for Charpy impact test was chosen as a reference as shown in Fig. 3.8. The load curve started from 0 N up to maximum value in less than 2 ms. Then the load decreases back to zero within 10 ms. This is the common shape of load-time curve produced from Charpy impact test.

From the PD simulation, the internal force of the whole specimen can be calculated, and load-time curves were produced. The load curves for V-notched, U-notched, and No-notch specimens are shown in Fig. 3.9a, b and c. All load curves of the PD simulation start at 0 N and raise to their maximum value. Then the load or internal force decrease after the numerical specimens fully penetrated. Although the shapes of the numerical curves are not as smooth as the experimental curve in Fig. 3.8, the start and the end of the curves have good similarity with the experimental curve, i.e. start and end with zero load. The suspected factor of the high fluctuations of the numerical curves is the discretization of the specimen, where the number of nodes is insufficient [20]. Low number of nodes causes the simulation of crack propagation to be jagged. Sufficient number of nodes can solve this problem but it will increase computational time or require high-performance computer.

The maximum load value of the U-notched specimen, i.e. ±14 kN is higher than the one for V-notched specimen, i.e. ±12 kN, similar to expectation based on experimental data. However, the maximum load value for No-notch specimen, i.e. ±9 kN is far from expectation since it has the lowest value among the three specimens.

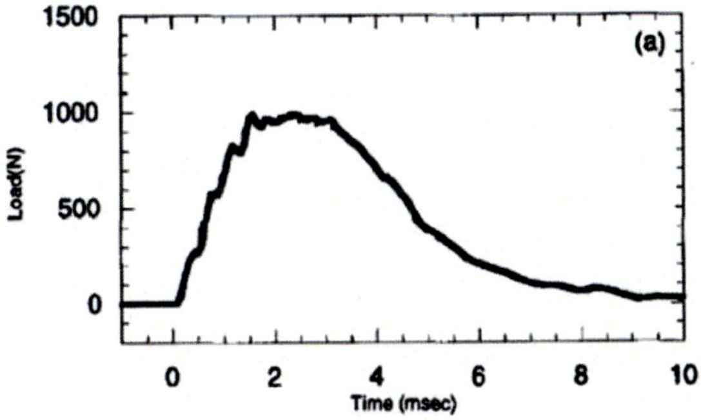

Fig. 3.8 Typical load-time curve shape obtained in PC Charpy impact test [10]

Fig. 3.9 PD simulation load-time curves of **(a)** V-notched specimen, **(b)** U-notched specimen, and **(c)** No-notch specimen

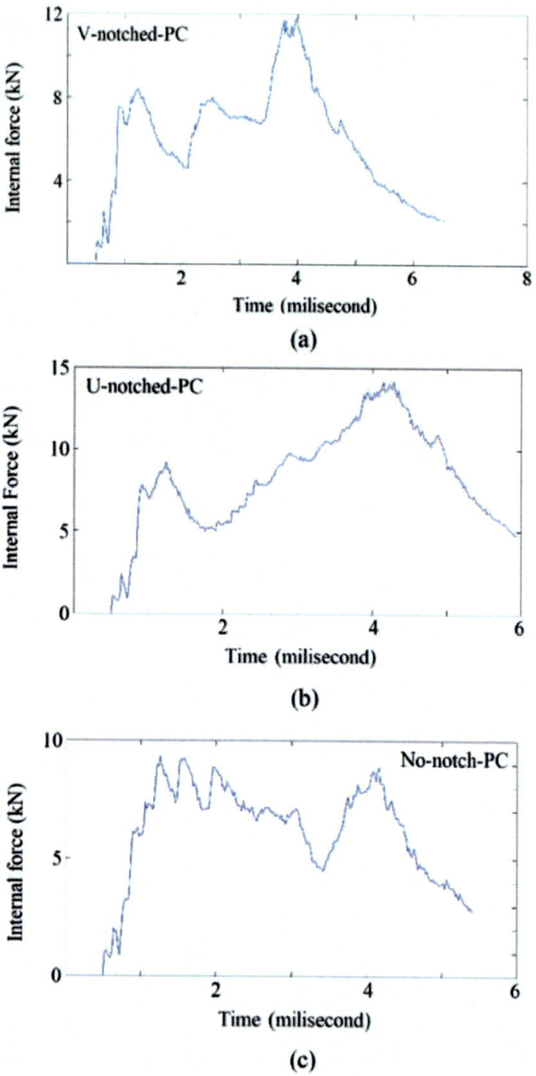

This error occurs due to the lack of plasticity properties in the material formulation of PC in the PD model. Polymeric specimen with no notch has a tendency to have wide plastically deformed region [11]. It is due to no stress concentration appears during the test.

In solving the above-mentioned shortcomings, firstly the number of nodes should be increased to reduce the fluctuation of the load-time curve. Secondly, plasticity formulation should be included in the material component of PD. This action is to improve simulation of impact fracture that includes significant plastic deformation.

Besides, the area under the curve should be analysed as well to calculate total force energy absorbed by the specimens.

3.3.2 *Post-fracture Specimen Shape Comparison*

Another comparison between numerical and experimental results is specimen shape after test is finished. Figure 3.10a, b and c show the shape of V-notched, U-notched and No-notch specimens respectively after experimental impact test completed. While Fig. 3.10d, e and f are the shape of V-nothed, U-notched and No-notch specimens respectively at the end of numerical impact tests. The red-coloured nodes denote the number of bonds that have broken for the nodes. The more intense the red colour of a node, the greater number of bonds have broken for the node.

In experimental V-notched specimen test, the crack propagates in straight direction, parallel to the impact direction. The fracture surface is smooth. In the numerical

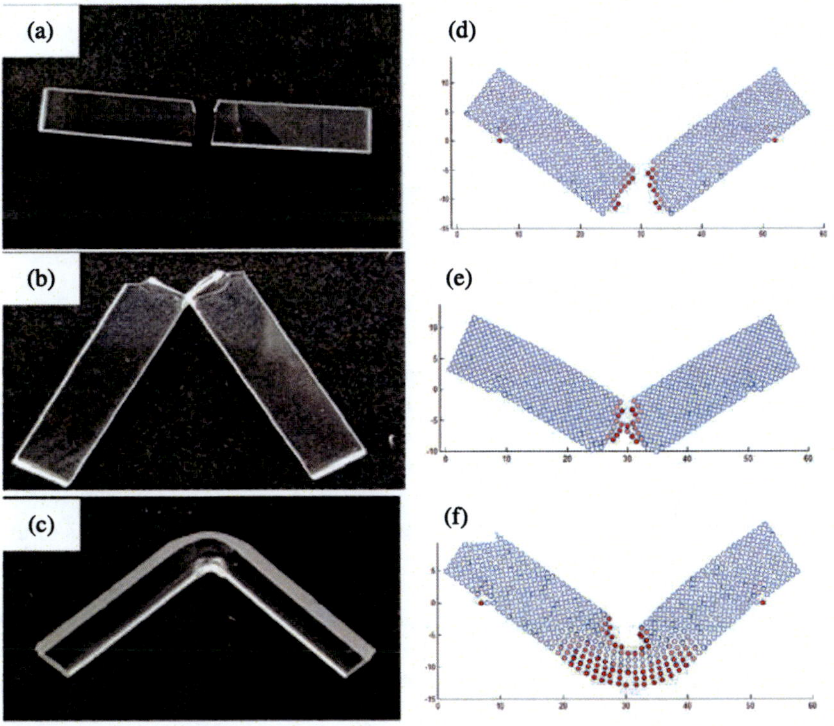

Fig. 3.10 Shape of specimens after impact test finished

specimen for the same test, the crack propagates in direction parallel to impact direction as well, but the fracture surface is a little rough. In experimental U-notched specimen test, the crack also propagates in straight direction, parallel to the impact direction. The fracture surface is rougher than the one in V-notched specimen. In the numerical specimen, the crack propagates in direction parallel to impact direction and the fracture surface is slightly rough. Finally, in experimental No-notch specimen test, the crack propagates in ambiguous direction since the specimen is not totally fractured. This is due to the hammer weight was not sufficient. The plastically deformed region is identifiable, i.e. at the location of impact. In the numerical specimen, the crack propagates in ambiguous direction as well. The fracture surface is rough, and the damaged or plastic region is big as shown in Fig. 3.10f. These behaviours are in good agreement with the previous about effect of notches [21].

The qualitative similarity between experimental and numerical results in term of shape of specimen after failure can be considered as good. This is because in the numerical result, the location of crack, the roughness of crack surface, bending, and plastic region are identical with the experimental result.

In order to improve the crack propagation direction and fracture surface roughness, the number of nodes should be increased to have detailed and more accurate crack propagation criteria [22, 23]. However, the discretization should consider the capability of computer since increased number of nodes means increased computational cost. Regarding the unbroken specimen, the experimental test should use suitable hammer weight to allow complete impact test, where the specimen split into two. Alternatively, for more accurate PD simulation, the weight of the hammer should be considered. Hence unbroken specimen can be achieved in simulation.

3.4 Conclusion

The PD model has successfully simulated the impact test on PC specimens. The different types of specimen, i.e. different notch shapes have different strength and fracture properties. The PD numerical model able to adequately predicts the strength and the load-time curve of the 3 types of specimen. The PD numerical model also sufficiently simulates the differences in term of crack direction, crack surface roughness, internal force of the specimen, and damage region. Several drawbacks of the current PD method were identified and methods of solving them were discussed.

Acknowledgements This research work is fully sponsored by Inisiatif Putra Muda grant provided by Universiti Putra Malaysia.

References

1. Foster JT, Silling SA, Chen WW (2010) Viscoplasticity using peridynamics. Int J Numer Meth Eng 81(10):1242–1258
2. Ha YD, Bobaru F (2010) Studies of dynamic crack propagation and crack branching with peridynamics. Int J Fract 162(1–2):229–244
3. Isiet M, Mišković I, Mišković S (2021) Review of peridynamic modelling of material failure and damage due to impact. Int J Impact Eng 147:103740
4. Yamamoto T, Odette GR, Lucas GE, Matsui H (2020) Confocal microscopy–fracture reconstruction and finite element modeling characterization of local cleavage toughness in a ferritic/martensitic steel in subsized Charpy V-notch impact tests. J Nucl Mater 283–287(2):992–996
5. Kunigita M, Aihara S, Kawabata T, Kasuya T, Okazaki Y, Inomoto M (2020) Prediction of Charpy impact toughness of steel weld heat-affected zones by combined micromechanics and stochastic fracture model—part I: model presentation. Eng Fract Mech 230:106965
6. Cao Y, Zhen Y, Song M, Yi H, Li F, Li X (2020) Determination of Johnson–Cook parameters and evaluation of Charpy impact test performance for X80 pipeline steel. Int J Mech Sci 179:105627
7. Azizi MA, Ariffin AK (2019) Peridynamic model for nonlinear viscoelastic creep and creep rupture of polypropylene. J Mech Eng Sci 13(4):5735–5752
8. Foster JT, Silling SA, Chen W (2019) An energy based failure criterion for use with peridynamic states. Int J Multiscale Comput Eng 9(6)
9. Seldén R (1987) Fracture energy measurements in polycarbonate and PMMA. Polym Test 7(3):209–222
10. Kayano Y, Keskkula H, Paul DR (1998) Fracture behaviour of polycarbonate blends with a core-shell impact modifier. Polymer 39(4):821–834
11. Graupner N, Kühn N, Müssig J (2021) Influence of sample thickness, curvature and notches on the Charpy impact strength—an approach to standardise the impact strength of curved test specimens and biological structures. Polym Test 93:106864
12. ISO 179–1:2010—plastics—determination of Charpy impact properties—part 1: non-instrumented impact test
13. Silling SA (2000) Reformulation of elasticity theory for discontinuities and long-range forces. J Mech Phys Solids 48(1):175–209
14. Silling SA, Bobaru F (2020) Peridynamic modeling of membranes and fibers. Int J Non-Linear Mech 40(2–3):395–409
15. Silling SA, Askari E (2005) A meshfree method based on the peridynamic model of solid mechanics. Comput Struct 83:1526–1535
16. Macek RW, Silling SA (2207) Peridynamics via finite element analysis. Finite Elem Anal Des 43:1169–1178
17. Swallowe GM (2013) Mechanical properties and testing of polymers: an A–Z reference. Springer Science & Business Media
18. Gonzalez DG, Rusinek A, Bendarma, A, Bernier R, Klosak M, Bahi S (2020) Material and structural behaviour of PMMA from low temperatures to over the glass transition: quasi-static and dynamic loading. Polym Test 81:106263
19. Irgens F (2008) Chapter 9: viscoelasticity. Springer-Verlag, Berlin, Continuum Mechanics
20. Azizi MA, Ariffin AK, Mohamed NAN (2015) The peridynamic model of viscoelastic creep and recovery. Multidiscip Model Mater Struct 11:579–597
21. Bura E, Seweryn A (2018) Mode I fracture in PMMA specimens with notches—experimental and numerical studies. Theoret Appl Fract Mech 97:140–155
22. Freimanis A, Paeglitis A (2017) Mesh sensitivity in peridynamic quasi-static simulations. Procedia Eng 172:284–291
23. Henke SF, Shanbhag S (2014) Mesh sensitivity in peridynamic simulations. Comput Phys Commun 185(1):181–193

Chapter 4
Enhancement of Mechanical Properties for AZ31B Quenching in Nano Fluid

M. M. Mubasyir, M. F. Abdullah, K. Z. Ku Ahmad, R. N. I. R. Othman, and A. H. Isahak

Abstract This paper focuses on the enhancement of the mechanical properties of AZ31B magnesium alloy. The experimental approach was constructed to determine the suitable factor combination of temperature, soaking time, and the medium containing nanoparticle on magnesium alloy surface structure to improve the mechanical properties using quenching in nanofluids. Twelve dogbone-shaped AZ31B-based specimens were tested. The diameter and thickness of the specimen were 6 and 10 mm, respectively. All specimens underwent heat treatment in a furnace at 260 and 350 °C and a soaking time of 30 and 60 min. The samples were then submerged into three different solutions: distilled water, distilled water containing 0.1 wt % wt carbon nanotube (CNT), and distilled water containing 0.1 wt % nano silica. The addition of CNT and nano silica improved the ultimate tensile strength (UTS) at 18.86% and 11.59% and the yield strength at 66.47% and 57.08%, respectively, control sample. Microscopy was performed to correlate the addition of nanoparticles in the heat treatment and the enhancement for mechanical properties by examining the depth region affected. These show promising industrial applications such as in automotive, aerospace, and even in the military to maintain low weight with high tensile strength. Therefore, this paper reports the effects of heat treatment and CNT and nano silica in enhancing magnesium alloy mechanical properties (UTS and yield strength) using the quenching method.

Keywords AZ31B · CNT · Nano Silica · Quenching · Microscopy

4.1 Introduction

Magnesium alloy is a lightweight alloy that is highly demanded in many industries such as aerospace, structure, and military defence because it is the lightest structural metal alloy [1]. The magnesium alloy density is far lower than other materials such

M. M. Mubasyir · M. F. Abdullah (✉) · K. Z. Ku Ahmad · R. N. I. R. Othman · A. H. Isahak
Department of Mechanical Engineering, Faculty of Engineering, Universiti Pertahanan Nasional Malaysia, Kem Sungai Besi, 57000 Kuala Lumpur, Malaysia
e-mail: m.faizal@upnm.edu.my

S. Abdullah et al. (eds.), *Structural Integrity Cases in Mechanical and Civil Engineering*, Structural Integrity 23, https://doi.org/10.1007/978-3-030-85646-5_4

as aluminium and steel, whose densities are approximately 35% and 77% higher, respectively, compared to magnesium [2, 3].

Magnesium alloy is one of the recommended metallic alloy materials for aircraft or vehicle applications. It has low weight and good mechanical properties, which reduces fuel consumption. However, compared with conventional materials, such as steel, not much research has been carried out on the relationship between magnesium alloys and impact loading [4], especially under high-velocity impact forces.

Hexagonal close-packed (HCP) is well known in magnesium's structure. In HCP, inorganic compounds, larger atoms (or ions) occupy positions corresponding approximately to equal spheres in close packing, while smaller particles are distributed among the voids [5]. For instance, the presence of voids in such material, when such materials were filled into such pores if the materials are essential to avoid structure collapses, like carbon nanotubes (CNTs) and nano silica that both have nano-material structure. CNT has unique properties that can produce solid materials and improve the mechanical properties of other materials [6–8]. The molecular nanotechnology of CNT and nano silica can fill the pores and voids in a structure and produce Van der Waals bonds [4, 9, 10]. The fill-up nanoparticles on to material were currently used quenching method surface roughness quenched faster. It depends on particle size, as mentioned [12], the study on nanoparticle deposition on the minimum heat flux. It satisfies front speed during quenching in water-based nanofluids. Inspired by the idea of using nanofluids as quenchants, CNT nanofluids were used as quenchants, and the preparation method played a vital role in determining the heat transfer rates during quenching [13].

The quenching method will improve the mechanical properties of magnesium alloy [11]. Magnesium is used for high-velocity impact applications because of its enhanced energy absorption and shock reduction. However, the current methods for improving its mechanical properties still could not reach the level of Hardox steel, a common material used for armour plating. Heat treatment such as quenching is sufficient to enhance the mechanical properties of magnesium alloys. This alloy has excellent potential for armour plating to reduce armour weight while improving the fuel consumption efficiency of armoured vehicles and providing high-velocity resistance. However, the structural distribution of AZ31B has low strength to withstand the impact deformation structure. A mixture of nanoparticles such as CNT and nano silica on magnesium alloy could demonstrate a high depletion of ballistic resistance. Therefore, implementing CNT and nano silica to AZ31B may increase the impact resistance. The results will be given the correlation of mechanical properties improvement between mixture nanoparticles on the magnesium alloy.

This material also has a disadvantage in ductility, which can be changed with nano-materials in its structure. This research investigated the reaction of magnesium alloy with the nano-material mixture using the quenching method. The effectiveness of the quenching method on magnesium alloy and the improvement in the mechanical properties of AZ31B by this method were analysed by looking at the factors that influence its performance in terms of CNT percentage, nano silica percentage, soaking time, and temperature.

4.2 Methodology

The frameworks of the methodology in this study are constructed in Fig. 4.1. Characterization of each process and factors were illustrated. The heat treatment process on the selected magnesium alloy specimen, AZ31B, varied in terms of the soaking time and temperature of the furnace based on the factorial design. Combining CNT or nano silica into distilled water to obtain a nanofluid medium for the specimen to be submerged is known as the quenching process, enhancing the properties of the specimen. A tensile test was carried out to determine the ultimate tensile strength (UTS) and study the enhancement of mechanical properties, which will later compare with the teat treatment and quenching in nanofluids. The elongation applied in each experiment was 2 mm/min. Theoretically, metal alloys that undergo heat treatment will achieve a higher tensile strength but, to some extent, will reduce their elongation due to attaining a much brittle property [14]. Thus, the factors listed in contributing to the specific mechanical properties for the experiment are tensile strength and yield strength.

4.2.1 Experimental Procedure

The material used for the experiments was magnesium alloy AZ31B. The chemical composition of the alloy is presented in Table 4.1.

Figure 4.2 presents the dog bone-shaped specimen for the tensile test, which was prepared based on ASTM E8. The cross-section and the length of the specimen were

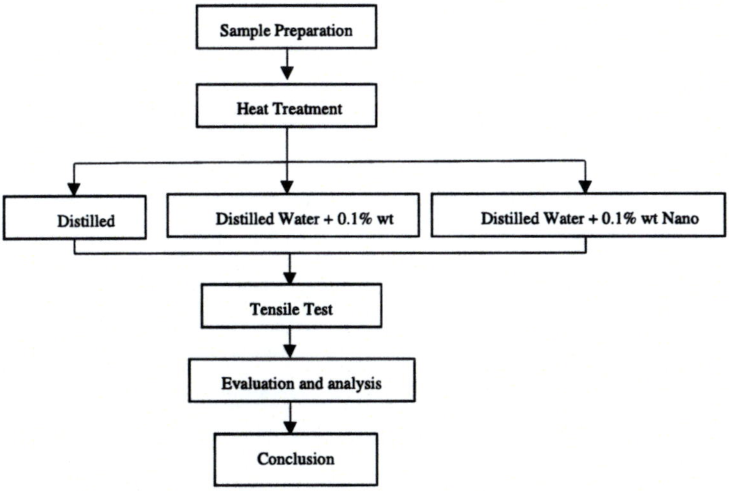

Fig. 4.1 Flow diagram of the methodology

Table 4.1 Chemical composition of Mg AZ31B alloy [3]

Al	Zn	Mn	Fe	Cu	Si	Ni	Mg
3.28	1.00	0.44	0.0029	0.000039	0.025	0.00023	Balance

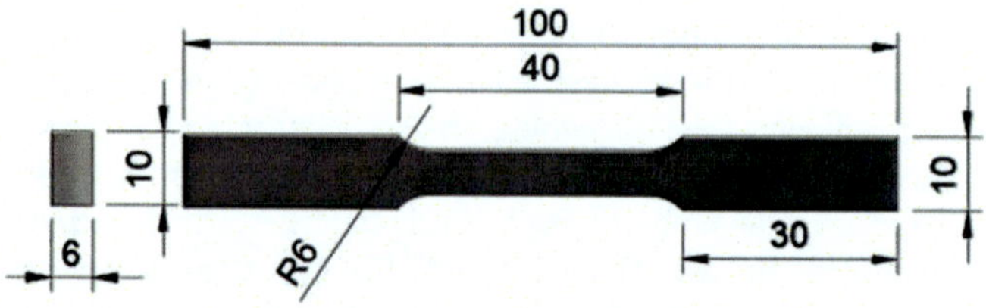

Fig. 4.2 Geometrical properties of the specimen used for tensile test [17]

6 mm × 10 mm and 100 mm, respectively. For the control sample, a tensile test on specimen AZ31B gave a value of 220 and 138 MPa for UTS and yield strength, respectively.

All specimens were classified into 12 specimens evaluated in two manipulated process factors from the heat treatment factor and three media that involves distilled water and nanocomposite to obtain nanofluids for the quenching process. Table 4.2 lists the categorizations of each specimen.

As mentioned in Fig. 4.1 and Table 4.2, the heat treatment process increases the magnesium alloy materials properties [11]. Focusing on the two factors to analyse the most suitable heat treatment process that enhances the specimen, especially in terms of energy absorption. The selected heat treatments were at 260 and 350 °C due to several reasons: the melting point of magnesium alloy AZ31B is 364 °C [9],

Table 4.2 Categorization of specimen

No. of Specimen	Temp (C°)	Soaking time (min)	Medium fluid composition (500 ml)
1	260	30	Distilled water
2	260	60	Distilled water
3	350	30	Distilled water
4	350	60	Distilled water
5	260	30	Distilled water + 0.1% wt CNT
6	260	60	Distilled water + 0.1% wt CNT
7	350	30	Distilled water + 0.1% wt CNT
8	350	60	Distilled water + 0.1% wt CNT
9	260	30	Distilled water + 0.1% wt Nano Silica
10	260	60	Distilled water + 0.1% wt Nano Silica
11	350	30	Distilled water + 0.1% wt Nano Silica
12	350	60	Distilled water + 0.1% wt Nano Silica

heating beyond the melting point will deform the structure of the specimen, 260 °C is the most suitable temperature for tempering treatment [4], hypothetically cooling it instantly will give in nanofluids might give potential outcomes. Later the treated specimens were submerged in a water-based nanofluid solution for quenching.

4.2.2 *Experimental Procedure*

As CNT and nano silica are hydrophobic, they are difficult to be dispersed directly [13]. Due to the delicate size of the nanocomposites, a certain amount of energy is required to breach the intermolecular forces between each particle. This will cause the nanocomposite to agglomerate even though it is mixed in distilled water. As a result, the nanofluid is not dispersed uniformly throughout the medium, which will affect the results. The nanofluid preparation underwent a certain procedure to ensure that the nano-materials composition in the medium is dispersed uniformly [15]. The quenching medium was prepared with 500 mL of distilled water + CNT/nano silica. Figure 4.3 shows the furnace used for heat treatment. Figure 4.4 displays the schematic diagram of the quenching process for the magnesium alloy, AZ31B in distilled water with the addition of either 0.1% wt CNT or nano silica as nanofluids.

The nanofluids were sonicated at 20 kHz for 15 min to get an even dispersion of CNT/Alumina in distilled water using a probe-type ultrasonic processor (QSonica 700 Sonicator) [16]. Quench probes of magnesium alloy were used in this study to study the effect of quenching in CNT/nano silica nanofluids. The chemical composition of the magnesium alloy was measured at the first stage. The quench probes were heated to near semisolid magnesium alloy temperature. After soaking for several minutes, it was quenched in three different media, as mentioned in Table 4.1.

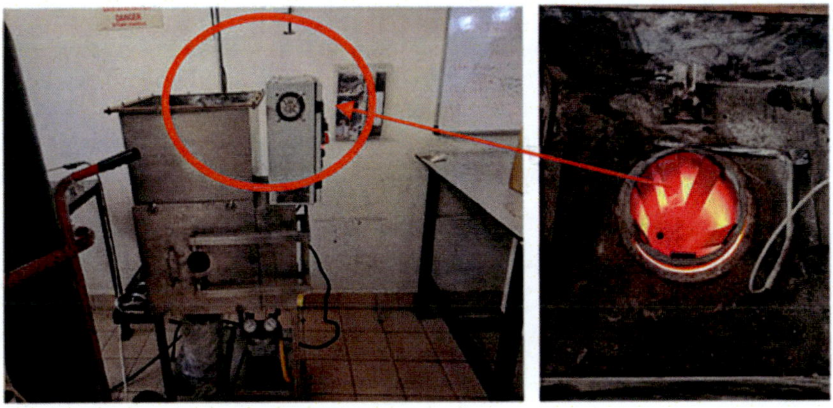

Fig. 4.3 Furnace heat treatment for specimen

Fig. 4.4 Schematic diagram of AZ31B quenching in nanofluid

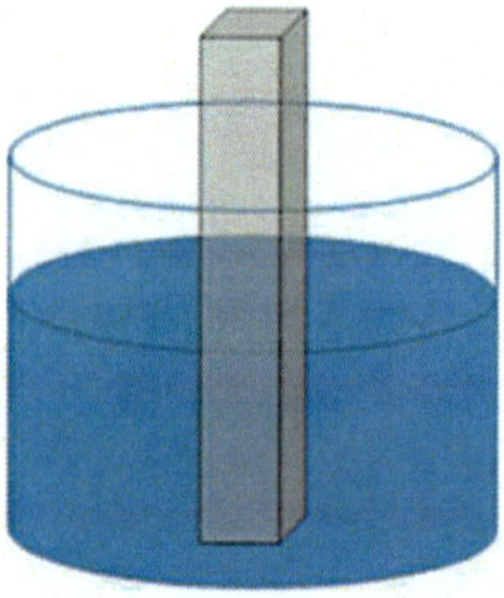

Instron the machine used to perform the tensile test on the specimen later analyse the correlation of the energy absorption, known as tensile energy absorption. Measurements were taken upon the capacity of a material to withstand a shock when subjected to sudden high tension. The tensile force applied to a material is graphed against the percentage elongation of the specimen due to the applied forces.

4.2.3 Preparation of Optical Metallographic

Optical microscopy was used to characterize structure grain boundaries, phase boundaries, and inclusion distribution and observe mechanical deformation [18]. Because the macro and microstructure of metals and alloys often determine the behaviour of the material, characterization of the effects of composition, processing, service conditions, and other variables on the macro and microstructure is frequently required [17, 18].

After the tensile test, each specimen was subjected to metallurgy analysis. Hot mould, grinding, polishing, and etching are the few steps before the specimen sample can be examined under the microscopy instrument. The macroanalysis of the specimen focused on the depth region of the specimen from the surface towards the affected region due to the heat treatment and presence of nanoparticles. It flowed lines of the specimen later relating to improving mechanical properties stated from the tensile test. Figure 4.5 shows the optical analysis for the specimen's cross-section

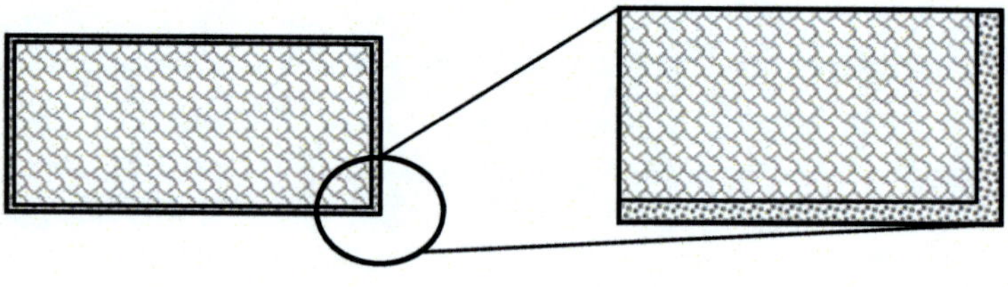

Fig. 4.5 Schematic diagram of optical analysis

to analyse the depth region and flowed line structure of the specimen. Solution DT software was used to analyse and measure the depth region and flowed line of each sample.

4.3 Results and Discussion

4.3.1 Tensile Test

Specimens 4, 8, and 12 are the sample under these conditions. The optical metallurgical analysis, which focuses on depth regions affected by the heat treatment and presence of CNT and nano silica, was around 55 and 22 μm, rather than that of the specimen soaked in distilled water (36 μm). CNT has a tendency of Van der Waals bonds to attach to the pores on the surface of the AZ31B [6, 7]. Meanwhile, nano silica contributes slightly to both UTS and yield strength increment but not as effective compared to CNT. Since nano silica is a hydrophobic material [18], it acts as an excellent coating to avoid corrosion for AZ31B. However, it does not excel in improving the mechanical properties compared to quenching in distilled water. The flowed line also displayed a significant difference with CNT ranging around 5–35 μm and nano silica meandering around 2–17 μm. Meanwhile, in distilled water without the presence of nanoparticle, the flowed line obtained around 3–47 μm. The presence of void in specimen 4 was more evident than in specimens 8 and 12.

Relating from the tensile test as scored specimens in ascending order 12, 4, and 8, respectively show a similar pattern compared with the depth region measurement affected by the quenching process in the selected medium. As for the analysis, flowed lines are more related to the capability of heat from specimen dissipated into the medium. Nevertheless, specimen 8 under the heat treatment and soaking time of 350 °C and 60 min displayed the most optimum outcome related to the diffusion depth of the nanoparticles from the surface of the AZ31B magnesium alloy when submerged in a 0.1% CNT quenching medium.

Specimens 1–4 were treated in distilled water, specimens 5–8 were treated in 0.1 wt % CNT in distilled water, and specimens 9–12 were treated in 0.1 wt % nano silica in distilled water. A shown in Fig. 4.6, the specimens with more than 250 MPa for UTS were treated in the CNT solution. The heat treatment of specimens at 350 °C for 1 h all passed the 8 value of strain as we analyse the following tables related to the specimen results. The potential approach of enhancing magnesium alloy with nanofluid becomes plausible (Table 4.3).

Based on UTS, specimen 8 scores the highest percentage differences at 18.86% with the heat treatment factor of soaking time 60 min and heat temperature of 350 °C when compared to the control sample. Regarding specimens 4 and 12 with the same heat treatment factor, only scored 11.93% and 11.59%, respectively. Thus, 0.1% wt CNT contributed to the magnesium alloy specimen enhancement in terms of ultimate tensile strength. Theoretically, the CNT solution enhances UTS.

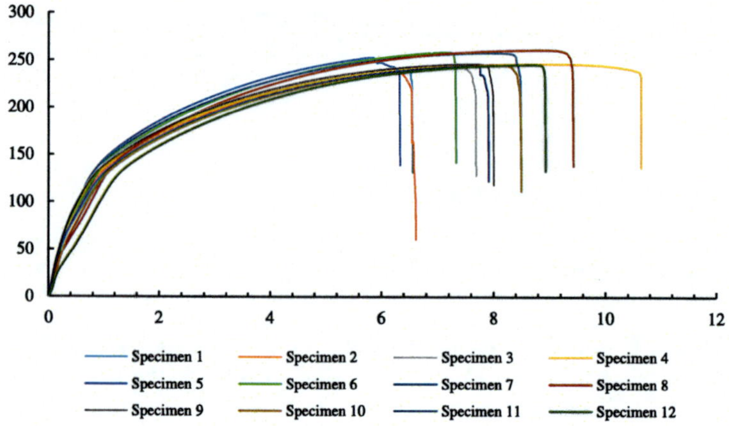

Fig. 4.6 Stress–Strain for all 12 samples

Table 4.3 Characterization of Specimen

Specimen	UTS (MPa)	Percentage of differences	Yield Strength (MPa)	Percentage of differences
Control sample	220	–	138	–
1	239.08	8.67%	210.390	52.46%
2	236.95	7.70%	208.518	51.10%
3	243.87	10.85%	215.270	55.99%
4	246.25	11.93%	216.704	57.03%
5	252.77	14.90%	222.435	61.18%
6	258.55	17.52%	227.527	64.87%
7	257.78	17.17%	226.850	64.38%
8	261.05	18.86%	229.726	66.47%
9	246.32	11.96%	216.769	57.08%
10	245.17	11.44%	215.747	56.34%
11	243.95	10.89%	214.674	55.56%
12	245.49	11.59%	216.029	56.54%

Based on yield strength, specimen 8 also scores the highest percentage differences at 66.46% for yield strength. Regarding specimens 4 and 12 with the same heat treatment factor, only scored 57.03% and 56.54%. Thus, 0.1% wt CNT contributed to the magnesium alloy specimen enhancement in terms of ultimate tensile strength. Theoretically, the CNT solution excels in enhancing the specimen ultimate tensile strength.

According to each respective medium, the repetitive pattern of towards best specimen can also be seen in Fig. 4.7, where between specimen 1 till 4 in distilled water

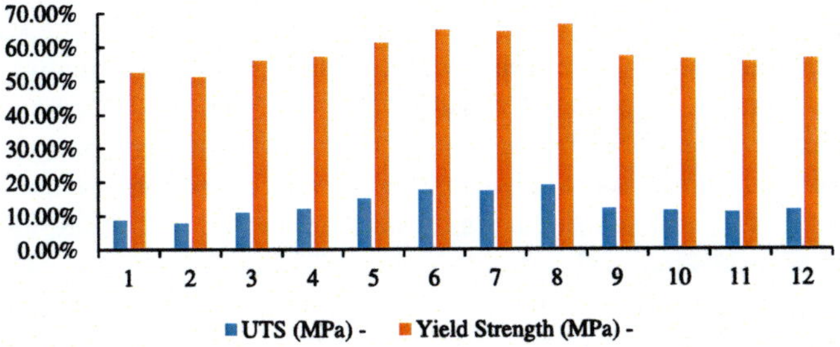

Fig. 4.7 Percentage differences of specimen compared to controlled sample

medium, specimen 4 scored better than the other specimen that undergoes different heat treatment on 2-factor heat treatment temperature and the soaking time. Similar to the other nanofluid media with CNT and nano silica where specimens 8 and 12 respectively display similar pattern which outstand the specimen according to the tier respective quenching medium. Thus, in terms of heat treatment finds confirms that magnesium alloy is best treated in 350 °C for 1 h of the soaking period based on the finding of this experiment.

4.3.2 *Optical Microscopy Preparation of Optical Metallographic*

Optical images contrast by adjusting the DCI of the optical microscope to have a clear view of the specimen depth region and flowed line affected by heat treatment. The specimens in Table 4.4 were taken from the heat treatment up to 350 °C and the soaking time of around 1 h to find the correlation of mechanical properties

Table 4.4 Metallurgical analysis of the sample for each medium

Specimen	Microscopy image	Depth region (μm)	Flowed lines (μm)
4	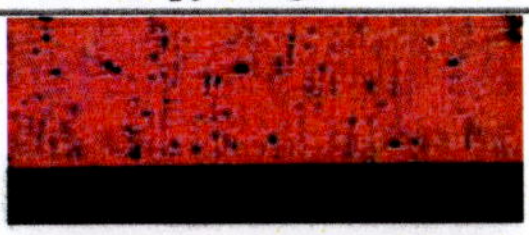	36	3–47
8		55	5–35
12	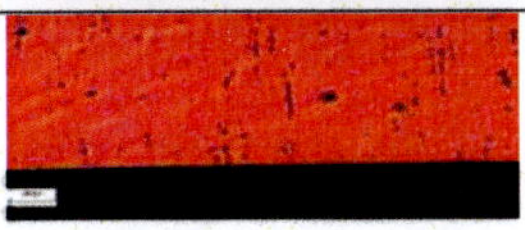	22	2–17

improvement and the structure analysis of specimen under optical microscopy. This is due to selecting the best specimen in terms of mechanical properties improvement from the 3 mediums, specimens 4, 8 and 12.

Specimens 4, 8, and 12 are the sample under these conditions. The optical metallurgical analysis, which focuses on depth regions affected by the heat treatment and presence of CNT and nano silica, was around 55 and 22 μm, rather than that of the specimen soaked in distilled water (36 μm). CNT tends to Van der Waals bonds to attach to the pores on the surface of the AZ31B [6, 7]. Meanwhile, nano silica contributes slightly to both UTS and yield strength increment but not as effective compared to CNT. Since nano silica is a hydrophobic material [18], it acts as an excellent coating to avoid corrosion for AZ31B. However, it does not excel in improving the mechanical properties compared to quenching in distilled water. The flowed line also displayed a significant difference, with CNT ranging around 5–35 μm and nano silica meandering around 2–17 μm. Meanwhile, in distilled water without the presence of nanoparticle, the flowed line was obtained around 3–47 μm. The presence of void in specimen 4 was more evident than in specimens 8 and 12.

Relating from the tensile test as scored specimens in ascending order 12, 4, and 8 respectively show a similar pattern compared with the depth region measurement affected by the quenching process in the selected medium. As for the analysis, flowed lines are more related to the capability of heat from specimen dissipated into the medium. Nevertheless, specimen 8 under the heat treatment and soaking time of 350 °C and 60 min displayed the most optimum outcome related to the diffusion depth of the nanoparticles from the surface of the AZ31B magnesium alloy when submerged in 0.1% CNT quenching medium.

4.4 Conclusion

Both the heat treatment process and preparation of nanofluid for the quenching medium affected the AZ31B specimen. The results of the tensile test of all specimens are presented in Fig. 4.6. The energy absorption is calculated based on the area under the curve of stress versus strain until the specimen failed, as shown in Fig. 4.7. Elongation and stress of each specimen affected due to the heat treatment and nanofluid exposure during the quenching process until specimen fail were analysed. The mechanical properties of each sample were affected by the ultimate tensile strength (UTS) and yield strength measured.

Acknowledgements The authors would like to express their gratitude and thanks to the Ministry of Higher Education of Malaysia via University Pertahanan Nasional Malaysia (research funding: FRGS/1/2018/TK03/UPNM/03/1) for funding and supporting this research.

References

1. Honsel GmbH, M.A. & Co KG Meschede.: Weight and cost saving with magnesium die castings. Proc.5th Int. Conf. Magnesium Alloys and Their Applications, 397–401 (2000).
2. Jones TL, DeLorme RD, Burkins MS (2007) Ballistic Evaluation of Magnesium Alloy AZ31B. Army Research Laboratory ARL-TR- 4077:1–14
3. Asgari H, Odeshi AG, Szpunar JA, Zeng LJ, Olsson E (2015) Grain size dependence of dynamic mechanical behavior of AZ31B magnesium alloy sheet under compressive shock loading. Mater Charact 106:359–367
4. Yu X, Li L, Li T, Qin D, Liu S, Li Y (2017) Improvement on dynamic fracture properties of magnesium alloy AZ31B through equal channel angular pressing. Eng Fract Mech 181:87–100
5. Jena PK, Kumar KS, Krishna VR, Singh AK, Bhat TB (2008) Studies on the role of microstructure on performance of a high strength armour steel. Eng Fail Anal 15:1088–1096
6. Li, Q., C.A. Rottmair, R.F. Singer.: CNT Reinforced Light Metal Composites Produced by Melt Stirring and by High Pressure Die Casting. Composites Science and Technology, 1–17 (2012).
7. Rahman M, Hosur M, Zainuddin S, Vaidya U, Tauhid A, Kumar A, Trovillion J, Jeelani S (2013) Effects of amino-functionalized MWCNTs on ballistic impact performance of E-glass/epoxy composites using a spherical projectile. Int J Impact Eng 57:108–118
8. Soliman EM, Sheyka MP, Taha MR (2012) Low-velocity impact of thin woven carbon fabric composites incorporating multi-walled carbon nanotubes. Int J Impact Eng 47:39–47
9. Guo P, Li L, Liu X, Ye T, Cao S, Xu C, Li S (2017) Compressive deformation behavior and microstructure evolution of AM80 magnesium alloy under quasi-static and dynamic loading. Int J Impact Eng 109:112–120
10. Tan,V. T. Ching.: Computational simulation of fabric armour subjected to ballistic impacts. Int J Impact Eng, 32: 1737–51 (2006).
11. Hamed Habibi Khoshmehr (2014) Ahmad Saboonchi: The quenching of silver rod in boiling carbon nanotube-water nanofluid. Int J Therm Sci 75:95–104
12. Kim H, Buongiorno J (2010) Nanoparticle deposition effects on the minimum heat flux point and quench front speed during quenching in water-based alumina nanofluids. Int J Heat Mass Transf 53:1542–1553
13. Babu, K., & Prasanna Kumar, T. S.: Effect of CNT concentration and agitation on surface heat flux during quenching in CNT nanofluids. International Journal of Heat and Mass Transfer, 54 (1–3), 106–117 (2011).
14. Brickman, D. B.: Pen cap failure analysis and prevention. American Society of Mechanical Engineering. (1997).
15. Radhamani AV, Lau HC, Ramakrishna S (2018) CNT-reinforced metal and steel nanocomposites: A comprehensive assessment of progress and future directions. Compos A Appl Sci Manuf 114:170–187
16. Paramsothy M, Chan J, Kwok R, Gupta M (2011) Addition of CNTs to enhance tensile/compressive response of magnesium alloy ZK60A. Compos A Appl Sci Manuf 42(2):180–188
17. Vander Voort, & Baldwin, W.: Metallography and Microstructures Handbook. ASM International, 9, 2733 (2004).
18. Qian Z, Wang S, Ye X, Liu Z, Wu Z (2018) Corrosion resistance and wetting properties of silica-based superhydrophobic coatings on AZ31B Mg alloy surfaces. Appl Surf Sci 453:1–10

Chapter 5
Comparison of Lattice Structure Configurations for Suitability in Turbine Blades Using Modal and Harmonic Response Analysis

Sajjad Hussain, Wan Aizon W. Ghopa, Salvinder Singh, Abdul Hadi Azman, Shahrum Abdullah, and Hafizan Kamaruddin

Abstract The aim of this paper is to investigate the modal characteristics and harmonic response of two lattice-based configuration called octet truss and cubic lattice in order to find out their suitability for use in turbine blades and other high speed rotating components. Solid model is used as reference. Lattice structures offer great flexibility and opportunities to the designers for providing high strength and lightweight structures compared to solid structures. Results indicate that lattice structures have natural frequency lower than solid. For cubic structure, natural frequencies are much lower which is undesirable as it limits the operating speed. Octet truss results are much better and more comparable to solid. Harmonic response results indicate that octet truss model's maximum deformation is 3.1 times less compared to the solid and von Mises stress is 3 times less compared to the solid model with similar dimensions and boundary conditions. However cubic lattice has 36 times higher deformation and 33 times higher von Mises stress compared to solid model. Hence it can be implied that octet truss structure can be a good replacement of solid rotating components such as turbine blades as they offer better dynamic response with lighter weight.

Keywords Lattice structure · Modal analysis · Harmonic response · Octet truss · Cubic lattice

5.1 Introduction

As the need for energy conservation increases, the need for the light parts/structures also increased [1]. Weight reduction is always desirable in dynamic applications as it increases the utility of the components. The production of light parts is important but at the same time, they must possess good mechanical properties [2]. Keeping

S. Hussain · W. A. W. Ghopa (✉) · S. Singh · A. H. Azman · S. Abdullah · H. Kamaruddin
Faculty of Engineering and Built Environment, Universiti Kebangsaan Malaysia, UKM, 43600 Bangi, Selangor, Malaysia
e-mail: waizon@ukm.edu.my

S. Abdullah et al. (eds.), *Structural Integrity Cases in Mechanical and Civil Engineering*, Structural Integrity 23, https://doi.org/10.1007/978-3-030-85646-5_5

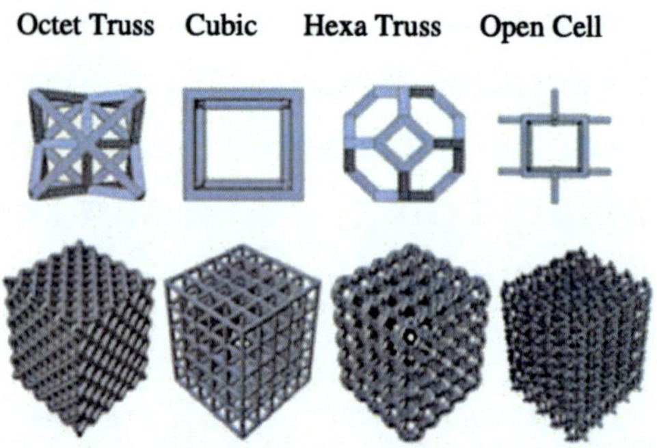

Fig. 5.1 Different configurations of lattice structures

in view these requirements, lattice structures were introduced. The reduction of parts' weight can be achieved using lattice structures. The lattice structure has high strength even when mass is reduced. Studies and research related to lattice structures have received higher attention due to their superiority over stochastic structure in producing lightweight high strength parts [3].

5.1.1 Lattice Structure

Lattice structure is basically a porous structure in nature that is formed by arranging unit cells of a monolithic material in space using different patterns. These patterns influence the mechanical performance of the newly formed lattice structure. Lattice structures offer great flexibility and opportunities to the designers for providing high strength and lightweight structures compared to solid structures. Due to these attributes, these structures are highly desirable in aerospace [4] and automobile industry [5] where thrust to weight ration and fuel consumption are of major importance.

Lattice structures are available in various types and configurations. The principle step involved in constructing lattice structures is the stacking of unit cells in x, y, and z directions [6]. The properties of the resulting lattice structure are controlled by these configurations [7]. Some common types of lattice configurations are shown in Fig. 5.1.

5.1.2 Applications of Lattice Structure

The choice of lattice is dependent on its application. Since each configuration has different properties, therefore it is quite important to select the most suitable type for a specific application. Lattice structure are gaining popularity in Biomedical,

Aerospace and Automotive industries due to their unique surface structure as well as their good mechanical properties.

In the aerospace and automotive industries, parts' weight is always a major area of concern as lighter parts are more efficient. Maloney et al. proposed a Nickel base micro-lattice structure to produce compact and light heat exchanger [8]. In 2018 [9], 25% weight reduction was achieved by constructing a pyramidal lattice core in interlocking structure for engine hood. Kulangara et al. [10] proposed a new design for spur gear using a honeycomb lattice structure with 19% volume reduction without compromising the strength.

Similarly in biomedical applications, the selection a specific lattice structure configuration depends on the desired function such as bone implants, dental implants and bone tissues. The porous nature of structures is important for bone ingrowth and stress shielding for biomedical implants.

Heinl et al. [11] analyzed two types of lattice structures for bone implant applications, diamond and hatched lattice structures. According to the results, the mechanical properties of the diamond structures were similar to trabecular bone, whereas the properties were between trabecular bone and cortical bone for a hatched structure.

5.1.3 Modal Analysis

Modal analysis is the basic and essential step of every turbine blade design. It gives the natural frequencies along with the mode shapes associated with those frequencies. It helps in validating any design as operational speed must not coincide with any natural frequency. Mode shapes are also equally important as they give information about the type of deformation to be experienced by the blade in that natural frequency zone. It also determines the type of balancing required (either rigid or flexible). Modal analysis can be performed experimentally as well as numerically.

In gas turbines, flow is highly unsteady, these periodic flow disturbances can induce high amplitudes of blade vibration under resonant conditions [12]. These excessive resonant stresses may lead to High-cycle fatigue failure which can eventually result in catastrophic engine failure. The need to predict resonant conditions at the design stage is now becoming more apparent with modern turbine designs, where increased blade loading and reduced axial spacing challenge the capabilities of conventional empirical design practices. The ability to forecast potential vibration problems before manufacturing can provide significant benefits in terms of reduced development costs and improved reliability [13]. All the blades in a turbomachine receive their major periodic excitation at a frequency equal to nozzle passing frequency. Since these forces are periodic, one has to consider several number of these harmonics in determining whether resonance takes place, when one of these harmonics coincides with any of the natural frequencies of any blade [14]. Modal analysis included determination of natural frequencies, mode shapes, forced response and damping characteristics. These characteristics can be determined experimentally and validated numerically. Experimental modal analysis can be performed either using bump test technique [15] or using vibration shaker [16].

During the operational life of turbine blades, vibratory stresses may be induced due to multiple reasons including unbalance, misalignment and mounting problems, with unbalance as the most common among them [17]. Hence it is inevitable to have proper understanding of unbalance response of turbine blades which is manifested in the form of vibration. Fatigue characteristics are also determined by using vibration analysis data. Hence vibration analysis is vital for durability. The vibration monitoring setup comprises of a sensor (accelerometer) and data acquisition system which is connected to analyzer. Depending upon the conditions, several techniques can be employed for vibration analysis [18] including Fast Fourier Transform (FFT), Short Time Fourier Transform (STFT), the Envelope Analysis (EA) and wavelet transform (WT).

Modal analysis is used for the prediction of failure in turbine blades as this determines the mode that has major contribution in the fatigue failure [19]. Rani et al. [20], in their work, used the modal analysis for identification of failure mode and crack location of 30 MW 1st stage steam turbine blade. In 2015, Yeon-Sun Choi and Kyu-Hwa Lee [21], used modal analysis and Campbell diagram to determine the root cause of a 1st row gas turbine blade failure. Nik Abdullah et al. [22] investigated the effect of cracks on natural frequencies of a beam and found that by increasing number of cracks, natural frequency reduces significantly that may lead to failure.

Experimental measurement of the forced response for turbine blade poses some challenges that include: the requirement of a large number of sensors for each engine stage, difficulties in dealing with multiple excitation frequencies, and sensors being located in the gas path [23]. It is much easier to perform the numerical analysis by employing the correct boundary conditions. It reduces both the cost as well as time. However, numerical analysis alone is not enough to validate the design as experimental analysis is always compulsory for the validation purpose. Numerical analysis reduces the number of iterations required experimentally and provides a direction for the experimental work.

Since lattice structures are gaining popularity with the passage of time and more industries are adapting these structures, it is important to evaluate these structures from all aspects. Modal analysis of these structures can give insights about the dynamic performance and suitability of lattice structures for high speed applications such as turbines, impellers and compressors. In present study, modal and harmonic response analysis of two lattice structure configurations is carried out and compared with solid model to determine the suitability of lattice-based models for dynamic applications such as turbines.

5.2 Methodology

Purpose of modal analysis is to identify and analyse the dynamic response of structures in terms of free vibration and forced vibration analysis. Every structure has

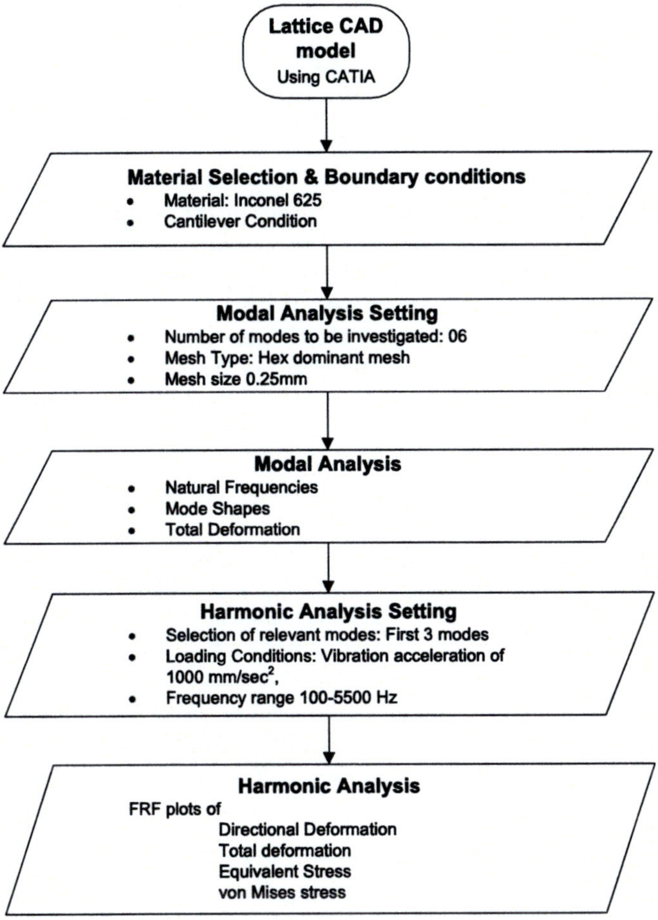

Fig. 5.2 Flow of the steps involved in this study

inherent natural frequencies at which the object can naturally vibrate. Energy is transferred from one form to the other with minimal loss (here vibrational to kinetic) at these frequencies resulting in resonance. High vibration amplitudes are undesirable for rotating machinery as they have adverse effects on operational stability, performance and life of the machine which may lead to catastrophic failure. Therefore, it is compulsory to investigate these frequencies at which the structure can vibrate erratically, experiencing higher vibration amplitudes.

Since lattice structures are gaining popularity with the passage of time and more industries are adopting these structures, it is important to evaluate these structures from all aspects. Modal analysis of these structures can give insights about the dynamic performance and suitability of lattice structures for high speed applications such as turbines, impellers and compressors. Figure 5.2 shows the flow of the steps adopted in this work.

5.2.1 2.1. Free Vibration Analysis

In the first step, geometry was selected, and CAD model was prepared. Two types of lattice structures, cubic and octet truss were used in this work. Solid structure was used as reference to determine the weight reduction variation in modal characteristics of each type of lattice from the solid structure. Inconel 625, an additive manufacturing material, commonly used for turbine blades was selected for this work. Natural frequency is dependent on stiffness and mass of the material as given in Eq. (5.1). Since for the same material, each geometry has different stiffness and mass, so the natural frequency will also change by changing the configuration of lattice structure.

$$f = \sqrt{(k/m)}/2\pi \tag{5.1}$$

Bounding box for all the configurations is the same and is shown in Fig. 5.3. All geometries are constructed using this bounding box. Lattice structure geometries used in this work are shown in Fig. 5.4. Unit cell size for both octet truss and cubic lattice is 4 mm. Thickness of each unit strut is 0.5 mm for both cubic and octet truss structures.

Modal analysis of the lattice structure configurations was performed in cantilever condition by applying fixed support at one end as blades are mounted on the rotor in cantilever condition. Modal analysis was performed based on an undamped free

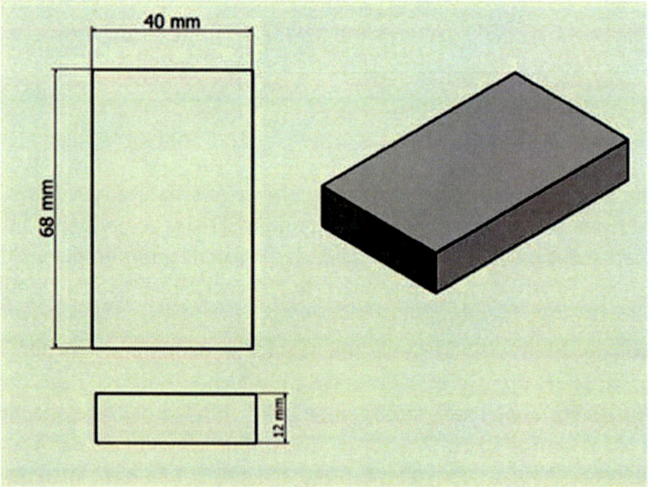

Fig. 5.3 Bounding box used for constructing lattice structure

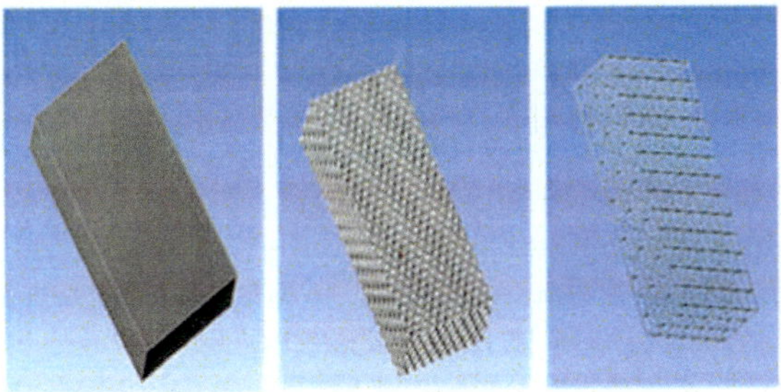

Mass	Solid	Octet Truss	Cubic
(g)	219.4	62.9	15.5

Fig. 5.4 Configurations used for analysis

Table 5.1 Boundary conditions for modal analysis

Boundary condition	Description
Material	Inconel 625
Young modulus	162 GPa
Density	8.44×10^{-6} kg/mm^3
Constraint	Fixed support at one end
Mesh element size	0.25 mm
No. of modes investigated	06

vibration system. Governing equation used is as following:

$$[M]\ddot{u} + [K]u = 0 \tag{5.2}$$

Both modal and harmonic analysis were performed using ANSYS workbench 2019 R1 Academic. Analysis was performed using normal environmental conditions (room temperature 22 °C).

There are various mode extraction methods available in ANSYS that include Block Lanczos, Supernode, Unsymmetric, Damped and QR damped. In this work, Block Lanczos method was employed due to its higher convergence rate. Type of meshing is hex dominant mesh with mesh size of 0.25 mm as shown in Table 5.1.

5.2.2 *Forced Vibration Analysis*

While modal analysis provides the information about the free vibration characteristics of the structure, harmonies response gives the forced vibration characteristics of structure.

For harmonic analysis, vibration in terms of acceleration is applied to the structure while fixed support is applied at the other end. Vibration was applied in radial directions (X and Z) of the structures. It is assumed that load is uniform. Harmonic analysis was performed on all three structures blades, i.e. solid structure, cubic lattice structure and octet truss structure using the same magnitude of vibration acceleration (10,000 mm/sec^2 in X and Z directions) and other boundary conditions. The harmonic analysis can be conducted in ANSYS using methods, Full Method or Mode Superposition method.

Mode superposition method was employed in this work as it uses the modal results before performing the harmonic response, so it is much faster as compared to the full method which does not use the modal history. The governing equation used in mode superposition method is given in Eq. (5.3).

$$(-\Omega^2[M] + j\Omega[C] + [K])\{\mathrm{x1} + \mathrm{jx2}\} = \{\mathrm{F1} + \mathrm{jF2}\} \tag{5.3}$$

Since most of the turbines operate below 5500 Hz, therefore frequency range was set from 100 to 5500 Hz. Clustering of results was selected around the natural frequencies as maxim deformation ad stress values are expected around the natural frequencies due to the excitation of respective modes. Loading condition for octet truss model is shown in Fig. 5.5. Same condition is also used for solid model and cubic lattice models.

Same boundary conditions are used for all models (solid, octet truss lattice and cubic lattice) as described in Table 5.2.

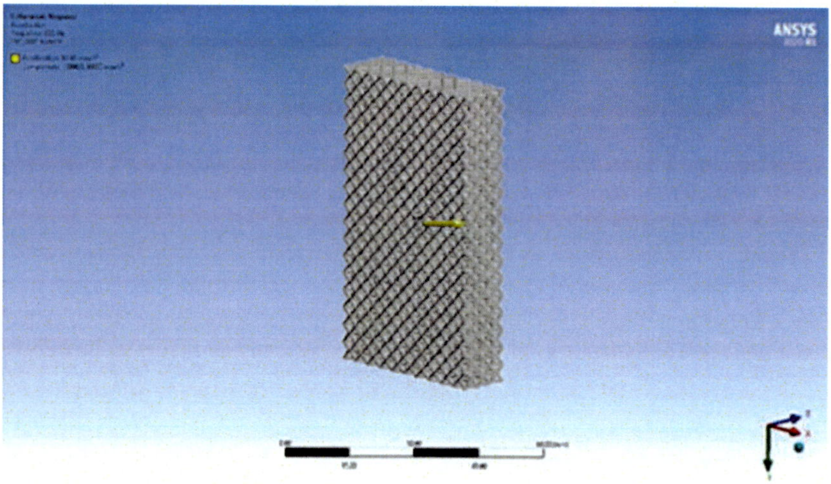

Fig. 5.5 Loading condition for harmonic response

Table 5.2 Boundary conditions for harmonic response

Boundary condition	Description
Material	Inconel 625
Young modulus	162 GPa
Density	8.44×10^{-6} kg/mm^3
Mesh element size	0.25 mm
No. of modes investigated	6
Constraint	Fixed support at one end
Frequency range	100–5500 Hz
Constraint	Fixed support at one end
Load	Vibration acceleration, 10,000 mm/sec^2 in Radial direction (X & Z)

5.3 Results and Discussion

Two types of analysis were performed on three models. In the first step, modal analysis was performed while using these results, forced response analysis was performed in the second step. Response was measured in terms of deformation (directional and total) and stress (normal and von Mises).

5.3.1 Modal Analysis Results

Figure 5.6 shows the modal analysis results for solid model obtained from ANSYS. As number of modes to be investigated was selected to six so the results indicate

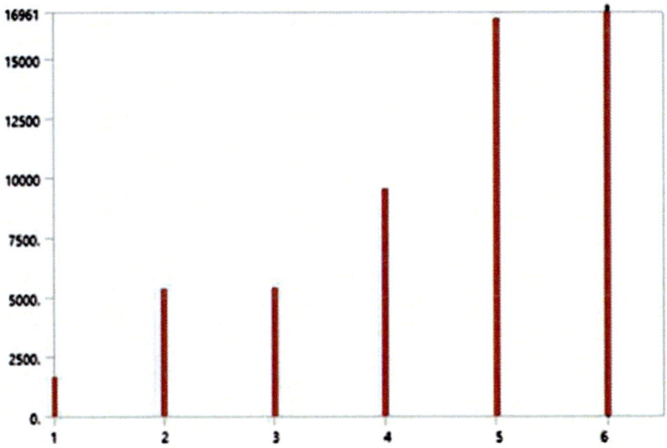

Fig. 5.6 Modal analysis results of solid model

the 1st six modes of the solid model. As per results, 1st mode occurs at 1678 Hz. This mode is flap wise bending mode 2nd mode occurs at 5343 Hz which is torsional mode. Other four modes are at quite high frequencies which are above the usual operational speed of turbines. Figure 5.7 shows mode shape results for 1st three modes of solid model respectively.

Mode shapes are associated with natural frequency. Figure 5.7 shows the 1st three modes of solid model. In 1st mode deformation is maxim at the top face uniformly while minimum at bottom where fixed support was applied, indicating that this mode is flap wise bending mode. In 2nd mode, the deformation pattern indicates the presence of torsion so this mode is torsional mode.

Modal Analysis results of octet truss lattice model obtained from ANSYS are shown in Fig. 5.8 modal Analysis results for octet truss structure indicates that 1st mode occurs at 918 Hz while 2nd and 3rd mode occur at 2517 and 3790 Hz respectively. Mode shapes results for octet truss lattice are shown in Fig. 5.9. Mode shape

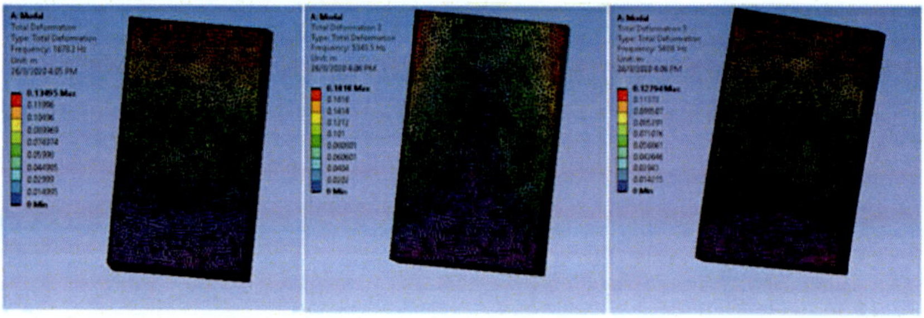

Fig. 5.7 Mode shapes of solid model

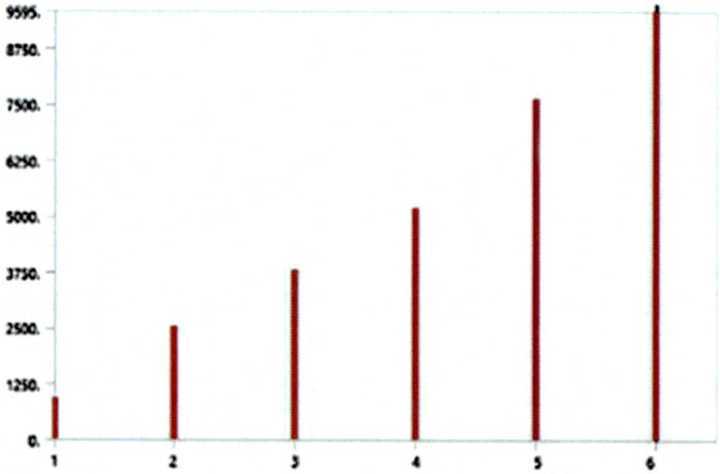

Fig. 5.8 Modal analysis results of octet truss lattice model

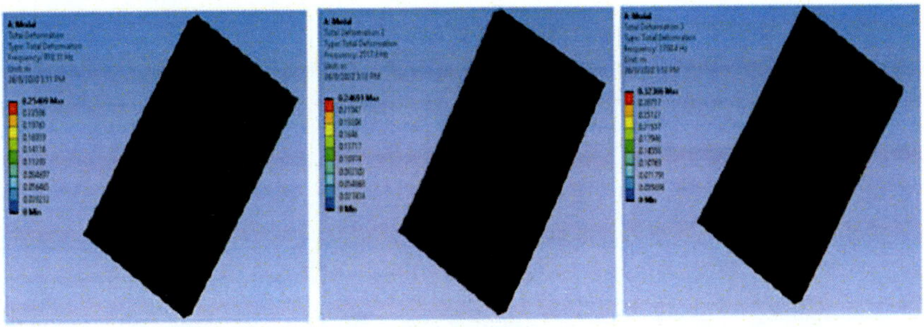

Fig. 5.9 Mode shapes of octet truss lattice model

results indicate that 1st and 2nd mode are flap wise bending modes while torsion is experienced in 3rd mode. Delayed torsional mode is always desirable in rotating machinery.

Modal analysis results of cubic lattice model are shown below in Fig. 5.10. As per results, 1st mode for cubic lattice structure occurs at 160 Hz which is quite lower as compared to solid and octet truss models. All six modes of cubic lattice model lie between 160 to 3000 Hz. Since higher modes have complex deformation pattern and are more prone to failures, therefore these results are not desirable for dynamic applications. Figure 5.11 shows the 1st three mode shapes for cubic lattice structure. Deformation is not uniform in cubic lattice model right from the 1st mode. Complex deformation patterns are observed in all other modes of cubic lattice model. Hence it can be implied that this type of lattice structure is not suitable for application in

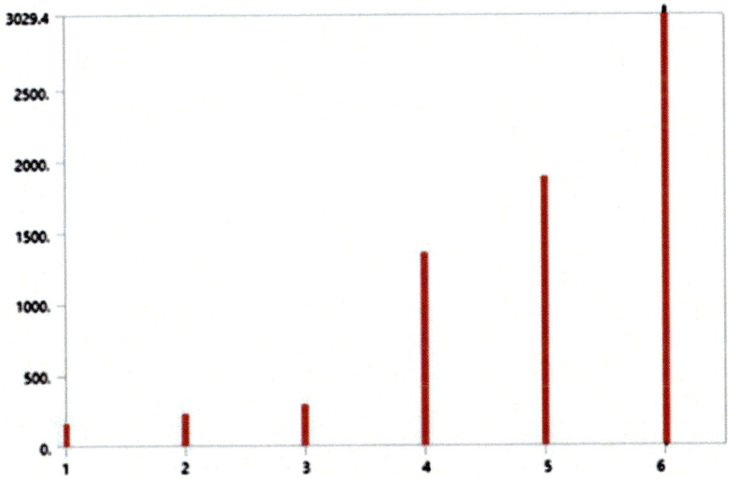

Fig. 5.10 Modal analysis results of cubic lattice model

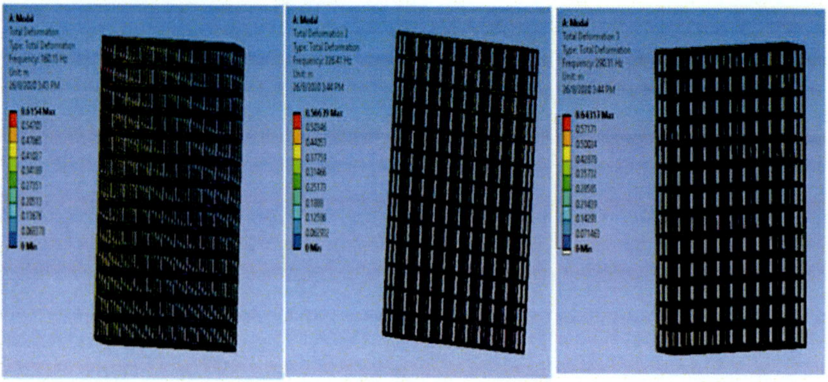

Fig. 5.11 Mode shape results of cubic lattice model

Table 5.3 Summary of modal analysis results

Mode	Natural frequency		
	Solid model	Octet truss lattice model	Cubic lattice model
1	1678	918	160
2	5343	2517	226
3	5419	3790	290
4	9508	5170	1365
5	16,700	7623	1882
6	16,961	9595	3029

rotating machinery. Summary of the modal analysis results for all three models is presented in Table 5.3.

5.3.2 *Harmonic Response Results*

After performing modal analysis, harmonic response analysis was performed to investigate the response of each model under external excitation force, which in this study was applied in terms of vibration acceleration. Since mode superposition method was selected, so modal analysis results were used. Frequency response of the models in terms of deformation and stress was measured.

Results indicate that for deformation on the X-axis, the maximum amplitude occurs at the third natural frequency, 5419 Hz with amplitude of 2.39×10^{-6} mm while at Y-axis, maximum amplitude occurs at the first natural frequency, 1678 Hz with an amplitude of 2.82×10^{-2} mm. For deformation on the Z-axis, maximum amplitude occurs at the third natural frequency of 5419 Hz with 2.7483×10^{-3} mm

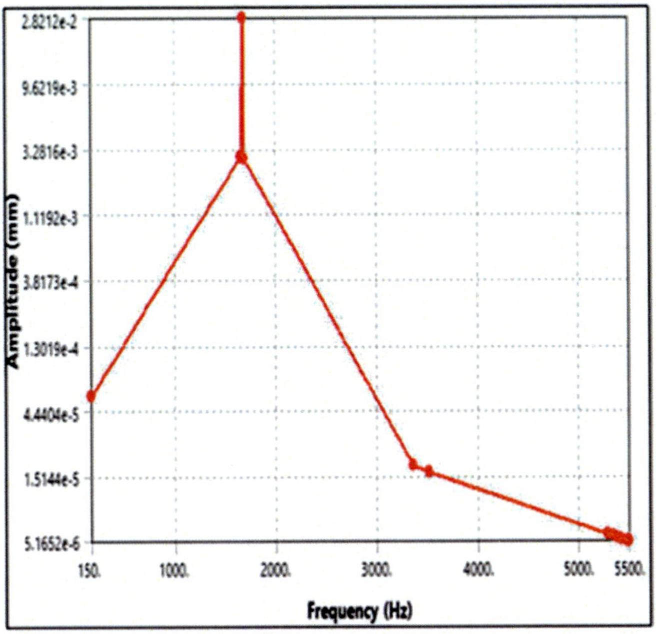

Fig. 5.12 Deformation in Y direction for solid model

amplitude. Hence the overall maximum amplitude occurs at 1st mode in Y direction as shown in Fig. 5.12. Similar loading condition were applied to octet truss model.

Directional deformation results for octet truss lattice model indicate that on the X-axis, maximum amplitude occurs at the second natural frequency of 2517 Hz with amplitude value of 1.1789×10^{-3} mm while on the Y-axis and the Z-axis can be observed maximum amplitude occurs at the first natural frequency of 918 Hz with amplitude values of 2.8497×10^{-6} an 9.0638×10^{-3} mm.

Hence maximum deformation occurs in Z direction at 1st mode as shown in Fig. 5.13. This result indicates that maximum deformation in octet truss lattice is 3.1 times less than solid model.

Directional deformation results of cubic lattice structures indicate that deformation in X axis, maximum deformation of 1.02 mm occurs at second natural frequency, 226.41 Hz while on the Y axis maximum amplitude of 0.74642 mm occurs at the fifth natural frequency, 1882.4 Hz. In Z axis, maximum deformation of 1.03 mm is observed at the first natural frequency, 160.15 Hz. Hence the overall maximum deformation for cubic lattice was observed in Z direction at 1st mode as shown in Fig. 5.14.

These deformation values are much higher compared to solid and octet truss models. Also, the natural frequencies of cubic lattice model are very low compared to solid and octet truss models. These low frequencies limit the operational speed of rotating machinery as it is always desired that natural frequencies are higher

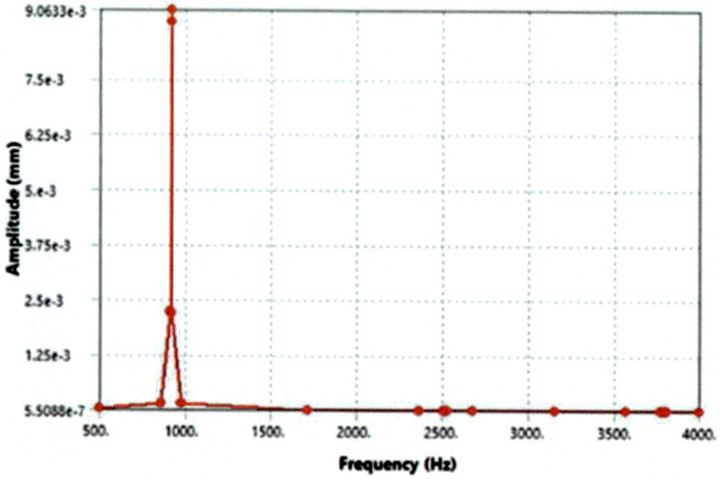

Fig. 5.13 Deformation in Z direction for octet truss model

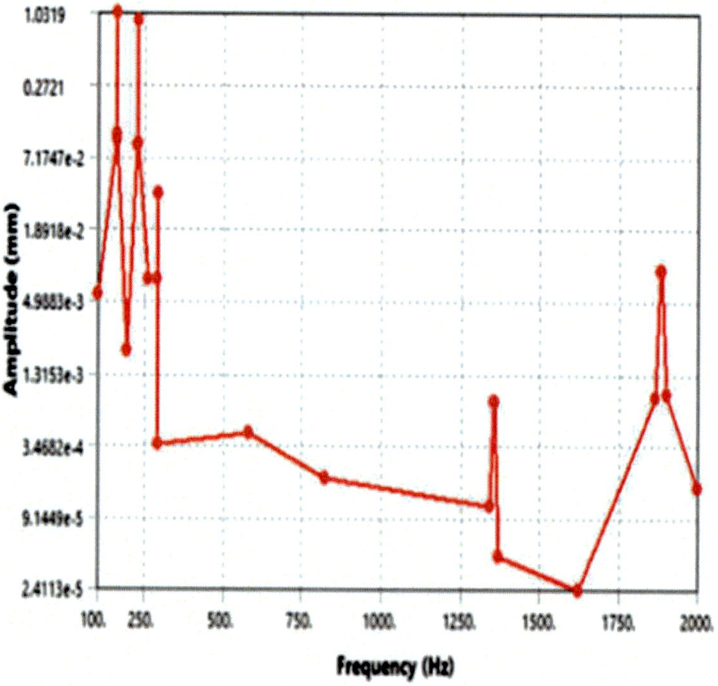

Fig. 5.14 Deformation in Z direction for cubic lattice model

than operational speed for rigid behavior. Although in some cases, operational speed is higher than natural frequency (Flexible modes) but usually only one 1st mode is experienced which is bending mode. But in cubic lattice number of modes are closely related so it is hard to find a safe operational zone.

Based on these results, it can be implied that cubic lattice model is not suitable to dynamic applications where rotation is involved and load is in the form of vibration. This is due to the fact that cubic lattice configuration is constructed in such a way that it can take unidirectional normal loads. But in rotating machinery, loads are not unidirectional so cubic lattice will not suit such applications.

Finally, frequency response function (FRF) was measured in terms of von Mises stress for all three models. Figure 5.15 shows the von Mises stress results for the first three modes for solid model. Von Mises stress results indicate that maximum stress of 1.79 MPa is observed at 2nd mode. This value is well below the endurance limit for Inconel 625 which is 244 MPa. Hence this model is safe under this loading condition.

For octet truss lattice, the highest value of Von Mises stress was observed at 1st mode with a magnitude of 0.6 MPa near the fixed support. This value is almost three times lower compared to solid model. This indicates that octet truss has better response to vibration compared to solid model. Figure 5.16 shows the von Mises stress distribution for first three modes of octet truss lattice model.

For cubic lattice, results indicate that maximum von Mises stress is experienced at 1st mode with a magnitude of 60.12 MPa. This value is very high compared to solid and octet truss models. It is almost 33 times higher as compared to the solid model. Figure 5.17 shows the von Mises stress distribution for first three modes of cubic lattice model. Summary of the results is presented in Table 5.4.

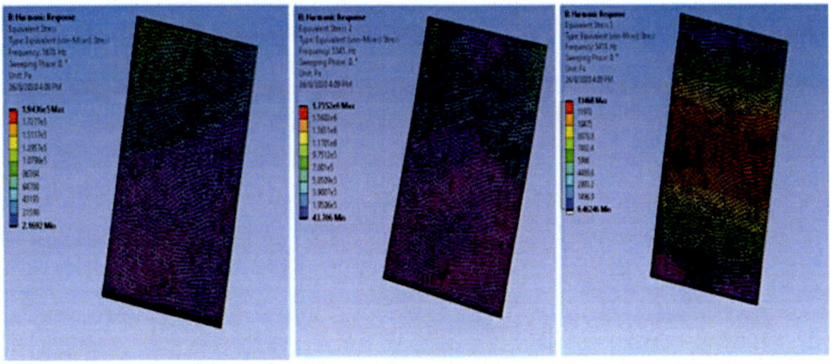

Fig. 5.15 Von Mises stress results for first three modes of solid model

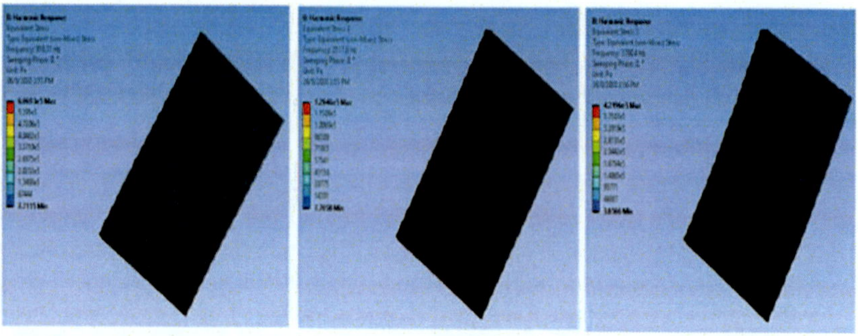

Fig. 5.16 Von Mises stress results for first three modes of octet truss model

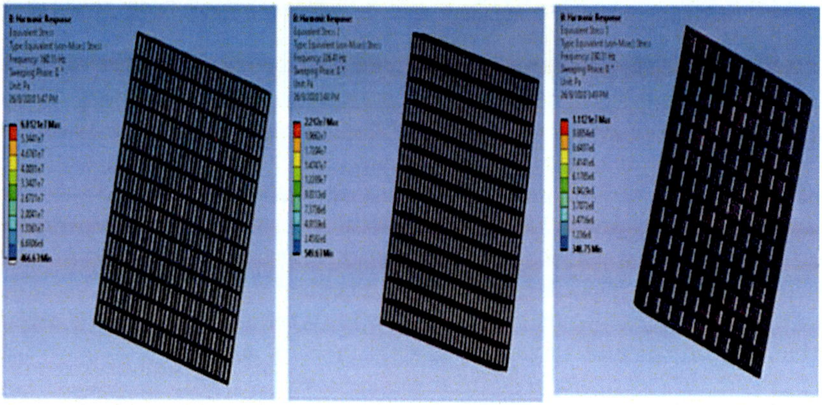

Fig. 5.17 Von Mises stress results for first three modes of cubic lattice model

Table 5.4 Summary of harmonic response results

Model	Maximum deformation (mm)	Maximum von mises stress (MPa)
Solid	0.0282	1.79
Octet truss lattice	0.00906	0.6
Cubic lattice	1.03	60.2

5.4 Conclusion

Two types of lattice structure configurations were used in this work to investigate their modal characteristics and harmonic response solid model was used as reference. Octet truss model was 3.33 times lighter compared to solid model of same dimensions. Modal analysis results indicate that both octet truss and cubic lattice models have lower natural frequencies compared to solid model. Solid model 1st six natural frequencies lie between 1680 to 16,691 Hz while those for octet truss natural frequencies lie between 918 to 9595 Hz and for cubic lattice, natural frequencies are much lower lying between 160 to 3000 Hz which is a drawback for cubic lattice model. Harmonic response analysis was performed by applying vibration excitation in terms of acceleration to the model while keeping one end fixed. Results indicate that octet truss configuration has lowest deformation and stress values among all three models. For octet truss lattice, maximum deformation observed is 3.1 times less compared to the maximum deformation of solid model while von Mises stress is 300% less compared to solid model.

Values of total maximum deformation and von Mises stress for cubic lattice model are almost 36 times and 33 times higher respectively to solid model. This implies that cubic lattice model does not possess acceptable modal properties and exhibits poor response to vibration. Hence it cannot be used for rotating machine components such as blades which are subjected to high vibrations. However, octet truss lattice structure exhibits good harmonic response to vibration excitation and performs even better than the solid. Hence it can serve as good replacement of solid in rotating components especially in turbine blades Thus It can be concluded that octet truss configuration is dynamically more stable compared to cubic configuration and can replace solid rotating components to achieve weight reduction without compromising dynamic properties.

Acknowledgements We would like to show our gratitude to Universiti Kebangsaan Malaysia for the research funding GUP-2018-078.

References

1. Manfredi D, Ambrosio EP, Calignano F, Krishnan M, Canali R, Biamino S, Pavese M, Atzeni E, Iuliano L, Fino P, Badini C (2013) Direct metal laser sintering: an additive manufacturing technology ready to produce lightweight structural parts for robotic applications: La metallurgiaitaliana
2. Chu C, Graf G, Rosen DW (2008) Design for additive manufacturing of cellular structures. Comput-Aided Design Appl 5:686–96
3. Ashby MF, Evans T, Fleck NA, Hutchinson JW, Wadley HN, and Gibson LJ (2000) Metal foams, a design guide. Elsevier
4. Wahab DA, Azman AH (2019) Additive manufacturing for repair and restoration in remanufacturing, an overview from object design and systems perspectives. Processes 7(11):802
5. Vasiliev VV, Barynin VA (2012) RazinAF: Anisogrid composite lattice structures development and aerospace applications. Compos Struct 94(3):1117–1127
6. Seharing A, Azman AH, Abdullah (2019) Gradient lattice structure bio mimicry design configurations for additive manufacturing. Int J Eng Tech 8:36–43
7. Beyer C, Figueroa D (2016) Design and analysis of lattice structures for additive manufacturing. J Manuf Sci Eng (138), 12
8. Maloney KJ, Fink KD, Schaedler TA, Kolodziejska JA, Jacobsen AJ, and Roper CS (2012) Multifunctional heat exchangers derived from three-dimensional micro-lattice structures. Int J Heat Mass Transf 55 (9–10):2486–93

9. Yin S, Chen H, Wu Y, Li Y, Xu J (2018) Introducing composite lattice core sandwich structure as an alternative proposal for engine hood. Compos Struct 201:131–140
10. Kulangara AJ, Rao CS, and Bose PS (2018) Generation and optimization of lattice structure on a spur gear. Mater Today Proceed 5 (2):5068–73
11. Heinl P, Müller L, Körner C, Singer RF, Müller FA (2008) Cellular Ti–6Al–4V structures with interconnected macro porosity for bone implants fabricated by selective electron beam melting. Actabiomaterialia 4(5):1536–1544
12. Bell DL, He L (2000) Three-dimensional unsteady flow for an oscillating turbine blade and the influence of tip leakage. J Turbo Mach 122(1):93–101
13. Moffatt S et al (2005) Blade forced response prediction for industrial gas turbine. Journal of propulsion and power 21.4:707–714
14. Madhavan S et al (2014) Vibration based damage detection of rotor blades in a gas turbine engine. Eng Fail Anal 46:26–39
15. Chaphalkar SP, Khetre SN, Meshram AM (2015) Modal analysis of cantilever beam structure using finite element analysis and experimental analysis. Am J Eng Res 4.10:178–185
16. Witek L (2016) Experimental modal analysis of the turbine blade. Zeszyty Naukowe Politechniki Rzeszowskiej, Mechanika 88(293):371–378
17. Yang W et al (2018) Research on unbalance response characteristics of gas turbine blade-disk rotor system. J Vibro-Eng 20.4:1676–1690
18. Neild SA, McFadden PD, Williams MS (2003) A review of time-frequency methods for structural vibration analysis. Eng Struct 25(6):713–728
19. Hou J, Wicks BJ, Antoniou RA (2002) An investigation of fatigue failures of turbine blades in a gas turbine engine by mechanical analysis. Eng Fail Anal 9(2):201–211
20. Rani S, Agrawal AK, Rastogi V (2019) Vibration analysis for detecting failure mode and crack location in first stage gas turbine blade. J Mech Sci Technol 33(1):1–10
21. Kim KK, Lee YS (2013) Modal characteristics according to the tip shape and assembly condition of the turbine blade. J Mech Sci Technol 27.11:3409–3417
22. Nik Mohamed NA, Mohamed Haris S, Alebrahim R, Sharifian S, Singh SSK (2012) Dynamic analysis of beam with multiple cracks using Transfer Matrix Method. Advances in Applied Mechanics Research: 7th Australasian Congress on Applied Mechanics, ACAM, 2012, pp 1091–1098
23. Lin H (2017) Effects of thermal and mechanical combined load on blade stress and fatigue life characteristic. In: IEEE International conference on mechatronics and automation (icma), pp 1348–1353

Chapter 6
Systematic Review: Overview on Trends and Future Opportunities of Additive Manufactured Lattice Structures

Asliah Seharing, Abdul Hadi Azman, and Shahrum Abdullah

Abstract This review analyses the trends and future opportunities of additive manufactured lattice structures and its fatigue behaviour. Additive manufacturing, a layer-by-layer process to fabricate components offers wide opportunities to produce complex lattice structures compared to conventional manufacturing techniques. The interest in lattice structure has increased with the breakthrough in additive manufacturing. Fatigue behaviour of lattice structure is important in ensuring structural integrity of the additively manufactured parts. A systematic review was conducted to analyse the trends in lattice structure and its fatigue behaviour in terms of articles published, citations, citations per article, journals, countries, field categories, and frequently used keywords. Based on Scopus's online database, the bibliometric analysis evaluates the trends in lattice structure research from 1965 to 2020. It can be concluded that the number of publications on lattice structures has increased significantly over the past decade and plays an important role in producing lightweight high-strength structures. Through keywords analysis, it is found that the term cellular structure or cellular solid and porous structure were commonly used until the early 2000, later, the term lattice structure became more popular with the introduction of metal additive manufacturing technology. Research patterns shifted from static analysis to dynamic analysis as its application is widen to various industries. Parameters such as unit cell topology, porosity, heat, and surface treatment have an influence on the fatigue behaviour of lattice structures.

Keywords Lattice structure · Additive manufacturing · Bibliometric review

A. Seharing · A. H. Azman (✉) · S. Abdullah
Department of Mechanical and Manufacturing Engineering, Faculty of Engineering and Built Environment, Universiti Kebangsaan Malaysia, 43600 UKM Bangi, Selangor, Malaysia
e-mail: hadi.azman@ukm.edu.my

S. Abdullah et al. (eds.), *Structural Integrity Cases in Mechanical and Civil Engineering*, Structural Integrity 23, https://doi.org/10.1007/978-3-030-85646-5_6

6.1 Introduction

Lattice structure, a porous structure [1] and also known as scaffolds [2], is used to produce high strength structures with minimal weight. Lattice structure exists in numerous types of topologies. Each topology offers a unique characteristic [3–6]. Additive manufacturing offers a wide opportunity to produce complex shapes and custom parts by offering design and geometry freedom. This results in high interests in lattice structure in the industrial applications, for example in automotive, aerospace, and biomedical fields [7].

Lattice structure is formed by repeating a single unit cell in the XYZ direction. The variety of unit cells that can be produced is limitless. Researchers have developed new lattice structure designs for the past few decades to identify and produce the optimum design for the desired mechanical properties and applications. The most common used lattice structures are octet-truss and cubic lattice structure. Octet-truss lattice structure offers the same strength in multiple directions of force, while a cubic lattice structure is proved to have the highest strength in unidirectional (z-direction). The parameters of a lattice structure consist of strut/beam, node, pore, and unit cells. Figure 6.1 shows an example of a cubic lattice structure with its parameters.

Cellular structure is widely present in nature and through biomimicry design has been adopted by engineers in various structures for the past three decades. Ashby and Gibson in their book [8] stated that even when the type of topology used is fixed in one type, the cells can be stacked in more than one way to form new structure, depending greatly on the creativity of the designer and the functionality required for the structure. With new types of topologies begin design, further investigation is required to understand its mechanical properties such as Young's modulus and fatigue behaviour and create more possibilities and opportunities in applying the lattice structure formed in various industries. Hence, the need for a systematic review

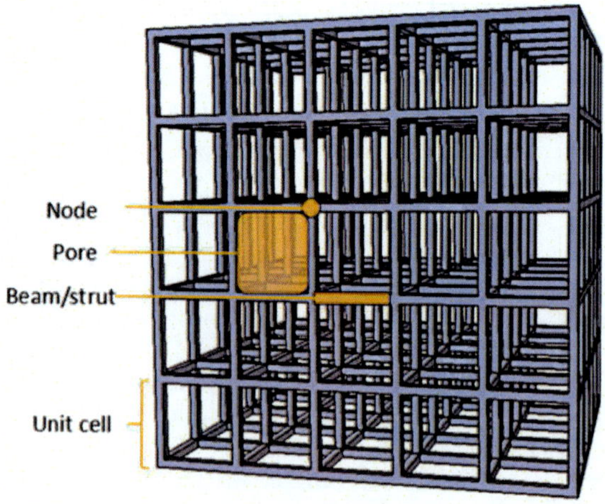

Fig. 6.1 Cubic lattice structure

for an overview on additive manufactured lattice structures publication trends and the challenges in lattice structure fatigue behaviour.

Bibliometric analysis is a quantitative analysis to analyse a specific research field limited to a length of time. The analysis that the authors presented is a systematic analysis to show the overview of chronological evolution in lattice structure research. To the authors' knowledge, there are many researches on lattice structure at different scales and focuses, but no bibliometric reviews were conducted since 1965. The primary purpose of this analysis is to investigate the current trends and progressions objectively and thoroughly by using the bibliometric method to obtain a perspective and analysis related to scientific publications on lattice structure.

6.2 Methodology: Systematic Literature Review

The steps in conducting a systematic review was proposed by Fink [9], which consisted of selecting the research question, gathering the article database and bibliographic content based on the keywords selected, screening of the criteria to include and excluding unrelated literature, implementation of a review criterion, and finally a synthesis of the findings. In the screening criteria, unrelated literature is analysed to be either included or excluded in the review. To implement a review criterion, a review protocol for the content analysis was implemented, this includes examination of the literature based on the keywords and database determined. The method used in this systematic review is summarized in Fig. 6.2.

6.2.1 Selection of Keywords

The selection of the research questions was made based on the topic of the paper regarding the publication trends and progress in the field of lattice structures, and its fatigue behaviour. All the data in this systematic analysis paper was searched and collected from the Scopus online database. The most important step in the data collection process is the selection of the keywords. In this case, since lattice structure is also known by other terms, five different keywords were used as a strategy to ensure there are no articles left behind. The keywords "lattice structure" OR "porous structure" OR "scaffold" OR "cellular solids" OR "cellular structure" were implemented during data collection, along with AND "additive manufacturing" OR "AM" OR "selective laser melting" OR "SLM" OR "3d printing".

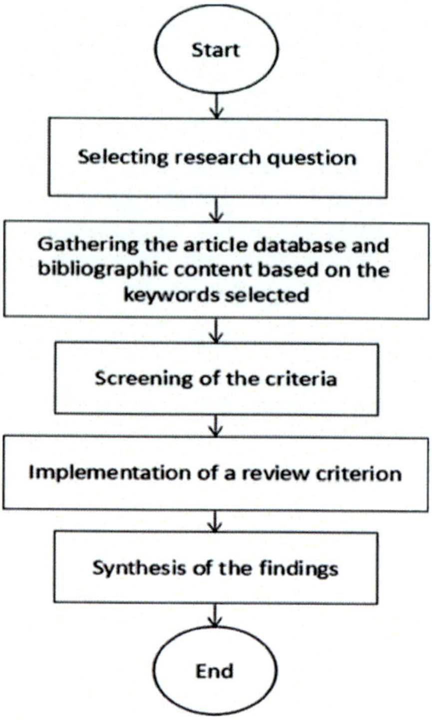

Fig. 6.2 Flowchart of the systematic literature review conducted

6.2.2 Application of Screening Criteria

Based on the method proposed by Fink and Rahito [10], there are a few review protocols that need to be followed in the screening criteria. The objective of this screening process is to identify the suitability of the articles to be included in the review. The process is divided into two sections which are (1) the fundamental background data of the publication such as author, publication type and year, time setting, and (2) the content and specific method in the study.

A second section of the review protocol describes the details of research trends in fatigue behaviour of lattice structures. In this review, it discusses based on the publications the manufacturing techniques, types of materials, types of lattice structures, progressivity of the structure, and type of mechanical testing used. The review focusses on lattice structure parameters influencing fatigue behaviour and the effect of manufacturing parameters, heat-treatment on additively manufactured lattice structure mechanical properties. The findings will improve the design and integration of additively manufactured lattice structures in the automotive and aerospace industry.

6.3 Results and Discussion

Based on the data collection method used, the results showed that 1,113 documents are closely related to the selected topic. This section will be discussed further based on the journal articles, citations, country-based, field and categories, source and publishers, and keywords that are highly mentioned.

6.3.1 Bibliographic Analysis: Publication Trends and Progress

The number of articles and the number of citations from 1965 to 2020 are shown in Fig. 6.3. The first article regarding lattice structure was published in 1965. The number remained stagnant until another article was published in 1970. Not a single article was published until 1984, where two articles were produced. The number of articles increased by one annually from 1984 to 1988. More articles were published in the 1990s and started to increase in the early 2000s. The number of articles increased rapidly in the 2010s until recently. This data showed that the topic was initially largely unexplored from 1965 to 2000, and interest increased starting from the year 2000 onwards. The trend in the number of citations throughout the years is approximately the same as the trend line for articles published, as shown in Fig. 6.4. The citation number did not achieve a high value (less than 50) until the early 2000s, which showed that the lattice structure's research contribution was still in the initial stage. The content gradually matured as the advancement of additive manufacturing technology progressed, notably in the year 2000 where the first metal electron-beam melting additive manufacturing machine was patented and licensed by ARCAM AB.

The 3D printing field changed from polymer to metal, from sintering to beaming or melting materials to produce high-quality and reduce time consumption production.

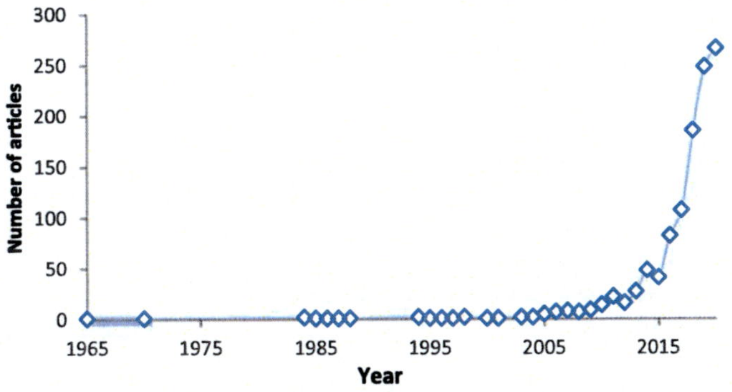

Fig. 6.3 Number of articles published from 1965 to 2020

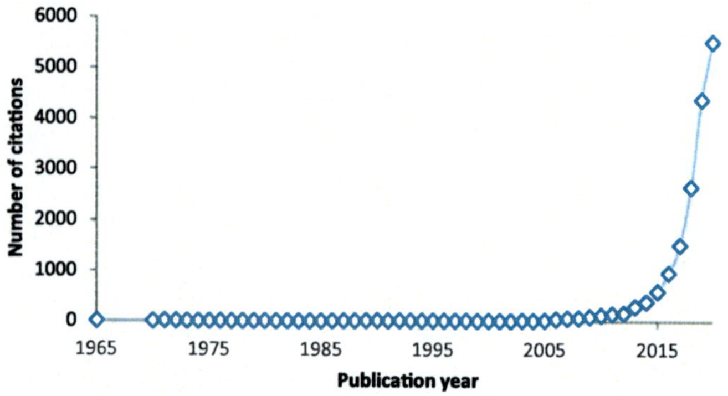

Fig. 6.4 Number of citations from 1965 to 2020

The number of citations increased rapidly after 2010, with 117 citations to 5,512 citations in 2020. The citation number increased growing since additive manufacturing is one of the main focuses in the industrial revolution 4.0 (IR4.0). Additive manufacturing matured as a technology, changing from mainly a rapid prototyping technology to producing final-user parts with reliable mechanical properties comparable to those manufactured using conventional manufacturing technology.

Figure 6.5 shows the top 20 countries based on the research publication in the selected topic. 241 journal articles published are written by researchers from the United States (US), which is 21.65% of the total number of articles. Followed closely by researchers from China with 226 journal articles. Researchers from the United Kingdom published 126 journals articles with the highest citation (35.8 citations

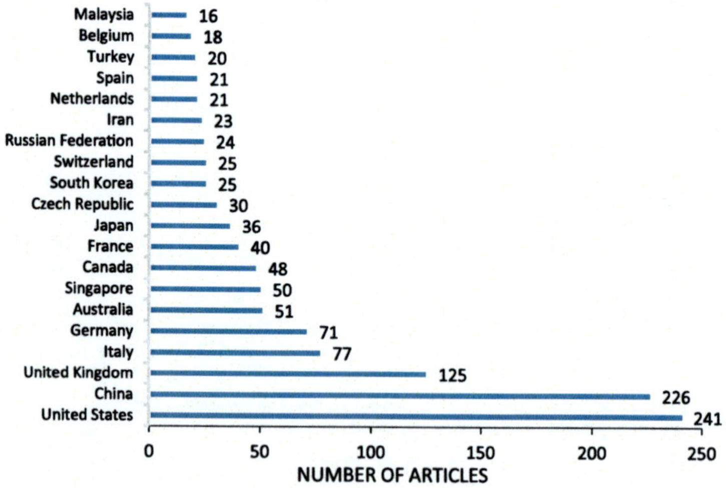

Fig. 6.5 Journal article publication on lattice structure according to countries

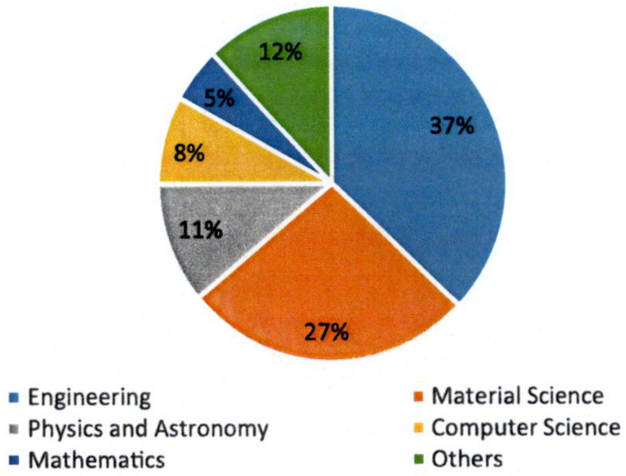

Fig. 6.6 Percentage of publications according to research fields

per article) compared to 13.3 citations per article for the US and 12.9 citations per article for China since 1965. Even though Germany ranked fifth in the chart, the publications started early in the 1970s. Researchers from Malaysia published 16 articles until the third quarter of 2020. Even though Malaysia only ranked in the 20th, the first journal [11] articles published regarding the study started in 2015 with 17 citations per article.

As shown in Fig. 6.6, most of the study's contribution is in the engineering and material science field with 37% and 27% respectively. At the early stage of the study, the applications used for lattice structure was not yet clear. However, the mechanical testing studied among researchers gradually led to increased focus on compression [12], tensile testing, and energy absorption aspects in understanding the characteristics of lattice structures. The focus was then turned to dynamic analysis [13] as its application widened to automotive and aerospace industries. Other contributions and applications are in biomedical [14] and health department [15]. The characteristic of the current research output for lattice structure is summarized in Table 6.1. The evolution of article contribution increased in quantity and quality.

Additive manufacturing journal only started to publish articles regarding lattice structure in 2014 and ranked second with 17.78 citations per article, as shown in Table 6.2. International Journal of Mechanical Sciences, Materials Science and Engineering A, and Journal of the Mechanical Behavior of Biomedical Materials have high citations per article with 34.41, 51.00 and 34.00, respectively. Acta Materialia started to publish literature regarding lattice structure in 2009, with 8 articles. After 11 years, the journal ranks the highest citations per article with 56.88 citations.

The main topic of a research article can be easily determined by the keywords used. The selection of keywords changes over the years. From 1965 to 1999, the number of articles in the selected field study is extremely low; hence the keywords information is unreliable. Starting from 2000 to 2010, the research topic became clearer with "lattice structure" and "cellular manufacturing" as the most frequent keywords used. The micro-structure and micro-architecture of the structure were recognized when the keywords like "porosity", "sandwich structures", "topology",

Table 6.1 Evolution of research articles in lattice structure from 1965 to 2020

Year	Article	Article (%)	Author	Authors per article	Cited reference	Cited references per article	Citation	Citations per article
1965–1969	1	0.1	1	1.0	5	5.0	13	13.0
1970–1974	1	0.1	1	1.0	34	34.0	30	30.0
1980–1984	2	0.2	6	3.0	14	7.0	20	10.0
1985–1989	4	0.4	5	1.3	69	17.3	25	6.3
1990–1994	2	0.2	13	6.5	10	5.0	50	25.0
1995–1999	5	0.4	12	2.4	8748	9.6	46	9.2
2000–2004	6	0.5	21	3.5	131	21.8	35	5.8
2005–2009	36	3.2	100	2.8	805	22.4	258	7.2
2010–2014	127	11.4	385	3.0	3053	24.0	1081	8.5
2015–2019	663	59.6	1772	2.7	18,429	27.8	10,075	15.2
2020	266	23.9	1135	4.3	10,905	41.0	5504	20.7
Total	1113		3451		33,498		17,137	
Average				2.9		21.0		13.7

" crystal structure", and "honeycomb structure" were highly mentioned throughout the decade.

However, the occurrence of keywords related to a fundamental topic shifted to the manufacturing process as the advancement of three-dimensional (3D) printing progressed. As shown in Fig. 6.7, more manufacturing-related keywords were written, which were "3D printers,", "selective laser melting", "additives", "melting", and "manufacture". The term "lattice structures" was selected compared to other synonyms because the configuration of porous structure were introduced [16] and the definition of lattice structure became more significant (as mentioned in the introduction).

The connections of the keywords were mapped and are shown in Fig. 6.8. The keywords were mentioned more than 52 times since 2011. The mapping shows the high occurrence was divided into four main clusters: blue, green, red, and yellow clusters. The red cluster mainly focuses on the structured progress and evolution when "lattice structures", "structural design", "topology", "shape optimization", "structural optimization", were included in the cluster. Other keywords such as "computer aided design", "finite element method" and "design" show the method used to identify the properties of the structure.

The green cluster focuses on the progress in the manufacturing field. "Selective laser melting" as the head of the cluster with a frequency of 263 as the main topic along with "melting", "fabrication" and materials used such as "aluminum alloys", "ternary alloys" and "titanium alloys". Additive manufacturing and lattice structure became closely related because of the high complexity of lattice structure design and the capability of additive manufacturing technology in producing complex forms.

Table 6.2 Top 20 productive journal article sources

Source title	Article (A)	Citation ©	C/A	Starting year
Materials and design	64	1927	30.11	2011
Additive manufacturing	59	1049	17.78	2014
Materials	28	305	10.89	2013
International Journal of Advanced Manufacturing Technology	25	228	9.12	2016
Proceedings of the ASME Design Engineering Technical Conference	25	136	5.44	2010
Rapid Prototyping Journal	20	169	8.45	2015
International Journal of Mechanical Sciences	17	585	34.41	2013
Materials Science and Engineering A	16	816	51.00	2011
Composite structures	13	186	14.31	2012
IOP Conference Series Materials Science and Engineering	12	22	1.83	2011
Journal of the Mechanical Behavior of Biomedical Materials	12	408	34.00	2014
Manufacturing technology	11	28	2.55	2017
Materials Science and Engineering C	11	262	23.82	2012
Advanced engineering materials	10	30	3.00	2010
CAD Computer Aided Design	10	176	17.60	2015
Proceedings of The Institution of Mechanical Engineers Part C Journal of Mechanical Engineering Science	10	18	1.80	2017
Extreme mechanics letters	9	53	5.89	2018
Materials Today Proceedings	9	27	3.00	2017
Metals	9	63	7.00	2017
Acta materialia	8	455	56.88	2009

Lattice structure is mentioned as a biomimicry structure [17]. biomimicry structure is defined as a design inspired by nature, and the main inspiration of lattice structures were living things such as the microstructure of butterfly wings [18], bamboo structure [19], and the microscale of the human bone. As shown in the blue cluster, "biomechanics", "bone", "cytology" and "cells" were in the same group as "porosity", "geometry" and "elastic moduli". Numerous researches were published on variety of topologies of lattice structure as bone replacement.

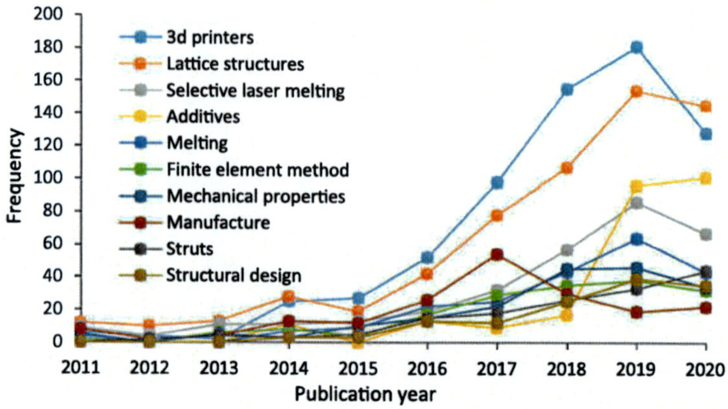

Fig. 6.7 Top 10 most frequent keywords used from 2011 to 2020

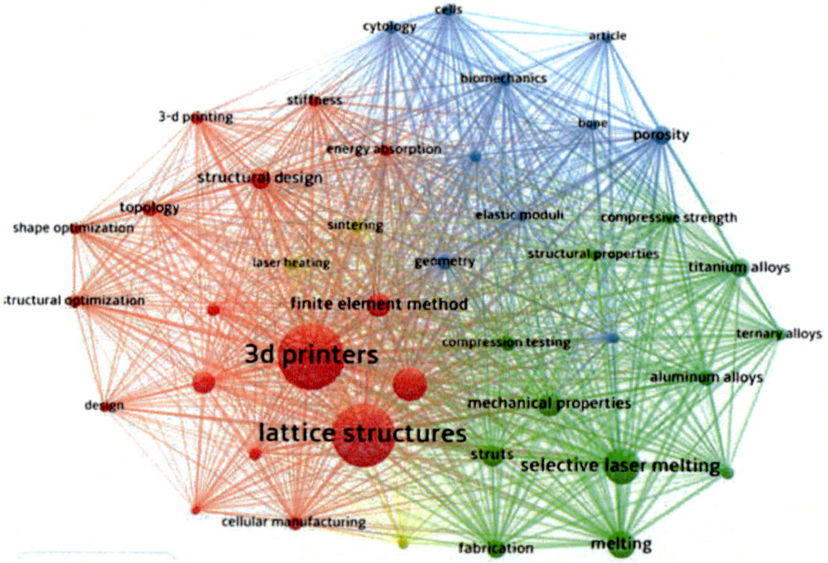

Fig. 6.8 The co-occurrence of index keywords visualization published from 1965 to 2020

6.3.2 Analysis on Research Trends in Fatigue Behaviour of Lattice Structures

Fatigue behaviour in lattice structure is broadly discussed in the literature. From the data collection result output, 234 documents discussed the fatigue behaviour of lattice structure. Figure 6.9 shows a summary of articles and citations from the year

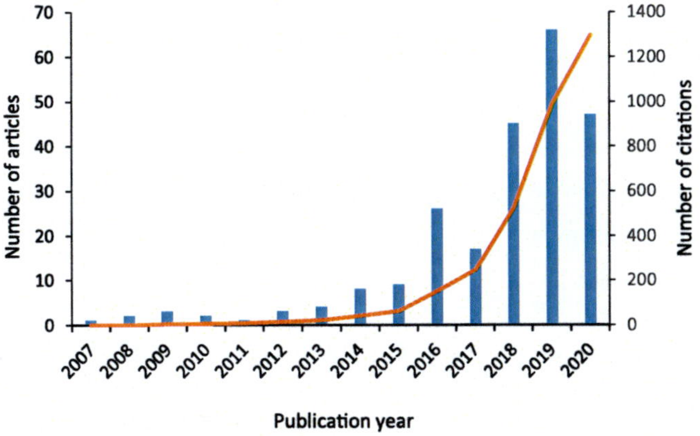

Fig. 6.9 Number of articles and citations from the year 2007 to 2020

2007 to 2020. The publications regarding the topic started in 2007 with only one article and increased annually.

The subject area of fatigue lattice structure is plotted in Fig. 6.10. The highest subject area is recorded from the engineering field with 184 publications, followed by the material science field with 135 publications since 2007, computer science

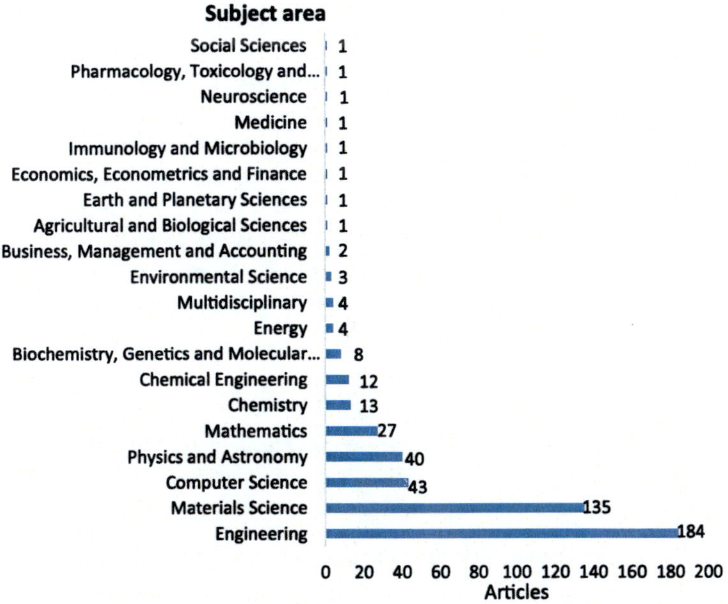

Fig. 6.10 Top 20 subject area of fatigue study in lattice structure

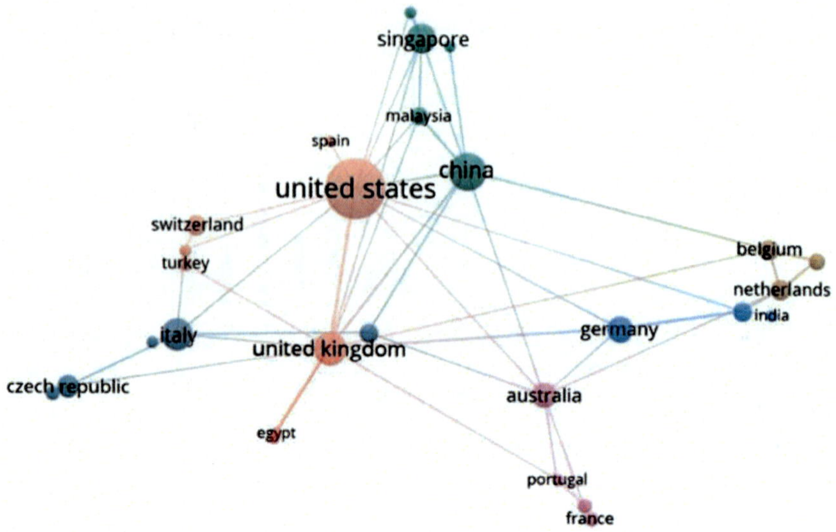

Fig. 6.11 International collaboration connections from 2007 to 2020

and physics, and astronomy with 43 and 40 publications in 13 years. Other non-engineering-related fields such as business, agricultural and economics published 2 and 1 research article. This shows that lattice structure offers various functionalities and can open huge opportunities in producing unique and high-strength low weight structure in various fields.

The collaboration of countries around the world is mapped and is shown in Fig. 6.11. The lines connected the nodes to other nodes represent the international collaborations between respected countries. The largest node with the highest number of publications and great connections is shown by the United States in lattice structure research. In addition, other countries also expanded their connections, such as Malaysia to China, Malaysia to United States, Malaysia to United Kingdom and Malaysia to Singapore. These collaborations not only extend and strengthen connection coverage but also managed to widen researchers' knowledge and improved their publications' quality.

6.3.3 *Lattice Structure Parameters Influencing Fatigue Behaviour*

In 2014, Yavari et al. stated that the type of cell topology affects fatigue properties the most compared to other topology parameters [20]. Choosing an optimum type of unit cell can increase the fatigue life up to ~0.6. The main factors that affect the mechanical properties of a structure inside the cell topology are the porosity, pore

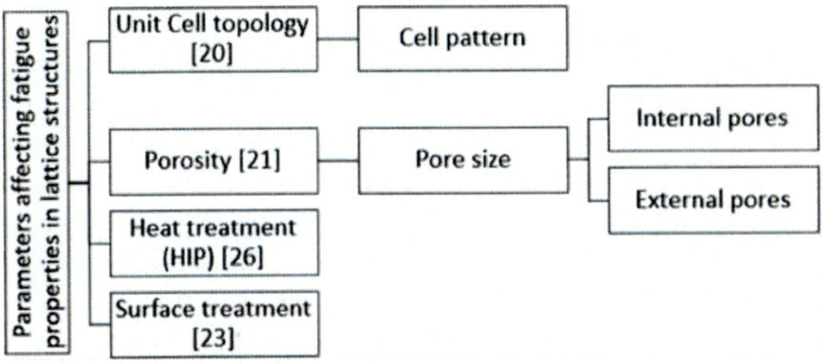

Fig. 6.12 Lattice structure parameters that influence fatigue behaviour

diameter, strut diameter, and strut shape [21]. Surface and heat treatment improve fatigue life by controlling the internal porosity. Metallic additive manufacturing can produce end products with micro internal porosity. Surface and heat treatment are used purposely to reduce internal porosity and prevent crack initiations. Figure 6.12 shows the summary of parameters affecting fatigue properties of lattice structures.

Yang et al. [22] conducted an experiment to compare fatigue resistance between as-built and sand-blasted gyroid lattice structure. Sand-blasted gyroid structure was recorded to show higher resistance to deform in the fatigue testing process which leads to higher fatigue life. Two types of gyroid lattice structures with normal and deformed surface were compared to study the influence of surface quality [23]. In low cycle fatigue, the normal gyroid show better fatigue life compared to deformed gyroid. However, in higher cycle value, the fatigue performance does not show any significant difference. Even though the effect of surface roughness in fatigue behaviour is not clear, it is significantly proven that the uneven surface grown by surface defects are the highest factor of crack propagation.

6.3.4 Effect of Manufacturing Parameters, Heat-Treatment on Additively Manufactured Lattice Structure Mechanical Properties

There are numerous studies on the effect of processing parameters and heat treatment process towards the mechanical properties of additively manufactured lattice structures. Großmann stated that the size of the melt pool highly affects the mechanical properties of lattice [24]. Tsopanos et al. described that the low laser power can produce lower strength lattice structure because of the unmelted powder in the strut cross-sections [25]. They conducted uniaxial compression tests and concluded that manufacturing parameters directly influenced the relative elastic modulus and plateau stress to relative density.

Robinson et al. conducted experiments to study the influence of heat treatment towards the energy absorption of AlSi10Mg of honeycomb structure [26]. Mechanical behaviour of heated (850 °C–920 °C) honeycomb structure was proven to increase when the ductility improved compared to annealed lattice and rolled lattice. Galati et al. heated lattice until 1050 °C and described that heated lattice change the status of brittle Ti6Al4V lattice to ductile [27]. Almost similar conclusions were made by Suzuki et al. that stated that heat treatment at 300 °C for two hours and 530 °C for 6 h improve ductility and reduce stress of EBM-manufactured lattice structure [28]. Sereshk et al. discussed in their paper and stated that with the right specific heat treatment, strength, crashing and energy absorption are significantly improved [29]. Besides heat-treatment, the build orientation during the manufacturing process also influences the fatigue behaviour of a structure. After heat treatment, the crack propagation behaviour is modified and the fatigue crack growth resistance of a material is improved [30]. However, due to the complex forms of additive manufactured structures such as topology optimized forms and lattice structures, this can limit the finishing operations and hence fatigue strength can be compromised [31]. Besides complex forms, geometrical errors in strut shape can also have negative impact of fatigue strength in the manufacturing of lattice structures using selective laser melting technology [32].

6.4 Conclusion

This bibliometric review summarized the trends of lattice structure research publications and its fatigue behaviour. The trends in lattice structure research is investigated based on the citations, subject categories, countries, journal publishers, and index keywords. In the era of IR 4.0, where industries are interested in smart technology such as additive manufacturing, this advancement opens a large opportunity in lightweight products, consequently increasing interest in lattice structure and how to integrate the structure in various industries. This can be observed in the analysis shown in this paper, where the number of articles and citations has largely increased since the development of metal additive manufacturing since the early 2000s. Lattice structure has been studied since 1965 and a major increase is observed in the last 10 years with the progress in additive manufacturing. With the increase in research in this field, the applications and the suitable use for the lattice structure have increased, such as biomedical health and engineering automotive. Fatigue behaviour and its influencing parameters have been determined, such as unit cell topology, heat and surface treatment and lattice structure porosity.

References

1. Zhou S et al (2016) Topological design and additive manufacturing of porous metals for bone scaffolds and orthopaedic implants: A review. Biomaterials 83:127–141
2. Lietaert K, Cutolo A, Boey D, Van Hooreweder B (2018) Fatigue life of additively manufactured Ti6Al4V scaffolds compression and compression fatigue load. Sci Rep December 2017, 1–9
3. Zhang XZ, Leary M, Tang HP, Song T, Qian M (2018) Selective electron beam manufactured Ti-6Al-4V lattice structures for orthopedic implant applications: current status and outstanding challenges. Curr Opinion Solid State Mater Sci 22(3):75–99
4. Maskery I et al (2016) A mechanical property evaluation of graded density Al-Si10-Mg lattice structures manufactured by selective laser melting. Mater Sci Eng A 670:264–274
5. Mazur M, Leary M, McMillan M, Sun S, Shidid D, Brandt M (2016) Mechanical properties of Ti6Al4V and AlSi12Mg lattice structures manufactured by Selective Laser Melting (SLM). In: Laser additive manufacturing: materials, design, technologies, and applications. Elsevier Ltd, 119–161
6. Azman AH (2016) Method for integration of lattice structures in design for additive manufacturing
7. Seharing A, Azman AH, Abdullah S (2019) Comparative analysis between the mechanical behaviour of gradient and uniform lattice structures using finite element analysis. J Eng Sci Technol 14(5):2779–2791
8. Ashby MF, Gibson LJ (1997) Cellular solids: structure and properties
9. Fink A (1998) Conducting research literature reviews: from the internet to paper, Fourth Edi
10. Rahito, Wahab D, Azman AH (2019) Additive manufacturing for repair and restoration in remanufacturing: an overview from object design and systems perspectives. Processes 7(802)
11. Hasib H, Harrysson OLA, West HA (2015) Powder removal from Ti-6Al-4V cellular structures fabricated via electron beam melting. Miner Met Mater Soc 67(3):639–646
12. Wang J, Wang C, Tang H, Liu H, Xiao L, Song W (2015) Mechanical behavior of open-cell rhombic dodecahedron Ti–6Al–4V lattice structure. Mater Sci Eng A (640):375–384
13. Zhang Y, Liu T, Ren H, Maskery I, Ashcroft I (2018) Dynamic compressive response of additively manufactured AlSi10Mg alloy hierarchical honeycomb structures. Compos Struct 195:45–59
14. Chen WM et al (2016) Lattice Ti structures with low rigidity but compatible mechanical strength: design of implant materials for trabecular bone. Int J Precis Eng Manuf 17(6):793–799
15. Lopez H et al (2010) Next-generation biomedical implants using additive manufacturing of complex, cellular and functional mesh arrays. Philos Trans R Soc A Math Phys Eng Sci 368(1917):1999–2032
16. Seharing A, Azman AH (2020) A review on integration of lightweight gradient lattice structures in additive manufacturing parts. Adv Mech Eng 12(6):1–21
17. Seharing A, Azman AH, Abdullah S (2019) Gradient lattice structure bio mimicry design configurations for additive manufacturing. Int J Eng Technol 8(1):36–43
18. Wilts BD et al (2017) Butterfly gyroid nanostructures as a time-frozen glimpse of intracellular membrane development. Sci Adv 3(4):e1603119
19. Tan T et al (2011) Mechanical properties of functionally graded hierarchical bamboo structures. Acta Biomater 7(10):3796–3803
20. Yavari SA, Ahmadi SM, Pouran B, Schrooten J, Weinans H, Zadpoor AA (2014) Relationship between unit cell type and porosity and the fatigue behavior of selective laser melted meta-biomaterials. J Mech Behav Biomed Mater 43(march2015):91–100
21. Han C et al (2018) Continuous functionally graded porous titanium scaffolds manufactured by selective laser melting for bone implants. J Mech Behav Biomed Mater 80:119–127
22. Yang L et al (2019) Compression-compression fatigue behaviour of gyroid-type triply periodic minimal surface porous structures fabricated by selective laser melting. Acta Mater 181:49–66
23. Yánez A, Paula M, Cuadrado A, Martel O, Monopoli D (2020) Surface roughness effects on the fatigue behaviour of gyroid cellular structures obtained by additive manufacturing. Int J Fatigue 138(April):105702

24. Großmann A, Felger J, Frölich T, Gosmann J, Mittelstedt C (2019) Melt pool controlled laser powder bed fusion for customised low-density lattice structures. Mater Des 181:108054
25. Tsopanos S, Mines RAW, Shen Y, Cantwell WJ, Brooks W, Sutcliffe CJ (2016) The influence of processing parameters on the mechanical properties of selectively laser melted stainless steel microlattice structures. J Manuf Sci Eng 132(August 2010):1–12
26. Jin N, Zhenyu Y, Wang Y, Cheng H, Zhang H (2020) Effects of heat treatment on microstructure and mechanical properties of selective laser melted Ti-6Al-4V lattice materials. Int J Mech Sci 190(15 Januari 2021):106042
27. Galati M et al (2019) Ti-6Al-4V lattice lattice structures produced by EBM: heat treatment and mechanical properties. Proc CIRP 88:411–416
28. Suzuki A, Sekizawa K, Liu M, Takata N, Kobashi M (2019) Effects of heat treatments on compressive deformation behaviors of lattice-structured AlSi10Mg alloy fabricated by selective laser melting. Adv Eng Mater 1900571:1–9
29. Reza M, Sereshk V, Shrestha R, Lessel B, Phan N (2019) Mechanical behavior of additively-manufactured gyroid lattice structure under different heat treatments. Solid Freeform Fabrication Symposium, 2263–2269
30. Ye H, Le F, Wei C, Ye K, Liu S, Wang (2021) Fatigue crack growth behavior of Ti-6Al-4V alloy fabricated via laser metal deposition: effects of building orientation and heat treatment. J Alloys Comp 159023:868
31. Maccioni L, Fraccaroli L, Borgianni Y, Concli F (2021) High-cycle-fatigue characterization of an additive manufacturing 17–4 PH stainless steel. Key Eng Mater 877:49–54
32. Mahmoud D, Al-Rubaie KS, Elbestawi MA (2021) The influence of selective laser melting defects on the fatigue properties of Ti6Al4V porosity graded gyroids for bone implants. Int J Mech Sci 193:106180

Chapter 7
Review on Coastal Liquefaction at Sabah Bays

Noor Sheena Herayani Harith, Isabella Boy Setanis, Ejria Saleh, and Tze Liang Lau

Abstract Soil liquefaction is a phenomenon that occurs when the soil loses strength. During the incident, especially when there was an earthquake, it caused the soil to behave like a liquid and caused the ground to lose the ability to bear its structure. Fortunately, none of the events occurred in Sabah province. However, in the gulf and coastal zones of Sabah and there is a need to identify the state of its soil structure that has potential to liquefaction especially areas close to the fault zone. The scope of the study will focus along the coastline of Sabah's bay. Liquefaction factor is determined using the Factor of Safety (FS) calculation where if the FS is more than 1.0, therefore no need for any liquefaction consideration, however if the FS value is less than 1.0, then it will be classified as liquified soil. At the end of the study, it can be summarized that one location is found to be fully exposed to liquefaction. Furthermore, other bay areas show the signs of liquefaction only at a depth of 0.9 m meters below the ground. Nevertheless, three other locations in West Sabah showed no signs of liquefaction. This study presents an easier method for evaluating liquefaction involving simple calculations, and the results obtained using the techniques are rather acceptable.

Keywords Soil liquefaction · Earthquake · Sabah bays area

N. S. H. Harith (✉) · I. B. Setanis
Faculty of Engineering, University Malaysia Sabah, Jalan UMS, 88400 Kota Kinabalu, Sabah, Malaysia
e-mail: sheena@ums.edu.my

N. S. H. Harith
Faculty of Science and Natural Resources, Natural Disaster Research Centre (NDRC), University Malaysia Sabah, Jalan UMS, 88400 Kota Kinabalu, Sabah, Malaysia

E. Saleh
Borneo Marine Research Institute, Universiti Malaysia Sabah, 88400 Kota Kinabalu, Sabah, Malaysia

T. L. Lau
School of Civil Engineering, Universiti Sains Malaysia, 14300 Nibong Tebal, Pulau Pinang, Malaysia

S. Abdullah et al. (eds.), *Structural Integrity Cases in Mechanical and Civil Engineering*, Structural Integrity 23, https://doi.org/10.1007/978-3-030-85646-5_7

7.1 Introduction

7.1.1 Sabah Fault Line and Seismicity

Sabah experienced earthquakes originating locally with magnitudes up to 6.0 on the Richter scale over the years. Besides the local earthquake, Sabah was also affected by large earthquakes located in the Southern Philippine and the Macassar Straits, the Sulu Sea, and the Celebes Sea, where locals testified that tremors were felt during the earthquake incident originated from these areas. In the circumstances when a greater magnitude of the earthquake occurs, there is a higher likelihood that dangers such as soil liquefaction will occur along with structural damage.

Sabah had its own earthquake series in 1900 and this is the first-ever known earthquake from a collection of various historical records. Since then, about 500 of low-to-moderate magnitudes have occurred in this region between magnitude of M_W 2.0–6.9. In Sabah seismic cases, the area is divided into several seismic sources that be characterized by earthquake concentrations including area sources; Ranau, Labuk and Lahad Datu, line source; Pitas, Perancangan, Pensiangan, Tawau and Kunak. The seismic source zone identified is shown in Fig. 7.1, with an average depth of 15 km.

Earthquake liquefaction in Sabah is a necessary study due to its active earthquake. Upon the occurrence of liquefaction by the earthquake, ground deformations can be categorized as primary deformation if located on or close to the ruptured fault. Mohd

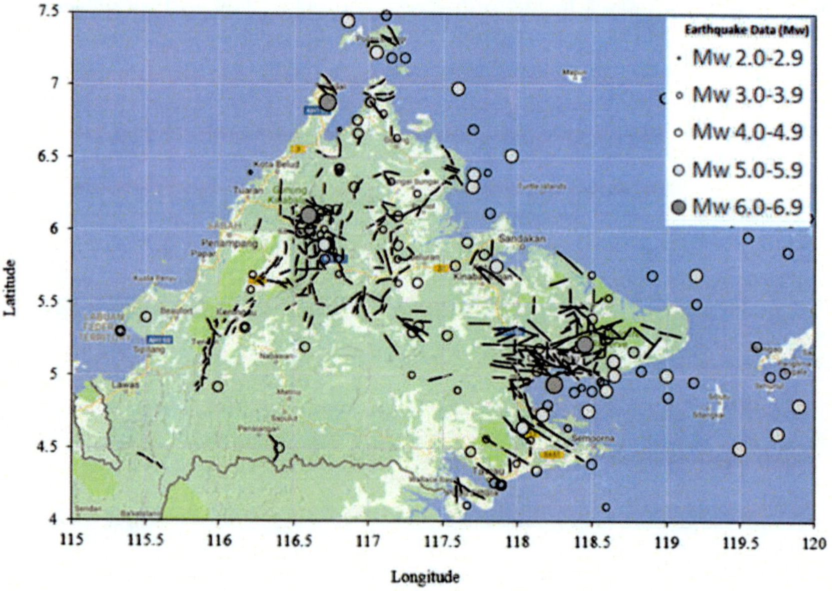

Fig. 7.1 The local seismic source and seismicity of Sabah

Fadhli et al. [1] has studied the effects of vibration on porosity of unsaturated soil with different water content is correlated to the earthquake events in the Ranau and Tawau areas. From their findings, vibrations, and seepage of liquids in the soil are problems that require major emphasis and attention to ensure geo-environment sustainability.

Fractured soil reduces the shear strength of intact soil and increases hydraulic conductiveness. In another discovery, Mohamad et al. [2] took a sample of the borehole that showed the presence of 7.2 m thick where the contents underneath were soft and consisted of clay with some organic matter. According to Sajid [3], excessive pores water pressure causes the ejection of soil through holes in the ground, thus creating sand boils. Hence, when the water pressure exceeds the level, it forces fine granular particles to move with each other as the load increased. In one case around Kota Kinabalu area, it faced post-construction settlement problems due to the construction on soft soil. The most effective way to reduce the disasters caused by the earthquake liquefaction is to estimate the hazard and to increase the information for future use in building design and construction.

Ahmadreza [4] noted that the most populous cities live within coastal area, i.e., Kota Kinabalu, Sandakan, Tawau, Lahad Datu and Semporna. As in case of famous liquefaction due to the earthquake can be seen in the previous Sulawesi earthquake in 2018 where the citizens live along the coastline. The liquefaction incident was caused by the coastal and submarine landslides as characterized by the flow of liquefied gravity [5].

7.1.2 Liquefaction Scaling Factor

In a study of soil liquefaction hazard on the coast of Terengganu, Hashim et al. [6] found that serious erosion on land deposits along the coastline. The soil consists of sandy with low plasticity and shallow groundwater. If a cohesionless soil is saturated, excess pore pressures may accumulate during seismic shearing and might lead to liquefaction. When seismic waves pass through the saturated soil layer by the rapidly applied loading, this might alter the fine granular structures and causes the fine coarse particles to collapse.

Liquefaction is a tendency of saturated soil that is not consolidated to behave like a liquid when introduced to the earthquake [7, 8]. Millea [9] states that if saturated granular material is subject to a continuous cyclic loading, the soil can achieve zero effective stress and will suffer a partial or complete loss of strength. The phenomena of deposited soil liquefaction are best described using the term reduction of soil shear strength due to the formation of pore water pressure in the soil structure specified as follows:

$$\tau = \sigma' tan\varphi \tag{7.1}$$

$$\sigma' = \sigma - u \tag{7.2}$$

Table 7.1 Magnitude scaling factor [11]

Earthquake magnitude	Magnitude scaling factor
5.5	2.20
6.0	1.76
6.5	1.44
7.0	1.19
7.5	1.00
8.0	0.84
8.5	0.72

Symbols indicate that, τ is shear strength, σ' is an effective normal stress, φ is an angle of internal friction of the soil, σ is the amount of normal stress and u is pore water pressure. Factor of safety (FS) is a measure of the soil's capacity to withstand liquefaction during the earthquake [10]. FS is calculated from the value of the Cyclic Resistance Ratio (CRR) per Cyclic Stress Ratio (CSR) multiply with magnitude scaling factors (MSF) (Eq. (7.3)) on different layers of soil as shown in the Eq. (7.4) [11].

$$\mathrm{MSF} = 87.2(M_w)^{-2.215} \tag{7.3}$$

$$\mathrm{FS} = \frac{CRR}{CSR}(MSF) \tag{7.4}$$

The symbol indicates that M_W is a moment magnitude and MSF can be determined using Table 7.1.

7.2 Methodology

In the preliminary study, borehole data, Sabah Geological Map and Malaysian Standard National Annex MS EN 1998–1:2015 [12] are referred. Figure 7.2 shows the methodological flow chart performed in this study. Peak Ground Acceleration (PGA) hazard map of Sabah at 10% probability of exceedance in 50 years return period is taken into consideration. Sabah Seismic Hazard Map has been developed to provide the most comprehensive and detailed information needed to help engineers design buildings, bridges, highways and services that can sustain earthquake loads. Peak Ground Acceleration (PGA) is fundamental in seismic design as it is directly associated with force, and buildings can be built to withstand at horizontal loads.

The soil liquefaction analysis modelling, described in Youd et al. [13], as well as recalculation proposed by Cetin et al. [14], will be used to estimate the potential liquefaction at Sabah bays. A simplified approach measures the resistance of the soil layer in terms of Cyclic Resistance Ratio (CRR) and Cyclic Stress Ratio (CSR). These

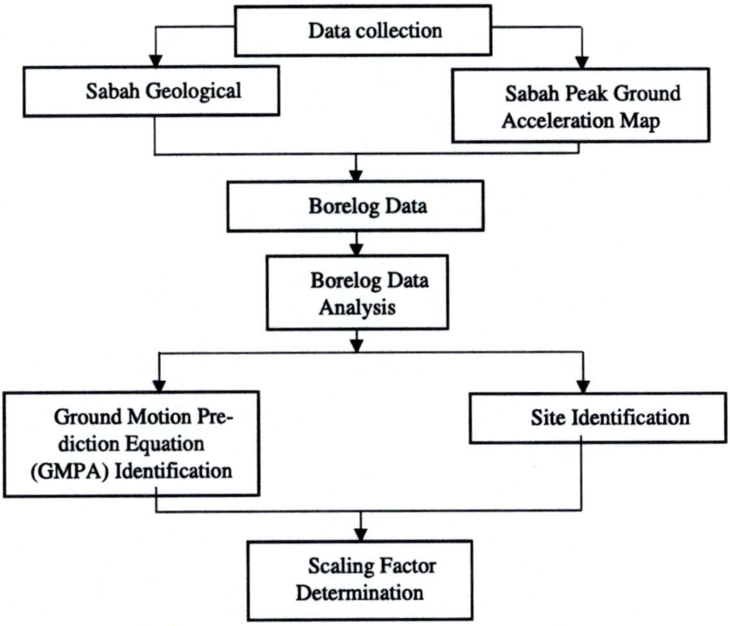

Fig. 7.2 Flow chart of methodology

are measures in assessing the factor of safety (FS) of a given soil layer that subjected to the liquefaction. If the liquefaction analysis shows the FS value is more than or equivalent to 1.0, there is no need for any liquefaction consideration. Nevertheless, if the soil layers reflect the FS less than 1.0, thus it is considered as the liquefying capacity.

The FS for the analysis of liquefaction in sixteen Sabah bays areas are to be analyzed starting from the borehole data analysis. Then, the modelling of liquefaction analysis should be determined at 50 km in the radius to closest fault zones. The site locations can be seen in Fig. 7.3 marked in red circle indicates the location of bore log data in the west, east and Southeast of Sabah. The magnitude of M_W 6.5 will be applied in this study. Then, the ground motion parameters will be calculated from the Ground Motion Prediction Equation (GMPE). The selected GMPE, including those by Frankel et al. [15], Toro et al. [16], and Silva et al. [17] will be used in this study. Figures 7.4 and 7.5 show the samples of bore log data analysis done in this study.

7.3 Result

The tendency of liquefaction from the 15 Sabah bays areas have been analyzed, and the following summary of abbreviations will be used: Kimanis Bay = BK, Gaya Bay = BG, Likas Bay = BL 1, Sepanggar Bay = BS 1, Usukan Bay = BU, Marudu Bay

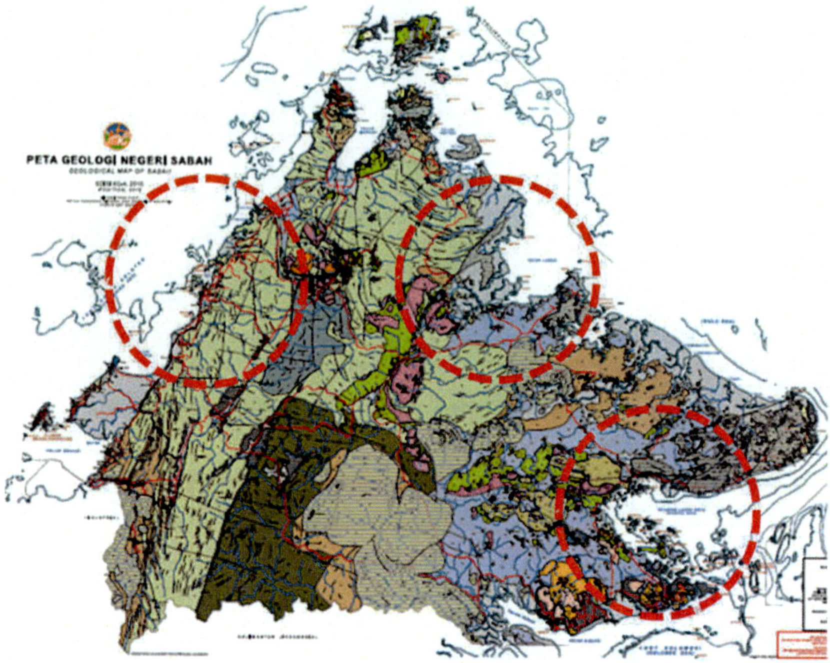

Fig. 7.3 Example of fault identification at 50 km site radius around at East Sabah (three red circle shows the study area) [18]

= BM 2, Paitan Bay = BP, Labuk Bay = BL 2, Darvel Bay = BD, Tawau Bay = BT 2, Keningau = KGU, Kota Kinabalu = KK, Penampang = PPG, Ranau = RNU, and Tambunan = TBN.

In Table 7.2, the parameters at each selected location site indicate that only one area (Marudu Bay) is susceptible to liquefaction. The PGA value on the site is 0.14 g and close to the Mesilou fault zone. Three locations at Kimanis Bay, Sepanggar Bay and Keningau Bay area showed no signs of liquefaction at PGA less than 0.04 g. However, other bay areas showed signs of liquefaction at a depth of more than 0.9 m meters below ground.

Figures 7.6 and 7.7 showed boring samples analyzed for FS in BM2 and KGU locations, respectively. The red line shows the FS limit of less than 1.0 or more than 1.0, respectively. BM2 sample showed the FS is in a range of 0.5 to 1.5 and depth from 1.0 to 18 m. Although at a depth 4.0 to 7.0 m has an FS of more than 1.0, however, overall, the BM2 sample has an FS value of less than 1.0, therefore it is considered as liquefied soil. KGU samples showed the FS value as a whole more than 1.0, thus there was no need for any liquefaction considerations.

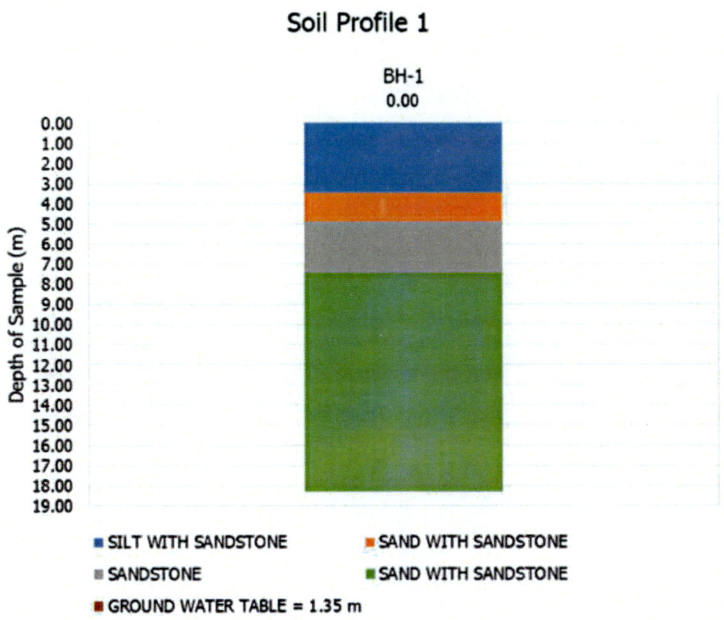

Fig. 7.4 Sample of soil profile 1 of current study

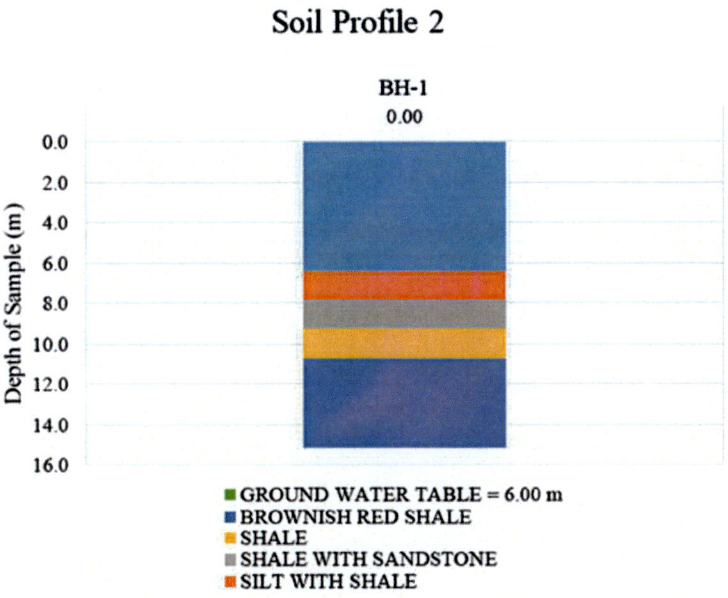

Fig. 7.5 Sample of soil profile 2 of current study

Table 7.2 Liquefaction factor of safety (FS) results at 15 locations along Sabah bays

Location	Model	Fault	PGA (g)	FS
Kimanis Bay	BK	Lobou-Lobou	0.02	>1.0
Gaya Bay	BG	Mensaban	0.03–0.04	<1.0 at 3 m depth
Likas Bay	BL1	Lobou-Lobou	0.04	<1.0 at 0.9 m depth
Sepanggar Bay	BS1	Lobou-Lobou	0.04	>1.0
Usukan Bay	BU	Mesilou	0.05	<1.0 at 2.0 m depth
Marudu Bay	BM2	Mesilou	0.14	<1.0
Paitan Bay	BP	Lobou-Lobou	0.05	<1.0 at 1.5 m depth
Labuk Bay	BL2	Lahad Datu	0.07	<1.0 at 4.0 m depth
Darvel Bay	BD	Lahad Datu	0.08	<1.0 at 3.0 m depth
Tawau Bay	BT2	Lahad Datu	0.08	<1.0 at 2.0 m depth
Keningau	KGU	Lobou-Lobou	0.03	<1.0
Kota Kinabalu	KK	Lobou-Lobou	0.04	<1.0 at 1.5 m depth
Penampang	PPG	Lobou-Lobou	0.04	<1.0 at 1.5 m depth
Ranau	RNU	Mensaban	0.15	<1.0 at 12 m depth
Tambunan	TBN	Lobou-Lobou	0.05	<1.0 at 1.5 m depth

Fig. 7.6 Factor of Safety (FS) at BM2 location (red line shows the limit of FS at 1.0)

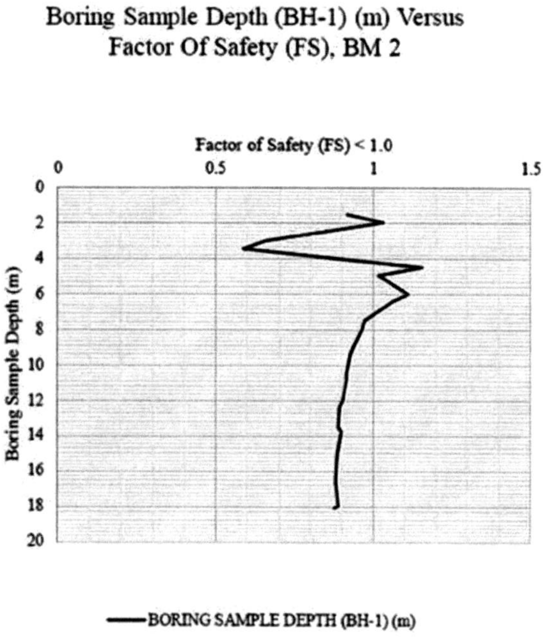

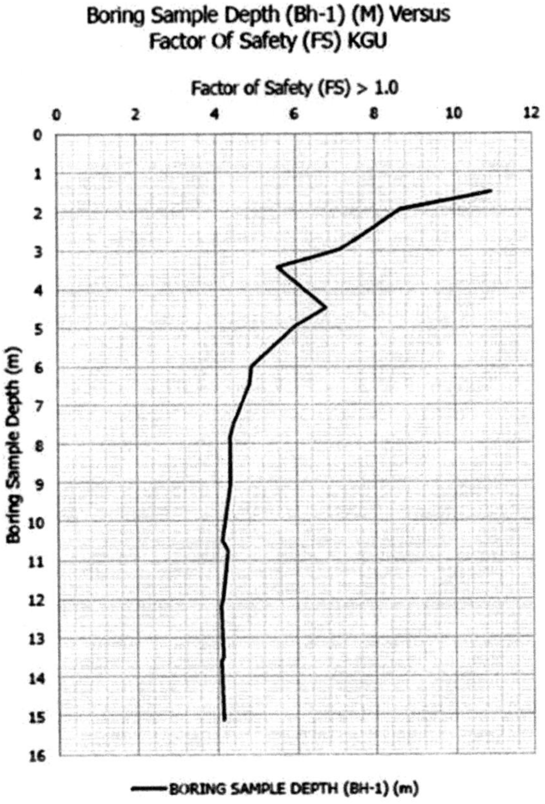

Fig. 7.7 Factor of Safety (FS) at KGU location (red line shows the limit of FS at 1.0)

7.4 Conclusion

The findings showed that the liquefaction results in terms of factor of Safety (FS) in selected Sabah's bay areas indicated that only one area (Marudu Bay) was exposed to liquefaction. The peak ground acceleration (PGA) value on site is 0.14 g and closest to the Mesilou fault zone. Three locations at Kimanis Bay, Sepanggar Bay and Keningau Bay areas showed no signs of liquefaction at low PGA of 0.04 g. In other bays, area showed signs of liquefaction at a depth of more than 0.9 m. This study presents an easier method of evaluating liquefaction involving simple calculations and the results obtained using the techniques are quite acceptable.

References

1. Mohd Fadhli AR, Noraliani A, Kamarudin A, Radzuan S, Mohd Zamri R, Zulkiflee I (2019) Impact of vibration on double-porosity unsaturated laterite soil with different water contents.In: MATEC Web of Conferences. 255:1–6
2. Nor Omar Md, Razali CE, Hadi AAA, Som PP, Eng BC, Rusli MB, Mohamad FR (2016) Challenges in construction over soft soil—case studies in Malaysia. In: Soft soil engineering international conference 136:1–8
3. Ali S (2001) Liquefaction. Liquefaction and Factors 1–15. https://www.academia.edu/37954099/Liquefaction_and_Factors. Last accessed 2020/11/01
4. Ashraf A (2015) Elemental distribution of core marine sediments in the coast of Sabah, Malaysia by using neutron activation analysis and inductively coupled plasma spectroscopy. PhD. Thesis, Universiti Putra Malaysia, Serdang
5. Sassa S, Takagawa T (2019) Liquefied gravity flow-induced tsunami: first evidence and comparison from the 2018 Indonesia Sulawesi earthquake and tsunami disasters. Landslides 16:195–200. https://doi.org/10.1007/s10346-018-1114-x
6. Hashim H, Suhatril M, Hashim R (2017) Preliminary study of soil liquefaction hazard at Terengganu shoreline, Peninsular Malaysia. In: IOP Conf Ser Mater Sci Eng 210(1):1–10
7. Rauch AF (1997) EPOLLS: an empirical method for predicting surface displacements due to liquefaction-induced lateral spreading in earthquakes. Ph.D. Dissertation. Virginia Polytechnic Institute and State University, United States
8. Herath P (2016) Calculating liquefaction potential of Northern Mississippi using shear wave data. PhD. Thesis, University of Mississippi
9. Millea MT (ed) (1990) Liquefaction mitigation technology. Report Number TN-1808. Office of Naval Technology, Virginia
10. Holzer TL, Noce TE, Bennett MJ (2009) Scenario liquefaction hazard maps of Santa Clara Valley, Northern California. Bull Seismol Soc Am 99(1):367–381
11. Song CR, Mikell N (2013) Earthquake and piping hazard assessment for Desoto, Tunica, and Tate County, Mississippi. PhD. Thesis, University of Mississippi
12. MS EN 1998 Part 1 (2017) Malaysia National Annex to Eurocode 8: Design of structures for earthquake resistance - Part 1: General rules, seismic actions and rules for buildings. Department of Standards Malaysia, Cyberjaya
13. Youd TL, Idriss IM, Andrus RD, Arango I, Castro G, Christian JT, Dobry R, Finn WDL, Jr, LFH, Hynes ME, Ishihara K, Koester JP, Liao SSC, Marcuson III WF, Martin GR, Mitchell JK, Moriwaki Y, Power MS, Robertson PK, Seed RB, Stokoe II KH (2001) Summary Report. Liquefaction resistance of soils: summary report from the 1996 NCEER and 1998 NCEER/NSF workshops on evaluation of liquefaction resistance of soils. In: Proceedings of the NCEER workshop on evaluation of liquefaction resistance of soils. Utah, 5–6 January, 1–40
14. Cetin KO, Seed RB, Der Kiureghian A, Tokimatsu K Jr, LFH, Kayen RE, Moss RES (2004) Standard penetration test-based probabilistic and deterministic assessment of seismic soil liquefaction potential. J Geotechnic Geoenviron Eng 130(12):1314–1340
15. Frankel A, Mueller C, Barnhard T, Perkins D, Leyen-decker E, Dickman N, Hanson S, Hopper M (1996) National seismic-hazard maps: documentation June 1996. Open- File Report 96–532. Geological Survey, Denver
16. Toro GR, Abrahamson NA, Schneider JF (1997) A model of strong ground motions from earthquakes in Central and Eastern North America N Best estimates and uncertainties. Seismol Res Lett 68:41–57
17. Silva W, Gregor N, Darragh R (2002) Development of regional hard rock attenuation relations for Central and Eastern North America. Internal Report November 1, 2002. Pacific Engineering, El Cerrito
18. Farhan Mohd Zainudin MS, Zubir N, Yan ASW, Amnan I, Among HL, Javino F (2015) Geological map of Sabah. Miner Geosic Malaysia (JMG)

Chapter 8
Ground Motion Prediction Equations for Sabah Region

Noor Sheena Herayani Harith, Nur Afifah Kassim, Samnursidah Samir, and Azlan Adnan

Abstract The reduction in earthquake wave amplitude is characterized by the ground motion prediction equation or GMPE. The collection of the GMPE is one of the essential steps in probabilistic seismic hazard assessments (PSHA). The calculation involves using multiple data such as geological, seismological, and site of interest. This paper aims to compile data catalogues in terms of magnitude and Peak Ground Acceleration (PGA) for the Sabah region and determine the suitable GMPE for low-to-moderate seismicity. The selected models were plotted and compared with the dataset of earthquakes for the period of 2004–2019. From this study, it can be concluded that two out of ten equations were found to fit the earthquake recorded.

Keywords Low-to-moderate · GMPE · Peak ground acceleration · Curve-fitting

8.1 Introduction

8.1.1 Background of Study

The vast majority of the world's earthquakes, around 90%, occur along this geologically unstable area of Southeast Asia. Plate boundaries are always moving in slow motion eventually creates tension that causes a significant earthquake shaking. The deformation of Southeast Asia was a combined result of continental collision and oceanic subduction. However, Malaysia is very fortunate despite being surrounded

N. S. H. Harith (✉) · N. A. Kassim
Faculty of Engineering, University Malaysia Sabah, Jalan UMS, 88400 Kota Kinabalu, Sabah, Malaysia
e-mail: sheena@ums.edu.my

N. S. H. Harith · S. Samir
Faculty of Science and Natural Resources, Natural Disaster Research Centre (NDRC), University Malaysia Sabah, Jalan UMS, 88400 Kota Kinabalu, Sabah, Malaysia

A. Adnan
School of Civil Engineering, Universiti Teknologi Malaysia (UTM), 81310, Skudai, Johor Bahru, Malaysia

S. Abdullah et al. (eds.), *Structural Integrity Cases in Mechanical and Civil Engineering*, Structural Integrity 23, https://doi.org/10.1007/978-3-030-85646-5_8

by countries with large earthquakes, but its position is stable in Eurasian Plate. Malaysia's location is also geographically outside of the ring of fire [1] (Fig. 8.1).

Earthquakes are one of the world's most devastating and frightening natural disasters. The unpredicted earthquake hazards can damage the surrounding people and buildings. The greatest threat in the earthquake occurrence was the collapse or fall of human-made and natural structures that caused the extensive loss of life and property. As a result, the unpredictable earthquake's effect may be considered in countries with a high risk of massive earthquakes and regions subjected to low-to-moderate seismicity.

Since the 1930s, strong ground motion networks have been established along the seismically active regions to understand the process of seismic activity [3]. Ground motion records were essential for engineers to understand the seismic behavior of

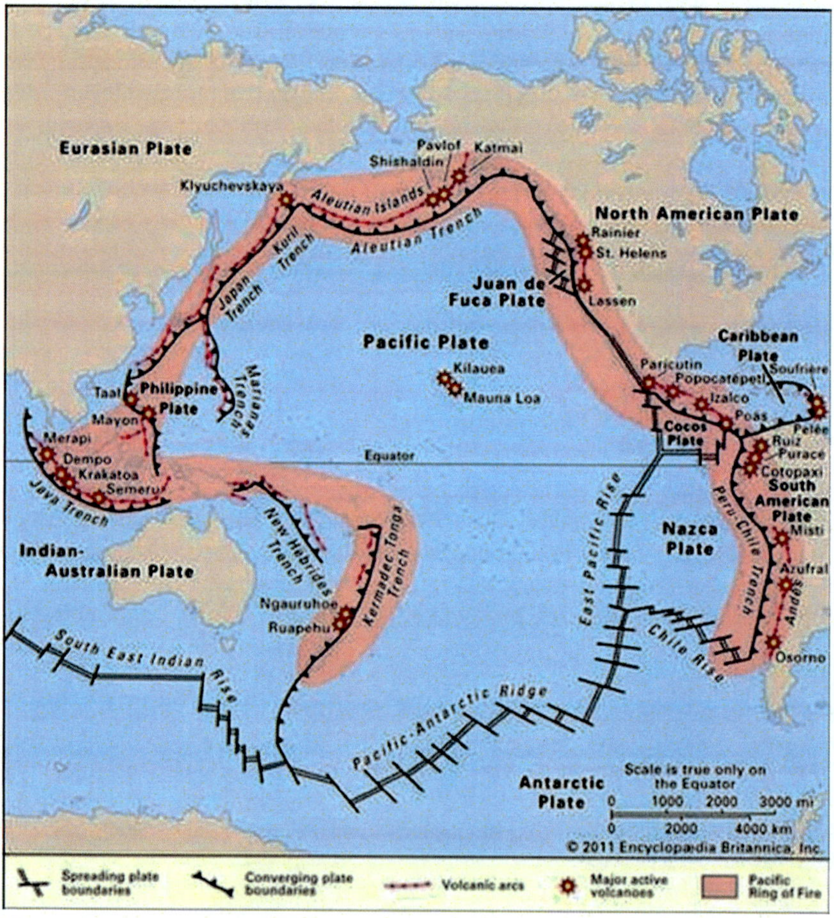

Fig. 8.1 Malaysia location within the ring of fire [2]

structures. The ground motion simulations have also been used to conduct a linear and nonlinear dynamic analysis of structures. Although ground motions are typical in earthquake engineering studies, the current substantial ground motion databases in Sabah are still limited.

According to the past researcher by Harith et al. [4], seismic hazard analysis is usually carried out to establish seismic design criteria for building and other critical structures such as nuclear power plants and dams. The analysis depends on the identification of the seismic sources and identifiable from earthquake catalogues. Thus, the essential Ground Motion Prediction Equation (GMPE) in seismic hazard studies for estimating the seismic sources' ground motions depends on earthquake magnitude, source-to-site distance, and local site condition. The chosen GMPEs must represent the region's geological and seismological attributes containing a wide range of magnitude and source-to-site distance.

Ground Motion Prediction Equations (GMPEs) are the mathematical equations that provide a means of predicting the level of ground shaking and its associated uncertainty at any given site or location, based on an earthquake magnitude, source-to-site distance, local soil conditions, and fault mechanism. These equations are typically developed by regression of the peak ground motion values to first-order variables such as magnitude and distance and possibly several other predictive variables [5]. The GMPEs are commonly developed for ground motion parameters such as Peak Ground Acceleration (PGA), Peak Ground Velocity (PGV), and Pseudo Response Spectral Acceleration (PSA) [6]. GMPE provides a fundamental tool for analyzing seismic hazards by describing the amplitude and attenuation of peak ground motion values from the events to the recording stations [5].

An essential step in any seismic hazard analysis study is selecting suitable Ground-Motion Prediction Equations (GMPEs) based on the region-specific parameters. These are widely used for predicting the level of ground shaking in seismic hazard analysis. Appropriate GMPE to calculate the ground motion in Peak Ground Acceleration (PGA) is a prerequisite for any seismic hazard analysis. Various Ground Motion Prediction Equations (GMPE) that become available made it possible to choose the GMPE that suits low-to-moderate seismic regions, where each model is compared to the corresponding records. The models were selected due to the criteria in magnitude and distance. Based on the previous study made by Liou [7], he stated that many GMPEs had been developed, and a comprehensive review of available GMPEs for spectral accelerations was provided during the past two decades.

8.1.2 Earthquake in Sabah

Sabah had rarely experienced a massive earthquake during the past few years but witnessed severe properties damaged. Five years ago, the earthquake magnitude had increased to M_W 6.0. This region has increasingly low-to-moderate seismic activities due to some moderately active faults around the area. Approximately 300 moderate earthquakes have occurred, ranging from a magnitude of M_W 4.0–7.0. In the history of earthquakes, the region already experienced a devastating earthquake with a magnitude of M_W 5.8 on July 26, 1976 centered in Lahad Datu, and just recently a M_W 6.0 on June 5, 2015 in Ranau. All signs indicate that it will continue to have the same problem in the future. The highest intensity of these earthquakes reached VIII degrees in Modified Mercalli Intensity (MMI) Scale equivalent with 3.4 to 6.5 g [8].

There is a need for reliable predictions of ground motion equations over the whole magnitude range in low-to-moderate seismicity regions. Hence, the knowledge of appropriate strong ground motion is essential to ensure structures like power plants, dams, or bridges. The GMPE that is available may not be suitable for accurately handling low-to-moderate earthquake conditions. However, to perform seismic hazard analysis for Sabah, it is fundamental to determine the GMPE using available ground motion records. Nowadays, it is broadly accepted to use GMPE to assess dynamic effects on structures through seismic hazard analysis. In this research, a few selected GMPE from the previous researcher worldwide needs to be reviewed and classified accordingly for low-to-moderate seismicity regions.

8.2 Methodology

8.2.1 Compilation of Earthquake Recorded

The Ground-Motion Prediction Equation (GMPE) was selected for rock sites in Peak Ground Acceleration (PGA). The PGA value of local datasets from strike-slip earthquakes consists of corrected and processed accelerograms. The distances between 10 and 1000 km with moment magnitudes in the range of 2.9 and 6.4 are compiled. The process of filtering earthquake time history is performed on the earthquake records before processing, according to Boore and Bommer [9] study. Original earthquake records are put through baseline correction and then plotted as a time history. Earthquake records are selected manually and damaged, or questionable records are excluded. Records with square waves due to the ground-motion values being too small are also eliminated. Any data collection in a short distance range will have a range of amplitudes because of the natural variability in the ground motion (due to source, path, and site variability).

At distances far enough from the source (depending on magnitude), some of the collection values will be below the amplitude cutoff and therefore excluded [10]. Figure 8.2 shows a sample of the ground motion recorded data where the first and second columns present the EW and NS component. The ground motion time history is recorded from the epicenter of the May 30, 2005 earthquake carrying a moment magnitude of 5.3 with a distance of 90.8 km from the seismic station. The time 0.0 s refers to the rupture initiation time, and the record has been baseline corrected and low-pass filtered at 0.13 Hz to remove high-frequency components contaminated by local geology and environmental noises.

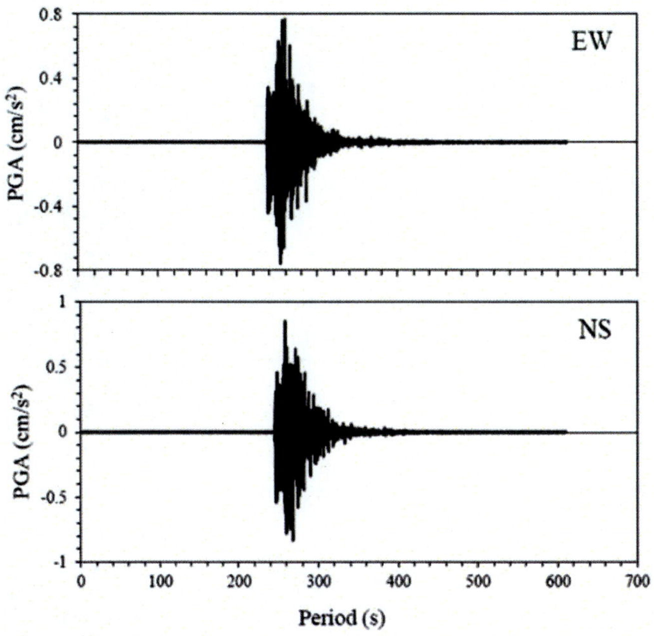

Fig. 8.2 Recorded time histories in East Malaysia due to the local-fault event occurring on May 30, 2005 ($M_W = 5.3$, Rhypo = 90.8 km)

The earthquake information was used as the parameters that focused on GMPEs for active seismicity determination in this study. Fortunately, the row of earthquake data at a different location for the Malaysian Meteorological Department (MMD) events has various stations throughout Malaysia. Liew et al. [11] stated that upon acquiring these data, they found that each seismic event recorded at multiple stations located at various distances spread out spatially throughout several regions such as Peninsular Malaysia, Sabah, and Sarawak. Each recording could be considered independent variables that can increase the data's size depending on its occurrence. The data provided by MMD occurred around Sabah from 2004 to 2019 with the records such as depth, magnitude, distance, and PGA to analyze that recorded data with the developed models. The distribution of all selected records with respect to PGA, magnitude, distance is shown in Figs. 8.3 and 8.4.

8.2.2 *Description of the GMPE Considered*

The earthquake originates deep below the ground surface, called a hypocenter and will create a seismic wave. The seismic wave waves then travel from the hypocenter to the ground surface. The earthquake wave amplitude will be reduced as it travels away

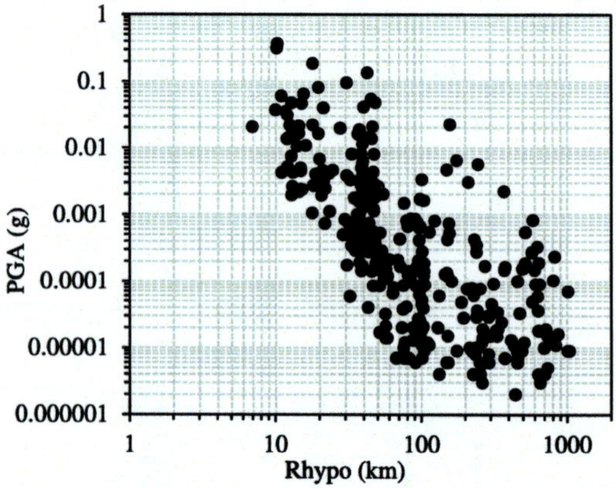

Fig. 8.3 Distribution of data used in terms of Peak Ground Acceleration (PGA) and distance

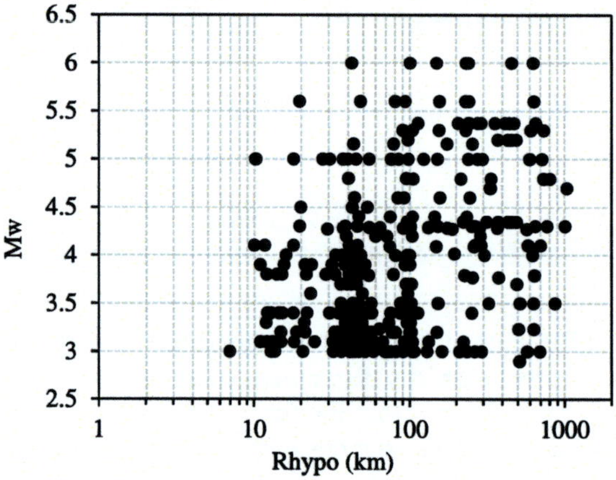

Fig. 8.4 Distribution of data used in terms of magnitude and distance

from its origin; thus, it is characterized by the ground motion prediction equation or GMPE. The majority of the widely used GMPEs offer the hazard in Peak Ground Acceleration (PGA). The seismic hazard analysis modeling should capture the uncertainties about earthquake input parameters such as the earthquake's magnitude and distance. The concept of determining a seismic hazard level has been prevalent in every country since first being introduced in the 1960s [12]. In the development of design ground motion, the critical part is in the seismic design of structures, where it is required to perform Probabilistic Seismic Hazard Analysis (PSHA). The aleatory uncertainties are handled with probability distributions with a solid GMPE to estimate earthquake ground motion parameters characterizing the earthquake source, propagation path, and geological condition.

In the seismic design of structures, the most critical part in developing seismic design is ground motion. Probabilistic Seismic Hazard Assessment (PSHA) requires a strong Ground Motion Prediction Equation (GMPE) to estimate earthquake ground motion parameters characterizing the earthquake source, propagation path, and geological conditions. Seismic hazard analysis is different in terms of the definition of seismic sources and GMPE. In the recent past, sufficient ground motion records from low-to-moderate magnitudes have become available to derive the equations. Comprehensive reviews and comparative studies of the past attenuation relations are made by Douglas [13], with each reporting a different mean and standard deviation at a given magnitude and distance. The selected GMPEs are compared with selected equations, and PGA estimates to investigate their sensitivity to the area assigned [14]. Many ground motion equations are necessary to illustrate the situation, especially for a region with no GMPE [15]. They are generally categorized according to global seismotectonic regimes, which are standard categories to differentiate GMPEs [13] (i.e., stable continental regions, subduction zones, and active regions with shallow crustal seismicity).

There are 10 GMPE that have been selected for comparative studies and reviewed. The description of each GMPE is summarized and listed in Table 8.1. The models were selected due to the criteria in magnitude from low-to-moderate ($2.9 \leq M_W \leq 6.4$) and the type of hypocenter distance. All the GMPEs are plotted into a regression analysis graph, PGA (cm/s^2 or gal) versus Rhypo (km), before selecting the most suitable model for low-to-moderate seismicity regions. The details for all equations can be found in Douglas [13] report.

Table 8.1 GMPE Model as described in Douglas [13]

GMPE Model (All models are cited from Douglas [13])	Symbol	Moment magnitude range, M_w	Hypocentral Distance range, Rhypo
Milne (1977)	M77	3.5–7.7	1–380
Blume (1980)	B80	2.1–7.6	0–449
Campbell (1981)	C81	2.7–7.8	3–480
Schenk (1984)	S84	2.5–6.5	2–600
Gaull (1988)	G88	2.6–6.9	2.5–175
Petrovski and Marcellini (1988)	PM88	3–7	8–200
PML (1988)	PML88	3–7	10–150
Stamatovska and Petrovski (1991)	SP91	3–8	10–500
Taylor Castillo (1992)	TC92	3–7.6	6–210
Dahle et al. (1995)	D95	3–8	6–490

8.3 Result

8.3.1 Classification of Ground Motion Prediction Equations

The ten GMPE models will be tested in their validity range (magnitude and distance). The range of magnitude considered here is between M_W 2.9 and 6.4 for comparison in terms of the PGA horizontal components' maximum value, as shown in Figs. 8.5, 8.6, 8.7, 8.8, 8.9, 8.10 and 8.11. The distances are limited to 10–1000 km, representing the near- to far-field earthquakes. PGA estimated by Ground Motion Prediction Equations for low-to-moderate seismicity regions is compared to actual records on rock sites. These actual earthquake recordings were plotted into different range magnitudes; $2.9 \leq M_W \leq 3.4$, $3.5 \leq M_W \leq 3.9$, $4.0 \leq M_W \leq 4.4$, $4.5 \leq M_W \leq 4.9$, $5.0 \leq M_W \leq 5.4$, $5.5 \leq M_W \leq 5.9$, and $6.0 \leq M_W \leq 6.4$ for comparison.

Some GMPEs show a good ability to predict and fit the observation data for example the equation of M77, B80, CS81, S84, and D95; estimate the data correctly at a range of magnitude less than M_W 5.4 except for magnitude at a range of M_W 5.5–6.4 which predicts a much larger PGA. However, it can also be seen that at a magnitude range of 5.5–6.4, these equations seem able to give a good fit if the PGA is larger at distance less than 100 km. The G88 model reflects the observation correctly at all ranges of magnitude.

PM88 and SP91 equations can predict the ground-motion records good for all ranges of magnitude; however, overpredict for more than 50 km. It is agreeable since the equations are for near-field earthquakes. The GMPE made by PML88 and TC92 predicts a larger PGA than the observed PGAs except for a magnitude between 4.5 and 4.9 that fits the data well.

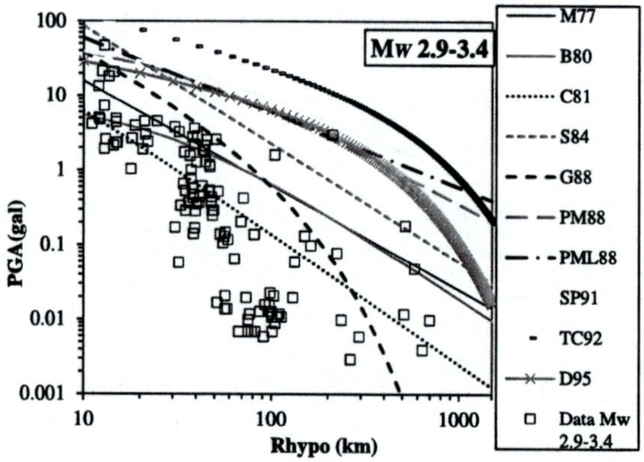

Fig. 8.5 Comparison of GMPEs attenuation curves and recorded PGA for magnitude between 2.9 and 3.4

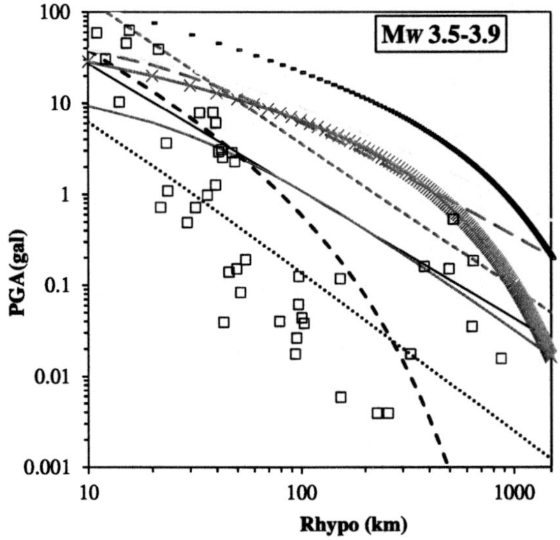

Fig. 8.6 Comparison of GMPEs attenuation curves and recorded PGA for magnitude between 3.5 and 3.9

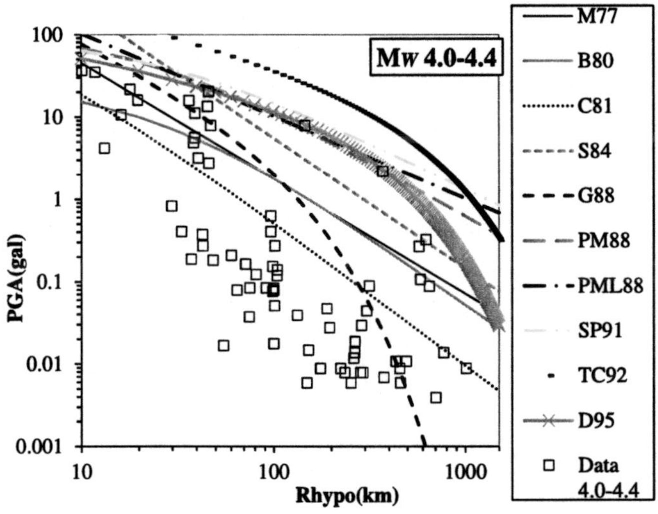

Fig. 8.7 Comparison of GMPEs attenuation curves and recorded PGA for magnitude between 4.0 and 4.4

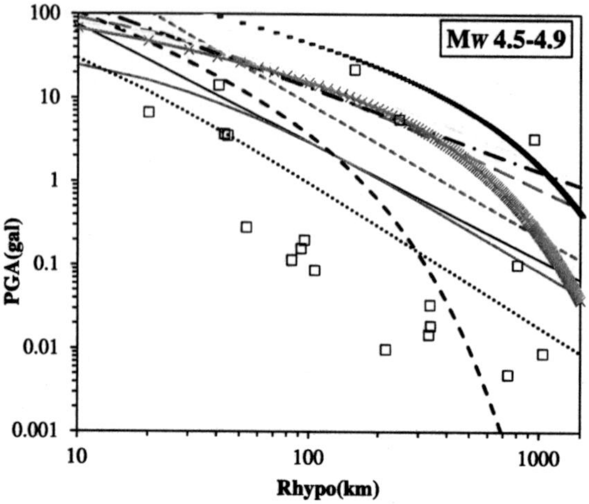

Fig. 8.8 Comparison of GMPEs attenuation curves and recorded PGA for magnitude between 4.5 and 4.9

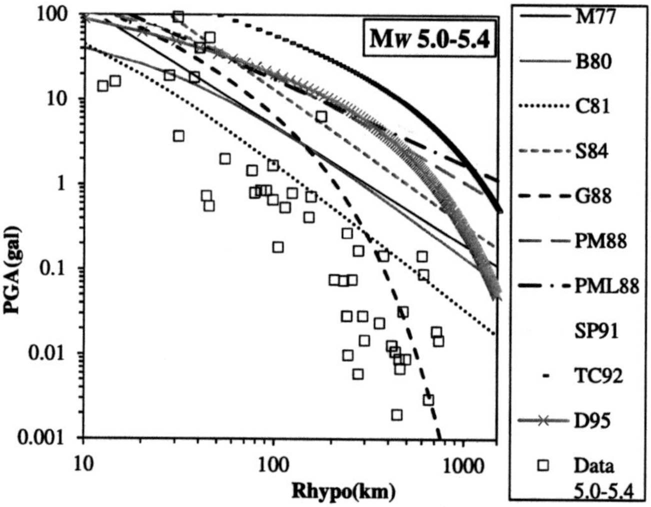

Fig. 8.9 Comparison of GMPEs attenuation curves and recorded PGA for magnitude between 5.0 and 5.4

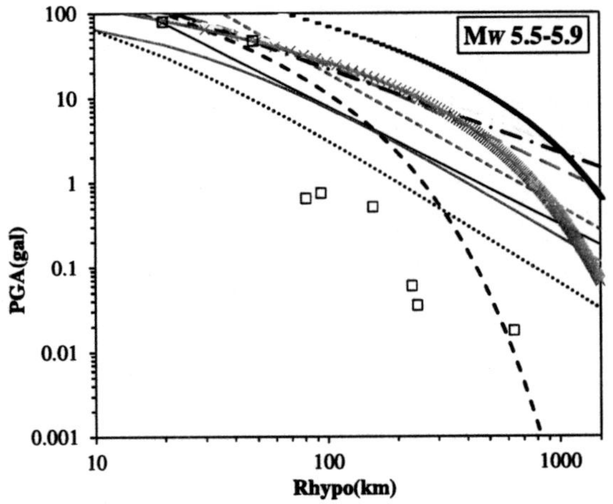

Fig. 8.10 Comparison of GMPEs attenuation curves and recorded PGA for magnitude between 5.5 and 5.9

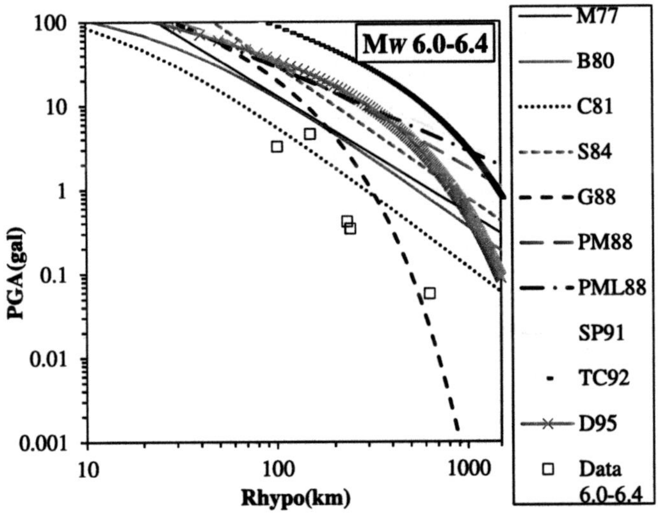

Fig. 8.11 Comparison of GMPEs attenuation curves and recorded PGA for magnitude between 6.0 and 6.4

8.4 Conclusion

In summary, this study shows that the observation recorded from small to large magnitude of earthquakes varies for Sabah's seismicity trend. Among the ten models, the G88 is identified as the best fitting model for distances of 10–1000 km and a magnitude less than M_W 6.4. Based on the hierarchy of curve-fitting GMPE, the model of G88 showed a perfect fit, followed by S84 show reasonable consistency with the earthquake data recorded. The models of M77, B80, and C81 perform roughly well only for magnitude at M_W 5.4 and below. The five other equations, PM88, SP91, PML88, and TC92 show the most inadequate fit for overpredicting the data throughout all ranges of magnitude for more than 50 km.

References

1. Sulaiman SS, Megat Mohamad Farish MML, Ismadi PNN, Moosom VS, Muhd. Yurizan ZA (2019) Is Malaysia located in the Pacific Ring of Fire? a legal perspective. In: International conference on law, environment and society. The European Proceedings of Social & Behavioural Sciences, pp 1±10
2. Encyclopaedia Britannica Homepage. https://www.britannica.com/place/Ring-of-Fire. Accessed 02 Dec 2020
3. Radiguet M, Cotton F, Manighetti I, Campillo M, Douglas J (2009) Dependency of near-field ground motions on the structural maturity of the ruptured faults. Bull Seismol Soc Am 99(4):2572±2581
4. Harith NSH, Adnan A, Shoushtari AV (2014) The compatible ground- motion prediction equations with East Malaysia for Shallow Crustal Earthquakes. J Civil Eng Res 4(3A):120±123
5. Kurzon I, Vernon FL, Ben-Zion Y, Atkinson G (2014) Ground motion prediction equations in the San Jacinto fault zone: Significant effects of rupture directivity and fault zone amplification. Pure Appl Geophys 171(11):3045–3081
6. Bhargavi P, Raghukanth STG (2019) Ground motion prediction equations for higher-order parameters. Soil Dyn Earthq Eng 118:98–110
7. Liao Y, Meneses J (2012) Engineering characteristics of ground motion records from the 2010 Mw 7.2 El Mayor-Cucapah earthquake in Mexico. Earthq Spectra
8. Leyu CH (2009) Seismic and Tsunami hazards and risks study in Malaysia. In: MOSTI (ed) Summary for policy makers
9. Boore DM, Bommer JJ (2005) Processing of strong-motion accelerograms: needs, options and consequences. Soil Dyn Earthq Eng 25:93–115
10. Boore DM, Atkinson GM (2008) Ground-motion prediction equations for the average horizontal component of PGA, PGV, and 5%-damped PSA at spectral periods between 0.01 and 10.0. Earthq Spectra 24:99–138
11. Liew MS, Danyaro KU, Mohamad M, Shawn LE, Aulov A (2017) Ground motion prediction equations for Malaysia due to subduction zone earthquakes in Sumatran Region. IEEE Access 5:23920±23937
12. Mcguire RK (2008) Probabilistic seismic hazard analysis: early history. Earthq Eng Struct Dyn 37:329–338
13. Douglas J (2019) Ground motion. Prediction equation, 1964–2019 https://www.strath.ac.uk/staff/douglasjohndr/. Accessed 02 Dec 2020

14. Cotton F, Scherbaum F, Bommer JJ, Bungum H (2006) Criteria for selecting and adjust- ing ground-motion models for specific target regions: application to Central Europe and rock sites. J Seismolog 10:137–156
15. Sabetta F, Lucantoni A, Bungum H, Bommer JJ (2005) Sensitivity of PSHA results to ground motion prediction relations and logic-tree weights. Soil Dyn Earthq Eng 25:317–329

Chapter 9
Fatigue Detection on Glass Fibre Reinforced Polymer Material Using Fiber Bragg Grating Sensor

Miminorazeansuhaila Loman and Mohd Hafizi Zohari

Abstract The effectiveness of monitoring systems for composite materials is improving owing to their increasing utilisation. Abrupt failure in composite requires an effective detection method and monitoring system. The fibre Bragg grating (FBG) sensor is one of the alternative sensors used for detecting and monitoring the structural health of an engineering structure. This study evaluated the applicability of the FBG sensor for fatigue damage monitoring in the composite. This study involved composite fabrication and experimental work. The glass fibre reinforced polymer specimens were fabricated using fibre glass and resin and made into flat workpieces. The workpieces were then utilised in a series of fatigue tests. Prior to the fatigue test, tensile tests were conducted to verify the ultimate strength of the material. Commencement of fatigue tests were recorded using the FBG sensor. Once the tests were started, the signals were acquired using the FBG sensor simultaneously. Data acquisition was continued during the fatigue test progression until the specimen failed. Results show the FBG wavelength shifted from its original position during tension loading and whenever the composite was released to its original position in the cyclic test. The FBG sensor seems a promising way to monitor fatigue damage and can be utilised in fatigue monitoring. Its wavelength shifts or changes is capable to monitor fatigue damage progression effectively.

Keywords Fatigue · Fibre bragg grating · Composite

9.1 Introduction

Composite seems to replace metal in many engineering structures such as aircraft [1, 2], automotive [3] and even in building construction and systems. It has a high strength-to-weight ratio compared to metal and effectively acts as an impact absorber in engineering applications [4]. Glass fibre reinforced polymer (GFRP) is a preferred type of composite utilised in engineering structures. Unlike metals and metallic

M. Loman (✉) · M. H. Zohari
Advanced Structural Integrity & Vibration Research, Faculty of Mechanical & Automotive Engineering, Universiti Malaysia Pahang, 26600 Pekan, Pahang, Malaysia
e-mail: miminorazean@ump.edu.my

S. Abdullah et al. (eds.), *Structural Integrity Cases in Mechanical and Civil Engineering*, Structural Integrity 23, https://doi.org/10.1007/978-3-030-85646-5_9

alloys that generally fail due to a single identifiable crack, which then initiates and propagates [5], GFRP fails in a brittle manner [6, 7]. Therefore, the strain state of composite must be monitored so that techniques to predict the probable failure can be developed [8].

In recent decades, monitoring structural health has been interesting because of economic savings and safety factors. Engineering structures may fail after a particular service life duration, which may harm people and the environment if it happens unexpectedly. Regular strain state observation of composites under various substantial situations, especially repeated loads, is the most challenging part as strain state is the main cause of catastrophic destruction on composite assemblies during service. Therefore, a successful fatigue monitoring system for composites primarily relies on understanding the fatigue damage processes related to the internal structure strain. Fatigue damage inspection employing visual means is complicated. Generally, indirect sensing is the leading approach for damage sensing in a localised area [9]. The indirect method includes ultrasonic, thermography, interferometers and acoustic emission. Unfortunately, several drawbacks exist in the existing approaches. None of these techniques provides direct observations in the micrometre range to cater to fatigue damage late detection problems [10]. Furthermore, the utilisation of ultrasonic in fatigue damage detection is difficult due to the parallel direction of the crack growth and the sensing waves transmission path [11]. The same study reported that this difficulty is further increased in monitoring thin composites structures of less than 5 mm.

Seeing that GFRP failure is commonly immediate and without warning, a proper understanding of its fatigue properties is necessary. This paper aims to assess the characteristic of the FBG signal in detecting fatigue damage for GFRP materials. The assessment was conducted at a cyclic load of 40% of the ultimate load, as the scope of this study. Therefore, discussion on crack length is not included as it requires more loading conditions to have a representative correlational analysis. This study intends to take advantage of the FBG specialities that can sense an early crack initiation sign. With a proper monitoring scheme, detection of early-stage fatigue damage is possible and thus avoiding sudden failure or incident.

9.2 FBG in Fatigue Damage Sensing

The damage mechanisms depend on the extrinsic and intrinsic factors including the direction of the reinforcement, mechanical behaviour of the material constituents, matrix and reinforcement ratio and layup sequence. Any presence of even a single internal defect or flaw that resulted from manufacturing can make a difference to material properties. Thus, predicting the fatigue properties of composite is more complicated because of its multiple failure modes [12]. Various sensors are used to detect fatigue failure [13]. Acoustic emission can detect fatigue failure in engineering components [14]. Papazian et al. [15] utilised multiple sensors for early-stage fatigue cracking characterization by various sensing techniques on a similar specimen. They found that the techniques can be used in fatigue failure monitoring of laboratory-scale experiments.

The existing method of strain gauge utilization in fatigue monitoring has weaknesses that may be overcome by replacing it with other types of sensors. Embedment of strain gauge into a structure for fatigue monitoring alters the mechanical properties of the composite material [16]. Fibre bragg grating (FBG) embedment does not change or alter any of the material properties of the structures where it is embedded [17]. A tiny size of FBG installed on a glass structure exerts minor effect on its look. Its impact is minimal because the size is small [18]. The technology of FBG is even adopted in the conservative aerospace [19].

Arena and Viscardi [20] reviewed the different novel strain measurement methods utilised. Their review involves a few techniques to sense strain and failure growth such as FBG, strain gauges, infrared thermography and digital image correlation. They presented that strain gauge or strain gauge-based extensometers have various weaknesses. They are very responsive towards electromagnetic fields, they do not save space and they are unsuitable for embedment in structure layups. Its low fatigue resistivity property makes strain gauge an unreliable fatigue strain monitoring sensor. FBG is not only lightweight, but it is also multiplex and has absolute measurement capability, making it ideal to handle strain and structural monitoring tasks. Its manufacturability from a single optical wire into possibly numerous gratings is also beneficial.

The best features of FBG in composite applications are that it can penetrate a dense sensing network with minor structural impact. FBG sensors have become a reliable, non-destructive, in situ tool not only to monitor but also to do diagnostics and control in civil structures. A comprehensive review on FBG monitoring by Sahota et al. [21] presents various interrogation methods of FBG strain sensors. FBG technology is a favourable sensor choice of optical fibre category because it is easily manufactured and possesses a relatively strong reflected signal. A periodic modulation of the index of refraction applied on the fibre core along the longitudinal direction produces functioning FBG [22]. That is, the fiber now acts as a dichroic mirror, reflecting part of the incoming spectrum. Equation (9.1), developed for vacuum, has to be adapted for silica, since the distances traveled by light are affected by the index of refraction of the fiber. The relationship between Bragg wavelength and core refractive index is given by

$$\lambda_B = 2\Lambda \mathrm{n}_{eff} \tag{9.1}$$

where Λ is the grating period measured as the distance between two adjacent grating planes, $\mathrm{n}_{\mathrm{eff}}$ is the effective core refractive index and λ_{B} is the Bragg wavelength. Rao et al. [23] showed that FBG sensors are better than conventional foil strain gauges in terms of sensitivity and durability. These characteristics make FBG sensors ideal for structural health monitoring of large structures.

9.3 Materials and Methods

To test the mechanical behaviour of GFRP, the workpieces were fabricated. Mechanical tests (i.e. tensile and fatigue testing) were then conducted. Signals generated from the loadings on the workpieces undergoing fatigue tests were acquired and analysed to obtain the association. Each procedure is explained as follows.

9.3.1 Workpiece Preparation

Workpieces made of GFRP to be used in mechanical and fatigue tests were prepared. In fabricating the polymer, epoxy resin was combined with four layers of fibre glass fabric to form a laminate sheet. Epoxy resin and hardener were combined in the ratio of 2:1. The fibre glass fabric was stacked alternately between the resin which was then clamped using a big wooden block. It was then left to compress well and dry. A panel was then formed as shown in Fig. 9.1 which was then cut into smaller pieces of rectangular shapes with the dimensions of 20 mm long, 5 mm wide and 2 mm thick, as shown in Fig. 9.2. In the scope of this study, a flat composite panel which was then cut with the aforementioned dimensions was fabricated. The hand layup method was

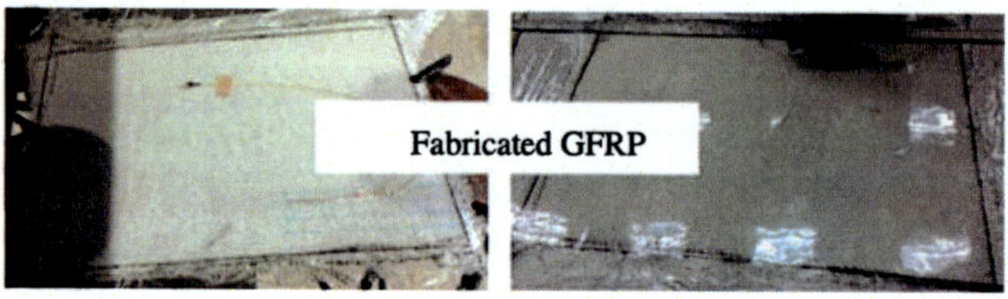

Fig. 9.1 Fabricated GFRP panel

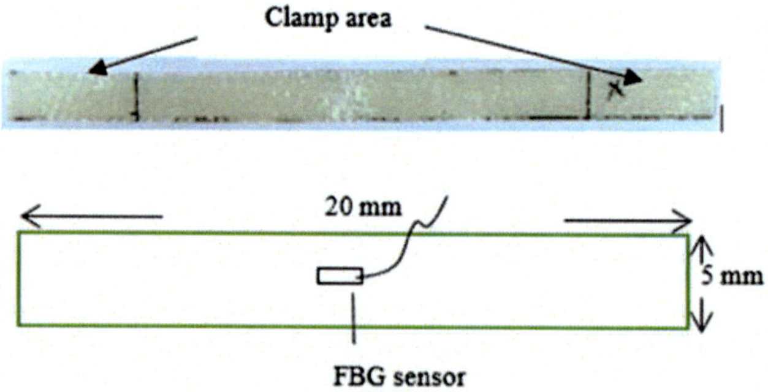

Fig. 9.2 Dimensions of workpiece for tensile and fatigue tests and position of FBG sensor during fatigue test

utilised to manufacture the GFRP composite panel. FBG sensor was attached to the specimen as in Fig. 9.2.

Each workpiece was bonded with an FBG sensor at the gauge length section. A total of eight workpieces were tested during the fatigue test the load ranges from 30 to 80% of the ultimate stress of the GFRP material. The FBG sensor was connected to the FBG interrogator system. The epoxy resin was mixed with hardener, and 1 mm-long FBG sensors with Bragg wavelengths of 1540, 1550 and 1560 nm were written on the same fibre optic. This wavelength shows the range of wavelengths that will shift during the test. The FBG sensor was aligned along the loading direction. All tests were performed on the Instron universal fatigue testing machine with a 25 kN load capacity. Load and displacement information was recorded in the incorporated Instron software by the built-in load cell and extensometer.

9.3.2 Tensile Test

Tensile test. Before the fatigue test was conducted, a similar workpiece and material were tested using the uniaxial tensile static test. The uniaxial static or tensile test had to be performed to verify the material's ultimate strength. The ultimate strength must be determined for any materials that are going to go through a fatigue test so that the workpiece does not fail because of reaching its ultimate strength. The average ultimate strength was set as a baseline parameter during the fatigue test where the maximum stress applied could not be higher than that. In this study, flat composite workpieces were subjected to constant amplitude strain and stress.

Fatigue test. For dynamic loading, a fatigue test was performed on the similar workpiece dimension of GFRP as a tensile test by referring to ASTM D3479 [24]. Different loads or stresses were applied to each specimen during the fatigue test. Tests were conducted on one loading condition three times. The fatigue test was started with minimum stress, and then the stress value was increased until close to its maximum or ultimate strength. The ultimate strength is the maximum stress that the material can bear before it fails. To ensure that the workpiece was not overly heated during the fatigue test, the Instron universal testing machine was applied with a 2 Hz frequency [25]. This frequency was selected as not too high to avoid overheating of the workpiece which might affect the FBG sensor as has been conducted by Kocaman et al. [26]. Fatigue tests were commenced in tension–tension mode of sinusoidal strain waveform at constant amplitudes. The workpieces were tested at one stress amplitude for 12,000 cycles. Figure 9.3 shows the experimental setup of the FBG sensor on the GFRP fatigue test specimen.

Data acquisition. In this study, FBG signal data were acquired during fatigue test progression. FBG wavelength was acquired using an FBG sensor with 1540 and 1550 nm wavelengths. The FBG sensor was bonded on the specimen gauge length area using a strong adhesive and left dried for one day. It was purposely

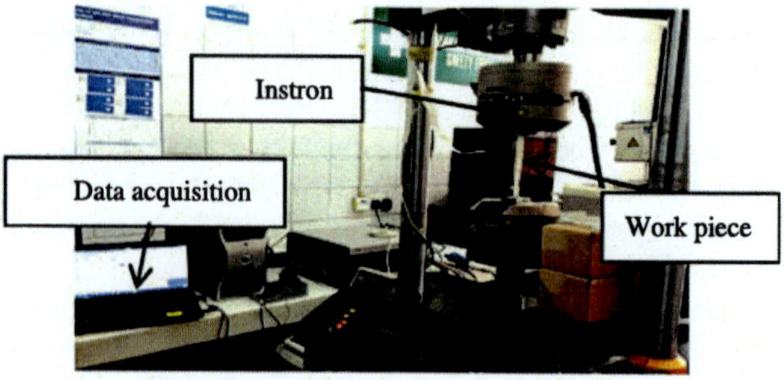

Fig. 9.3 Schematic of fatigue test and FBG data acquisition rig set up

done to ensure that it bonded well to the workpiece surface. The FBG sensor was connected to the FBG interrogator system, including an optical spectrum analyser (OSA), circulator, light and power source which were connected to software for data acquisition, display, record and further analysis.

9.4 Results and Discussion

Signal analysis

Wavelength. At the early stage of the fatigue test, the wavelength is 1548 nm as shown in Fig. 9.4. Here, the original wavelength is single and does not split or shift.

As fatigue test was commenced, the wavelength moved or shifted to the left and right following the loads undergone by the workpiece. FBG data were recorded in wavelength as they simply represented the changes that happened to the workpiece's inner structure. Generally, if the workpiece material was in compression mode, then the wavelength would shift to the left. When the load was in tension mode, then the wavelength oppositely shifted to the right. Figure 9.5 illustrates selected wavelength

Fig. 9.4 Original FBG wavelength before fatigue test

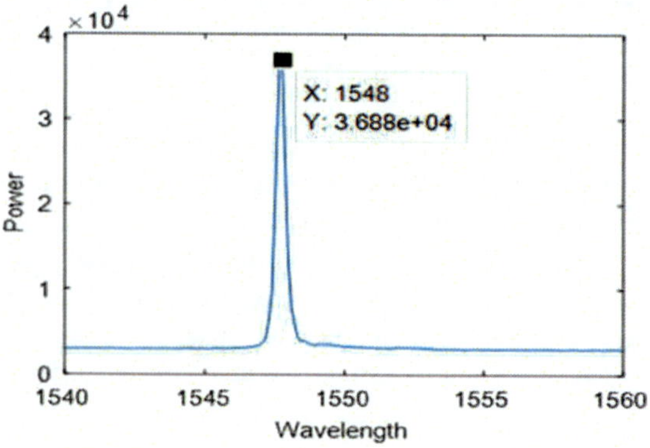

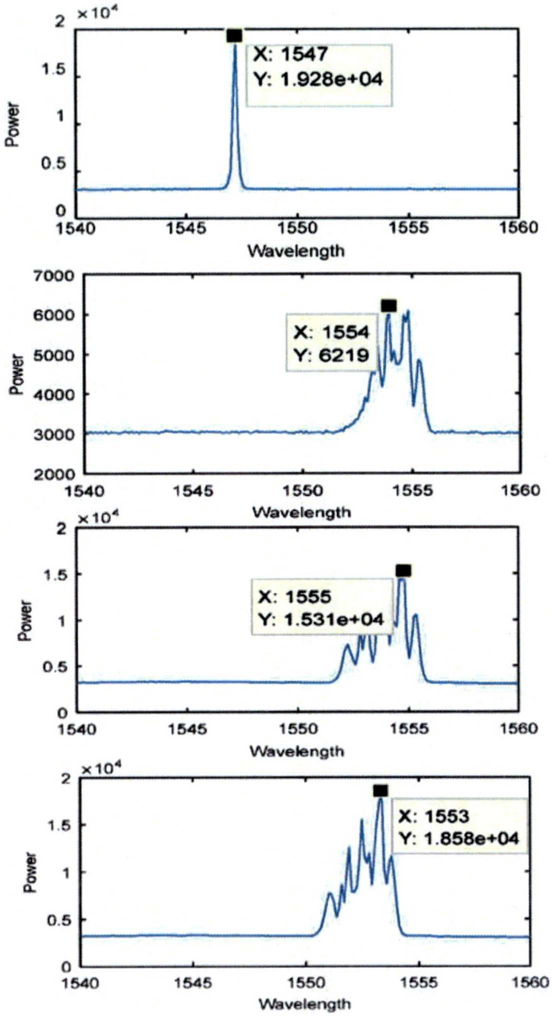

Fig. 9.5 FBG wavelength shifts of workpiece at 40% ultimate stress during a fatigue test

spectra acquired from the workpiece which experienced 40% loads from the ultimate stress by the FBG sensor during fatigue test progression. Initially, as soon as the fatigue test was started, the FBG spectra within the wavelength range of 1540 nm and 1560 nm were recorded. Figure 9.5 shows that wavelength shifts as the fatigue test progresses. These data were picked randomly after a certain duration at different intervals. However, not all figures show the complete left–right shifts of the wavelength in each data point. At an early stage of the fatigue test on the workpiece, FBG exhibited a wavelength value of 1547 nm. After a certain duration, its wavelength peaked at 1554 nm. A few peaks can be seen in this figure. At another moment during the test, the peak wavelength shifted to 1555 nm, and the last spectrum showed a

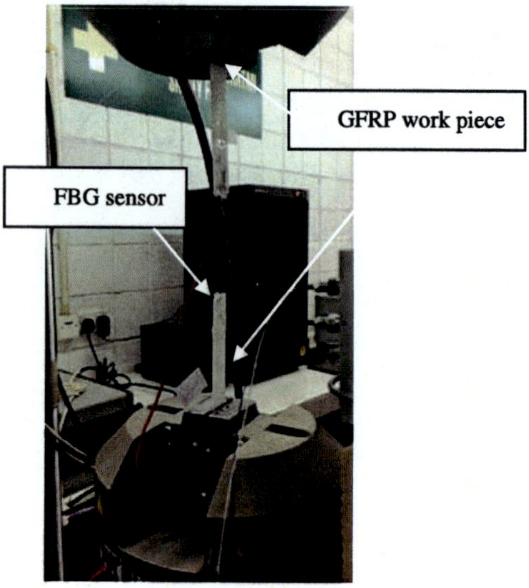

Fig. 9.6 Fractured workpiece after fatigue test with FBG sensor still intact

wavelength of 1553 nm. A perfect bonding between the FBG sensor and the workpiece surface can produce a smooth spectrum without any spectrum splitting. Xue et al. [27] supported this finding. Even though a few peaks appeared that could possibly be due to a slight decrease in bonding between the FBG sensor and surface of the composite workpiece during the cyclic loading, the wavelength shift is still noticeable. The FBG sensor was still intact and attached to the workpiece for the entire test duration and even after the workpiece already fractured as in Fig. 9.6. This finding exhibits successful fatigue damage sensing capability with a considerably wide sensor positioning which is about 10 cm from the exact failure location.

Data Analysis. The FBG wavelength spectrum was recorded multiple times during the fatigue test to monitor the condition of the FBG sensor when subjected to high-cycle fatigue loads. The result is illustrated in Fig. 9.5, where the x-axis represents the increment of the number of data, n. Such a recording was purposely carried out to understand the characteristics of FBG signal under fatigue loading. The integrity of the FBG obviously remains for the entire test duration. The spectra became more intense as the specimen approached fatigue failure with the increased loading cycle. More wavelengths of the same length or intensity were observed at the close time intervals. This phenomenon can be seen in Fig. 9.5 which clearly shows that the FBG sensor can detect the changes that occurred in the specimen structures and translate them into wavelength changes.

Upon commencing the fatigue test on the workpiece, the FBG wavelength was recorded simultaneously. Figure 9.7 shows the distribution of the wavelength shift or spectrum during fatigue test progression. Each dot in the figure represents a spectrum that has a wavelength value at a specific time. This wavelength value varies along the time duration depending on the loads given during the cyclic fatigue test. Figure 9.7 clearly illustrates that the FBG sensor recorded the condition of what was happening

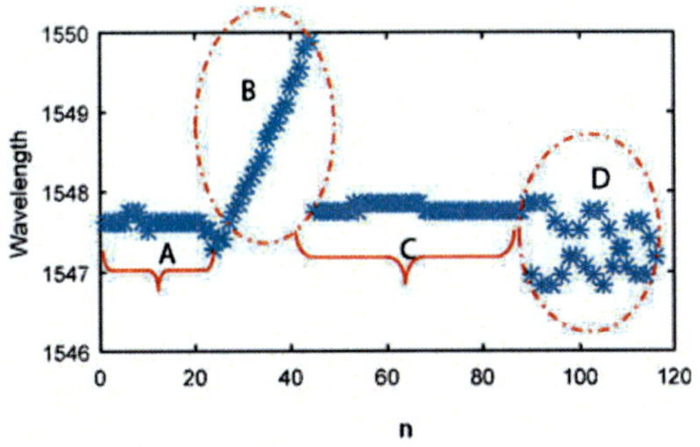

Fig. 9.7 Intensity of FBG wavelength for workpiece that undergone 40% of ultimate strength load during fatigue test

to the workpiece for the entire fatigue test. This result is consistent with the findings in [26]. FBG can monitor the structural health over quite a long duration by recording the changes experienced by the structure. The wavelength originated at less than 1548 nm then shifted to a higher wavelength of more than 1548 to 1550 nm at $n =$ 45. After that, the wavelength shifted to the lower value that was close to 1457.5 nm for a certain duration and lastly distributed at less than 1547 nm before the workpiece failed.

In Fig. 9.7, the wavelength is plotted versus n, where n is the summation of n_i for i $= 0$ to N, and N is the number of data. n_0 is the number of data before the fatigue test was started at $t = 0$ s, and n_N is referred to the final second of the fatigue test. As soon as the fatigue test started at A, the wavelengths were shifting the higher and lower values within the range of 1547–1548 nm. This might represent that the FBG was exhibiting the tension and compression experienced by the workpiece when fatigue loading was applied. However, the wavelength shifts were tremendously increasing without going back to their original intensity at B. This might imply that the plastic deformation started to happen where the FBG wavelength was deformed like the workpiece where it was bonded to. The FBG wavelength then went back to the normal wavelength. At this section, it did not deform the workpiece because, at this time, the workpiece might have lost its elasticity. Lastly at D, the workpiece experienced pre-failure where the FBG started to crack and finally failed. The FBG sensor was still in good condition and did not damage or break even after the test was completed. This result demonstrates the robustness of applying FBG in the structural health monitoring system, specifically one that involves an extreme environment.

Based on the presented FBG graphical wavelength historical data, the proposed monitoring approach is feasible for the various structural health monitoring systems. This is because the level of failure and respective prevention strategies varies between dynamic systems. Prevention of any plastic deformation may be regarded as crucial for a particular system. Where in other structural components, the similar warning requirement could be extended until crack initiated before a total failure occurred.

9.5 Conclusion

This study reveals that the FBG sensor is capable of detecting physical changes inside composite components. The workpiece under tension and compression loadings are reflected in the FBG wavelength shifts. It was proven by the data acquired on the GFRP workpieces that had undergone fatigue test. Although the wavelength shifts did not directly address fatigue failure in this study, it may be possible if further analysis is performed. The research objective is achieved successfully as the results clearly show the connection between the FBG wavelength shifts and cyclic loading applied to the composite plate workpieces along the fatigue test. The after test check-up demonstrates that the FBG remained on the composite workpieces and mostly did not degrade. Since the indication of the fatigue crack initiation is not clear until the crack size reaches a critical limit, this monitoring approach offers early fatigue damage detection. The critical level should be avoided because of the rapid crack propagation until a complete fracture is often too late for any preventive action. A more interesting part to be highlighted is this monitoring approach's capability in fatigue failure indication for the studied brittle material.

References

1. Gude M, Hufenbach W, Koch I, Koschichowa R, Schulte K, Knoll J (2013) Fatigue testing of carbon fibre reinforced polymers under VHCF loading. Proc Mater Sci 2:18–24
2. Adam TJ, Horst P (2014) Experimental investigation of the very high cycle fatigue of GFRP [90/0]s cross-ply specimens subjected to high-frequency four-point bending. Compos Sci Technol 101:62–70
3. Gardyński L, Caban J, Barta D (2018) Research of composite materials used in the construction of vehicle bodywork. Adv Sci Technol Res J 12(3)
4. Patel M, Pardhi B, Chopara S, Pal M (2018) Lightweight composite materials for automotive-a review. Int Res J Eng Technol (IRJET) 5(11)
5. Suresh S (2001) Fatigue of materials, 2nd edn. Cambridge University Press
6. Ferdous W, Manalo A, Peauril J, Salih C, Reddy KR, Yu P, Schubel P, Heyer T (2020) Test and modelling the fatigue behavior of GFRPR. Composites–effect of stress level, stress concentration and frequency. Eng Sci Technol Int J 23(5):1223–1232
7. Tanimoto T, Amijima S (1975) Progressive nature of fatigue damage of glass fiber reinforced plastics. J Compos Mater 9:380–390
8. Kocaman ES, Akay E, Yilmaz C, Turkmen HS, Misirlioglu IB, Suleman A, Yildiz MT (2017) Monitoring the damage state of fiber reinforced composites using an FBG network for failure prediction. Materials 10(32):1–19
9. Glisic B, Chen J, Hubbell D, Streicker B (2011) A comparison between Bragg-gratings long-gauge strain and temperature sensors and Brillouin scattering-based distributed strain and temperature sensors. In: Proceedings of SPIE 7981. https://doi.org/10.1117/12.881818
10. Garcea SC, Spearing SM, Sinclair I (2012) Exploring the fundamental of fatigue in composites: opportunities using x-ray computed tomography imaging. Air Force Research Laboratory annual report, University of Southampton, United Kingdom
11. Mouritz AP, Townsend C, Shah Khan MZ (2000) Non-destructive detection of fatigue damage in thick composites by pulse-echo ultrasonics-ultrasonic and SEM evaluations. Compos Sci Technol 60(1):23–32

12. Colombo C, Libonati F, Vergani L (2012) Fatigue damage in GFRP. Int J Struct Integr 3(4):424–440
13. Wang P, Takagia T, Takenob T, Miki H (2013) Early fatigue damage detecting sensors—a review and prospects. Sens Actuators A Phys 198:46–60
14. Karimian SF, Modarres M, Bruck HA (2020) A new method for detecting fatigue crack initiation in aluminum alloy using acoustic emission waveform information entropy. Eng Fract Mech 223(106771):1–12
15. Papazian JM, Nardiello J, Silberstein RP, Welsh G, Grundy D, Craven C, Evans L, Goldfine N, Michaels JE, Michaels TE, Li Y, Laird C (2007) Sensors for monitoring early stage fatigue cracking. Int J Fatigue 29:1668–1680
16. Belhouideg S, Lagache M (2018) Effect of embedded strain gage on the mechanical behavior of composite structures. J Modern Mater 5(1):1–7
17. Liu B, Zhang S, Jianping HE (2019) Deformation measurement of glass structure using FBG sensor. Photonic Sens 9(6)
18. Guo X, Wang B, Ma Z, Wang Z (2019) Testing mechanical properties of rock bolt under different supports using fiber bragg grating technology. Sensors 19(4098). https://doi.org/10.3390/s19194098
19. Claire D, Silvia T, Ivan G, James K, Travis N (2012) High-strain fiber bragg gratings for structural fatigue testing of military aircraft. Photonic Sens 2(3):215–224
20. Arena M, Viscardi M (2020) Strain state detection in composite structures: review and new challenges. J Compos Sci 4(60). https://doi.org/10.3390/jcs4020060
21. Sahota JK, Gupta N, Dhawan D (2020) Fiber Bragg Grating sensor for monitoring of physical parameters: a comprehensive review. Opt Eng 59(6):060901–060935
22. Werneck MM, Allil RCSB, Ribeiro BA, de Nazaré FVB (2013) A Guide to fiber bragg grating sensors, current trends in short and long-period fiber gratings. In: Cuadrado-Laborde C (ed) IntechOpen, pp 1–24
23. Rao MB, Bhat MR, Murthy CRL, Madhav KV, Asokan S (2006) Structural health monitoring (SHM) using strain gauges, PVDF film and fiber bragg grating (FBG) sensors: a comparative study. In: Proceedings of national seminar on non-destructive evaluation,7–9 Dec, Hyderabad
24. ASTM D3479/D3479M-19 (2019) Standard test method for tension-tension fatigue of polymer matrix composite materials. ASTM International, West Conshohocken, PA
25. Xue G, Fang X, Hu X, Gong L (2018) Measurement accuracy if FBG used as a surface-bonded strain sensor installed by adhesive, 57(11):2939–2946
26. Keulen CJ, Akay E, Melemez FF, Kocaman ES, Deniz A, Yilmaz C, Boz T, Yildiz M, Turkmen HS, Suleman A (2016) Prediction of fatigue response of composite structures by monitoring the strain energy release rate with embedded fiber Bragg gratings. J Intell Mater Syst Struct 27(1):17–27
27. Kocaman ES, Akay E, Yilmaz C, Turkmen HS, Misirlioglu IB, Suleman A, Yildiz M (2017) Monitoring the damage state of fiber reinforced composites using an FBG network for failure prediction. Materials 10(32):1–19

Chapter 10
Biomechanical Overloading Factors Influencing the Failure of Dental Implants: A Review

Muhammad Ikman Ishak, Ruslizam Daud, Ishak Ibrahim, Fauziah Mat, and Nurul Najwa Mansor

Abstract The increasing popularity of dental implants has led to an increase in the number of late implant failures. Although the failure of dental implants is rare, it produces a challenging clinical situation. Thus, the scope of causes that is detrimental to dental implants and peri-implant tissues is important to be explored. Inadequate forces resulting from occlusion may cause complications that leading to implant failure. It is found that the mechanical-related contributing factors (biomechanical overloading) significantly affect the implant persistence as compared to the biological factors. The present review, therefore, emphasises the causes of the overloaded dental implants and technical complications in clinical scenarios. A comprehensive search was performed via ScienceDirect, Scopus and PubMed databases using the related keywords. The literature indicates various factors could trigger biomechanical overloading and promote the occurrence of implant failures: parafunctional habits, implant diameter, length, thread shape and material and implant-abutment connection. The relationship between these factors and implant longevity is still controversial, with further randomised clinical trials and advanced quantitative assessments are required to help elucidate this issue.

Keywords Dental implant · Overloading · Complication · Failure · Fracture

10.1 Introduction

The use of endosseous dental implants for the treatment of patients with partial or complete edentulous arch regardless of the maxilla or mandible has secured high popularity since decades ago due to great effectiveness in restoring the oral function, aesthetic, and comfort. The high success rate of dental implants makes it highly demanded and being one of the undeniable teeth restoration techniques. The traditional approach utilising denture exhibits a major vertical bone level disuse atrophy after long use because of the reduced blood supply which leads to the accelerated

M. I. Ishak (✉) · R. Daud · I. Ibrahim · F. Mat · N. N. Mansor
Faculty of Mechanical Engineering Technology, Universiti Malaysia Perlis (UniMAP), Kampus Alam UniMAP, Pauh Putra, 02600 Arau, Perlis, Malaysia
e-mail: ikman@unimap.edu.my

S. Abdullah et al. (eds.), *Structural Integrity Cases in Mechanical and Civil Engineering*, Structural Integrity 23, https://doi.org/10.1007/978-3-030-85646-5_10

bone loss. Dental implants are reported to record a high survival rate which is about 95% and above, on average, after several years of loading based on numerous clinical follow-up studies [1, 2].

Despite reporting a high success rate, the complications that resulting in implant failures such as the loosening and/or fracture of the prosthesis/abutment screw and implant body do occur. Albeit they are infrequent, the impact of the implant failure is really significant and frustrating not only for patients but also for clinicians as it normally involves the loss of both prosthesis and implant parts [3]. According to Chrcanovic et al. (2018), the prevalence of dental implant fractures is reported to be merely about 1.0–1.5% based on a few review studies [4]. Some reports revealed that the low percentage of implant failures could associate with the short period of follow-up, besides a limited number of patients and implants involved [2].

Implant fractures commonly constitute significant implant failures and hence require maintenance or even complete implant removal for the worst-case scenario. In comparison, the posterior dental implants regardless of their location in the maxilla or mandible establish a greater failure rate than those placed anteriorly. This has caught attention in many studies evaluating the causes and treatments [3]. Moreover, the implant-supported fixed single crown was reported to be the most affected among the other types of teeth restoration. According to Gargallo Albiol et al. (2008), the onset of complications appears at the first 2–3 years after the functional load of the implant [5].

Osseointegrated dental implants exhibit different fashions of biomechanical responses towards occlusal force due to the absence of the periodontal ligaments [6]. It is thus believed that dental implants could be more susceptible to occlusal overloading as compared with natural teeth. The periodontal ligaments in the natural teeth provide physiological and functional adjustments towards occlusal overloading owing to a well-oriented position of the ligaments in resisting the axial force. The occlusal overloading factors may negatively influence the short- or long-term performance of implants which are often considered as one of the possible reasons for marginal bone loss and failure of the prosthesis and/or implant. This is supported by a general consensus that early implant failure may be associated with overloading [7]. The occurrence of peri-implant bone resorption without eventual implant failure could primarily be related to biological complications. Therefore, it is a necessity to control implant occlusion within the allowable physiological range and hence offer an ideal implant load for long-term implant survival. In addition, many past reports indicated that the persistence of dental implants is dependent on the occlusion [8].

Tissues infection such as peri-implantitis is an example of biological complications found in many clinical studies which also regarded as the contributor to the loss of a dental implant [3, 8]. Although the biological-related causal aspect may lead to the marginal bone loss, as aforementioned, its impact on implant failure, however, is less significant and unclear.

The objective of the present study was to investigate the possible causative overloading factors leading to implant complication and failure based on the current clinical literature as well as establish the need for future research. The relationship between overloading and implant failure is clarified during the bone healing stage after implantation and subsequent formation of osseointegration.

10.2 Materials and Methods

The present literature study was conducted through the use of Scopus and PubMed databases. The following keywords, “occlusal overloading”, “dental implant complication”, “dental implant technical complication”, “dental implant fracture” and “dental implant failure” were utilised during the search. A limitation was set for the articles published in the years, 1992–2020 only. The inclusion criteria for this study were English language clinical reports and literature reviews. In vivo and in vitro studies that were not related to dental implant failures were excluded. Thus, a total of sixty-nine articles were selected for this study.

Consideration was only given to modern endosseous dental implants irrespective of system type, implant body design, and time of onset of the complications. Zygomatic implants were excluded in the study although those fixtures were widely used in teeth restoration particularly in the posterior jawbone region. A dental implant system was counted to have a high tendency to experience failure if at least a fracture line was running through the implant body, abutment and/or prosthesis. Besides, the loosening of the prosthetic components, microfracture of the surrounding bones and marginal bone loss were also taken into account in the review.

10.3 Results

The results of this review study were presented in several sections based on the most common biomechanical overloading factors contributing to implant failure. Findings and conclusions from relevant clinical reports, human studies, animal studies, experimental works and previous reviews were included and presented.

10.3.1 Parafunctional Habits

A number of retrospective studies reported that patients with parafunctional habits such as clenching, bruxism, and ice-chewing are greatly exposed to occlusal overloading when compared with those without. There was about 20–35.9% of patients may generate forces of such value to initiate microfracture of the peri-implant bone tissues with concomitant bone loss and implant failure due to bruxism [9, 10]. This is parallel with other earlier investigations where the resulting restorative complications were found in patients with parafunctional habits and posterior-region implant treatment [11]. In a different study, 90% of the fractured implants were observed in the cases having cantilevered prosthesis and parafunctional habit [12]. However, a contradictory finding was found in a 1–10-year prospective study where no association exists between implant failure and bruxism [13].

Considering this factor, the odds ratio of dental implant loss for bruxers in association to non-bruxers is about 2.71, for a population of 98 bruxers and matched non-bruxers [10]. Besides, according to Chitumalla et al. (2018) through a 5-year retrospective study, it was reported that the survival rate of dental implants for patients with bruxism habit had been decreased through years which was 90, 87, 85, 75 and 72% after 1, 2, 3, 4 and 5 years, respectively [14]. The findings were in agreement with another recent clinical study, which stated that bruxism could be a significant etiologic factor resulting in biological and mechanical complications, supported by 80% of implant fractures observed in patients with bruxism [3]. The data was almost similar to the one published by Pommer et al. [15], where 89.6% of the fracture cases belong to the group with parafunctional activity [15]. The relationship between implant failure and parafunction has also been corroborated by other earlier investigators that underscoring the contribution of occlusal overloading in peri-implant disease and subsequent failure [16].

10.3.2 Cantilever Extension

The presence of a cantilever in the implant-supported fixed restorations should be avoided as it may induce undesirable occlusal overloading forces especially in the posterior regions [1, 8, 17, 18]. The extended portion is believed to leave a hinging effect resulting in considerable compressive stress on the implants. As far as the length of the cantilever is concerned, the number and location of implants thus play a vital role. A recent numerical assessment exhibited that the case with a short cantilever (5.5–8.5 mm) and six implants promoted a more encouraging mechanical stresses than the case with a long cantilever (up to 18.5 mm) and four implants [19]. This corroborates the data reported by other studies which have exhibited that fixed dental prostheses with cantilever arm ≥8 mm established crestal bone resorption, resulting in peri-implantitis [20]. In addition, an earlier investigation has suggested the average length of cantilever extension of the mandible and maxilla were 14.4 and 10.9 mm, respectively [21]. A current retrospective clinical study described a contradictory outcome in which cantilever extension of any size may lead to towing forces and subsequently increases stress in the bone and implant, resulting in crestal bone loss and parts failure [3]. Moreover, the mesial cantilevers were claimed to be notably favourable as compared with distal cantilevers [22]; however, this does not support the opinion of the later authors who indicated that the distally located cantilever parts were more promising [23, 24]. Besides, no difference in marginal bone levels has also been shown between partially edentulous patients with cantilevers and those without, which is evident through several recent meta-analyses, systematic reviews and long-term follow-up studies [25–28]. The finding is valid regardless of the implementation of distal or mesial cantilever with a limitation imposed by the cantilever design.

In terms of implant position, the increase in the distance from the centre of the prosthesis to the implant axis will increase the forces accordingly. This could lead to screw loosening and prosthetic parts fracture. It is suggested to have a centrally position contact and the posterior implants should be placed axially in relation to the opposing antagonist to reduce occlusal overloading [29].

10.3.3 Prosthesis Morphology and Contact

The morphology of the prosthesis particularly the cusp inclination exhibits an essential role in determining the direction of occlusal forces. Chewing force comprises vertical load component mainly, however, horizontal movement of the mandible and the inclination of the dental cusps generate unfavourable lateral forces [1]. The forces can be directed more axially if a flat cusp and shallow anatomy are involved. This is because the inclined cusp may originate an undesirable bending motion and transfer more forces to the implants than the flat cusp [8]. A cusp with a steep inclination could also increase the stress intensity in the prosthesis [30]. In several animal studies, the lateral forces are simulated by artificially designed cusp inclines as some animals (i.e. dogs) are unable to execute lateral movements [25]. The non-axial or lateral forces are not perpendicular to the occlusal plane and are considered to disrupt the bone–implant interface [31]. This is in agreement with in vitro works where non-axial loads contribute to high-stress magnitude at the cervical region of the bone [32, 33].

The non-passive fit of the prosthetic structure is also one of the causes of implant fracture [34]. Schwarz (2000) suggested that if the prosthesis screw loosens frequently or fractures although it has already been satisfactorily tightened, it could be due to the insufficient precision fit of the prosthetic structure [35]. The presence of misfits between implants and crowns has been attributed to the occurrence of overloading which, therefore, needs to be addressed [2].

10.3.4 Implant Materials

Other critical parameters to avoid occlusal overloading include the design of dental implants. This covers the type of material constituting the prosthetic components especially the retaining screw and implant body [2, 3, 36]. To date, the most common implant materials used are ceramics and titanium [37]. Metal fatigue was reported to be an important cause of the technical complications. Based on a risk factor analysis for implant fracture in a retrospective study involving 2670 patients, titanium grade 3–4 implants exhibited a 72.9% less tendency of fracture than the turned, grade 1 implants [4]. It is also noteworthy that abutment screws made of grade 1 titanium are more prone to deformation and fracture than titanium alloy. Therefore, a greater value of torque can be applied to the titanium alloy abutment screws and implant body [38]. The use of a full zirconia (polycrystalline ceramics) or an alloy with titanium, which is described as titanium zirconium (TiZr) is also of interest in the current implant manufacturing.

Yttria-stabilized zirconia (YSZ) is reported to have a better aesthetic than titanium alloy in reference to soft tissue colour [39]. Table 10.1 depicts a comparison of mechanical properties of titanium, alumina, and zirconia.

The retaining screws used to secure the prosthesis to the implant body are commonly made of either titanium or gold. Reduced screw loosening and improved clamping force are demonstrated by implant bodies with gold or gold-coated screws in comparison to those with titanium screws [34, 40]. This could be due to their greater elastic modulus than titanium that allowing an acceptable micro motion to dissipate force to the implant fixture. However, in terms of strength, titanium screws are stronger than gold with the occurrence of metal fatigue may easily lead a gold screw to fracture as compared to titanium. The incidence of galling is typically promoted by titanium retaining screws whereby being one of the major shortcomings of the material. According to Al Jabbari and the colleagues, this phenomenon has contributed to an excessive friction between the two mating surfaces (retaining screw and implant body) that resulting in a localised welding [41]. This is also known as the adhesive wear mechanism where it can be harmful to both retaining screw and implant body threads.

New emerging materials such as polymers and polymeric reinforced composites (polyetheretherketone (PEEK)) have also been studied for possible implant applications [42, 43]. The main concern in regard to a dental implant material is its biocompatibility as the implant is in exact contact with living tissues. From a mechanical point of view, the implant material should have properties that are close to the contiguous bones to inhibit minimal bone adaptation stress due to the stress shielding effect. Peri-prosthetic fracture is one of the complications observed which occurs because of a considerable bone degradation. The comparable value of Young's modulus between PEEK (3–4 GPa) and the bones could leave an insignificant stress shielding effect. Nevertheless, if the unmodified PEEK is employed, it may offer very limited inherent bioactivity and osseointegration. Thus, there is a necessity to investigate further the role of PEEK as the evidenced performance of the material is still controversial and it has not extensively been used clinically.

Table 10.1 Comparison of mechanical properties of titanium, alumina and zirconia [44]

Mechanical properties	Titanium (Grade 4)	Alumina	Zirconia
Compression strength (MPa)	840–1080	4100	2000
Young's modulus (GPa)	104.1	380	210
Fracture toughness (MPa $m^{-1/2}$)	>70	4	7–10

10.3.5 Implant Diameter

The overall strength of the bone–implant interface is predominantly influenced by the characteristics of load transfer that can be accounted for by the implant macro-design or geometry. The width or diameter of the implant is one of the factors, which has received prime attention [1, 45]. The ideal diameter for a promising implant performance is highly dependent on the quality of bone support. Insertion of right implants and their behaviour in the bones are crucial to prevent any potential hazards such as periodontal ligament damage and implant exposure especially for the areas with low bone quantity [46]. Osteogenic distraction or bone grafting is commonly performed to increase the bone volume. The procedure, however, requires high cost, prolonged treatment and healing times, membrane exposure and unpredictable bone resorption. Literature reported that Type I and II bone quality (based on the Lekholm and Zarb classification) exhibit inconsiderable effects by the implant diameter, whilst Type III and IV show a contradict result [47]. However, complications in Type II bone were still observed particularly for the implants placed posteriorly, therefore, suggesting wider implants to increase functional surface area.

Implants with large diameters tend to fracture more hardly than those with small diameters, especially when placed in the posterior regions [17, 34]. This is evidenced by a low implant failure rate as stated in many past clinical assessments. The results of one retrospective study involving the placement of 10,099 implants demonstrated that no implant fracture was found in implants of wide diameter (4.20–5.00 mm). The highest implant fracture rate occurred among implants with a narrow diameter (3.00–3.50 mm) with 1.35%, followed by the implants with a standard diameter (3.70–4.10 mm) with 0.34% [4]. This seems to be parallel with an earlier meta-analysis of 16 studies wherein the mean failure rate of implant diameters less than 3.00 mm (25%) is significantly greater than that of wider implants (13%) over the same period of follow-up [48]. The wide implants of 5.0-mm diameter or more are recommended for the restoration in the molar region [49, 50]. An implant with a diameter of 5.0 and 6.0 mm is three- and sixfold stronger than that with a diameter of 3.75 mm, respectively. In comparison, the width of the implant is slightly smaller than the width of almost every natural tooth. The larger the width, regardless of implant or tooth, the lesser the high level of stress distributed to the neighbouring bones [51]. Several reports, nevertheless, drawn contrary opinions and conclusions [52] and a few even failed to confirm the influence of implant width [3, 53].

As the narrow-diameter implants tend to fracture easily, they must carefully be inserted into the bone based on the manufacturer's suggestion. In some analyses, the implants with small diameters recorded satisfactory medium- and long-term survival rates which varying from 96.9 to 100% within a period of 1–7 years follow-up [54, 55]. Another indication of narrow implants apart from being used in the anterior regions is as a reliable alternative of the replacement for the missing molars especially for the case with medium horizontal bone atrophy [56]. With respect to implant diameter, the definitions of narrow, regular, and wide implants also remain unclear and need to be decided to serve as a reference for future implant dentistry.

10.3.6 Implant Length

The implant longevity is not only secured by the role of implant width alone in regard to the implant macro-design. The length of the implant is also associated with the overall implant surface area. Among the factors considered to choose the appropriate implant length are patient and site-specific local anatomical conditions. The lower implant failure rate is reported in the anterior mandible which is possibly due to a higher bone level and density as compared to other areas [57]. Initial bone-implant contact is important in determining the functional surface area and this majorly depends on the bone density. Type I bone promotes approximately 80% bone-implant contact, while Types II, III and IV exhibit a lower percentage [58].

Longer dental implants are normally desired in the compromised bone region (Type IV) [59], to avoid improper occlusion, moment and off-axis loads, parafunction, or high stress [60] that could lead to complications. The regions with limited bone volume and poor bone quality such as the posterior maxilla usually require an advanced sinus bone augmentation, whilst for the posterior mandible, nerve repositioning surgery is expected for the insertion of longer implants. As these procedures may induce adverse effects namely sinus perforation, and mandibular canal violation and paresthesia [61], thus, shorter implants are suggested as a dependable alternative [62].

As far as peri-implant bone loss is concerned, the short implants (5–8 mm) showed less bone loss in the third- and fifth-year follow-up when compared to longer implants in vertically grafted bone irrespective of atrophic arches, based on a systematic review of 13 randomised controlled clinical trials published between 2005 and 2016 with a minimum follow-up period of 12 months [63]. However, both implant groups presented an insignificant discrepancy of marginal bone resorption in the first year of evaluation. According to Chrcanovic et al. (2018), the probability of fracture was increased by 22.3% for every 1 mm increase in implant length [4]. The high possibility of fracture with longer implants could be due to the fact that wider-diameter implants which claimed to possess a lower tendency to fracture are not usually long. A recent consensus report on the implant length indicated contrasted findings wherein the survival rate of short implants (86.7–100%) was slightly lower than that of longer implants in the augmented bone (95–100%) for a period of 1–5 years follow-up [64]. Albeit the short implants disclose a greater variability, still, they are claimed to be a rational option for sites with limited bone height.

The length of the implant has a direct relationship with the crown to implant (C/I) ratio which is assumed to have an impact on occlusal overloading. A high clinical C/I ratio which is greater than 1 to 1 (2 to 3 C/I ratio) is expected to leave catabolic results, however, a number of researchers reported counterintuitive findings. Instead, the high C/I ratio resulted in an anabolic change in the bone [25, 65, 66]. Besides, the treatment of cases with interocclusal distance using short implants has been successful, similar to those using longer implants with bone grafting [67]. There is also confusion in relation to dental implant classification based on length in order to define short, regular, and long implants that needs to be further studied.

10.3.7 Implant–Abutment Connection

The type of connection between the implant and the abutment is one of the main anatomical features tested in the regions of implantation [3]. The lack of perfect fit at the implant–abutment interface can lead to fracture especially fracture of the abutment screw. Another factor related to the implant–abutment connection that also causes the fracture is the absence of passive adaptation between the abutment and the prosthesis [34]. The stability of the implant-abutment connection is affected by several factors—a geometrical feature of the connecting part of the implant body platform, abutment screw head design and diameter and preload.

Three main designs of the connecting part of the implant body platform are internal, external and one-piece type. The internal design comprises a submerged connecting part into the implant body in the form of conical, clearance fit and combined connections. Whilst, the external design is represented by the connecting part located above the implant platform typically in protruded hexagonal shape [68]. A perfect seal between the implant body and the abutment is greatly demanded to inhibit bacterial leakage and colonisation. The subsequent bacterial contamination at the interface is expected to initiate peri-implantitis and marginal bone resorption. A past review on the survival and complication rates of fixed partial dentures demonstrated that the cumulative incidence of the connection-associated complications such as screw loosening and fracture was about 7.3% [69]. The contribution of implant–abutment design in implant performance has also been confirmed by many recent review studies [68, 70].

Each design of the implant–abutment connection affects the load transfer path differently. Inadequate ones may put the peri-implant bone at high risk of overloading. The adoption of internal morphology has displayed a greater stability and superior load distribution relative to the external connection. Longer internal wall attachment that belongs to internal connection provides a firm structure to oppose micro motion inducing complications such as joint opening. Also, the abutment screw is believed to shield stress better in that interface type due to the dispersion of lateral loads deeply in the implant body [71]. Low risk of marginal bone loss is promoted by internal fixation as shown in short- and medium-term assessments. The reason behind this is that the mismatched implant and abutment diameters (Morse taper) have led to the offset of the interface horizontally away from the crestal bone region. This view is supported by Sasada and Cochran (2017) who write that an average bone loss of 1.5–2.0 mm was recorded in the first year of loading for the external-hexagon butt-joint implants, which is comparatively greater than that for the internally connected implants [72].

Some disagreements were found in regard to the role of the internal connection whereby the thin lateral implant wall at the connecting part could result in a greater absolute strain level [73, 74]. Subsequent stress shift from that area tends to increase the risk of peri-implant bone resorption or eventual implant fracture. A few studies reach different conclusions, finding no difference in fracture strength for implants with external fixation in comparison to the ones with internal fixation [75]. Together, the recent studies outlined that the internal connection may be more advantageous, however, further well-designed and long-term evaluations are required.

10.4 Challenges in Securing Dental Implant Stability

The investigation on dental implant failure due to biomechanical overloading was reviewed in Sect. 10.3 in accordance with prominent factors which are parafunctional habits, cantilever extension, prosthesis morphology and contact, implant body materials, diameter and length, and implant-abutment connection. The review revealed that biomechanical overloading is crucial and valuable to be considered because it provides insights and knowledge enhancement in the understanding of dental implant failure. To probe the contributing factors of technical complications reported in clinical practices, a critical grasp on the related aspects is a must.

The fundamental understanding of the patient's related factors or physiological alterations is significant to predict any possibilities of implant complications. Substantial amounts of clinical studies have shown a positive relationship between bruxism and implant failure. The complex bruxism loading characteristic and behaviour still create a challenge in the experimental testing including the computational analyses to obtain an accurate quantified result. Many of which have considered such loading or occlusal force in the basic form with major simplifications imposed such as inadequate magnitude, single principal loading direction and time-independent. The correct and detailed force features representing bruxism in terms of value, duration, frequency and direction are desired to precisely quantify its impact. Strain gauge transducer can be used to record bite force with that of bruxers is shown to be 827 N with a standard deviation of 620 N [76]. Moreover, the design of the prosthesis and occlusion as well as the number of implants should be constant throughout the analysis to avoid great variation in conclusions.

Another challenge associated with the perseverance of dental implant stability is the prosthesis or crown design. It is fundamental to know that the steep or highly inclined cusp needs to be avoided due to the existence of adverse lateral force components. Also, the quality of implant-crown attachment must be a concern to ensure a passive fit or contact is achieved. Study on this area is still limited in which further evaluations would be advantageous.

The selection of a particular implant design or geometry could significantly affect the stability and performance of the implant system. Therefore, it is also a must to place a major emphasis on the implant features especially those closely related to the bone contact such as implant–abutment connection, implant body material, diameter and length. The implant–bone interface should be maintained to secure a long-term satisfactory result. In comparison, as said, natural teeth sustain better mechanical stresses upon loading due to the presence of the periodontal ligament relative to the implant. The implant deflects about 10- to 100-fold lower wherein the strong attachment with the bone tissues tends to result in bone resorption and degradation of the osseointegration. Osseointegration is a process at the microscopic level which is important for the direct functional and structural bond between living tissue and surface of the load-bearing implant. Poor osseointegration may affect dental implant stability as the therapy is substantially depending on the quality of osseointegration. Good implant stability can be defined as the freedom of an implant structure from clinical mobility and the ability of an implant to withstand axial and lateral loadings [77]. However, it has also been reported that the implants may experience micromotion especially those imposed by early or immediate loading. Excessive micromotion could result in fibrous tissue formation around the implant and being a causative agent for the failure of osseointegration and subsequent implant loss [31]. In contrast, some suggested that low-frequency micromotion could encourage bone growth with a range of 50–150 μm is favourable for osseointegration [78]. There seem to be sufficient evidence emerged to support early-loaded implants as they can achieve success rates that are comparable with the ones of delay-loaded.

Dental implants are highly prone to destructive effects by the tensile and shear forces, in contrast to the compressive force. Thus, the resolution of a single resultant occlusal load into the three force components in the solid screw-threaded implants at the interface is substantially influenced by the implant macro-design. Furthermore, in implant treatment planning, it is desired that the stresses may evenly be distributed within the implant complex and surrounding tissues. The low value of stress is predicted to decrease relative stiffness and strain differences between the implant and bone. The occurrence of stress shielding phenomenon that is caused by a considerable discrepancy in the elastic modulus of the bone and implant urges the host bone to largely transfer physiological and parafunctional loadings to the implant. This triggers the bone remodelling process to arise in which the bone will more degrade rather than regenerate and encounter a reduction in mass and quality.

Irrespective of the biomechanical factors, marginal or peri-implant bone loss is the primary characteristic sign that precedes the fracture of implants. The crestal portion of the bone–implant interface normally sustains high stresses and the stress value gradually reduced towards the apical third. In terms of bone adaptation, several parameters have been suggested to relate it with the loading namely strain rate, history-related factors, energy density, frequency, gradients and strain value. Of these, strain value provides more convincing evidence as observed in many types of research. Through Frost's mechanostat theory, there are four zones of microstrain—disuse atrophy (less than 200 μ-strain), steady state (from more than 200 to less than 2500 μ-strain), physiologic overload (from more than 2500 to less than 4000 μ-strain), and pathologic overload (4000 μ-strain or more) [31]. The disused atrophy and pathologic overload zones are the most extreme ones and are predicted to decrease bone mass and volume, resulting in considerable bone resorption to surgical failure. In certain cases, dental implants, nevertheless, are also observed to experience catastrophic failure without preceded by proof of peri-implant bone loss.

Adequate pre-treatment and prosthetic planning are vital to minimise the possibility of prosthetic parts fracture and must be reliant on the biomechanical algorithm. As definitive load-bearing evaluation for the host site is still limited, this has piloted for costly and invasive approaches to seek solutions. Biases against several biomechanical overloading factors are replete wherein some of them are not even evidence-based. Distribution of mechanical stress and strain at the bone–implant interface are believed to play a huge role in peri-implant bone loss and therefore, may also help to enlighten the prevalence of peri-implantitis in which on the rise. Patients who suffered from implant failure signs are normally subjected to several treatment method possibilities depending on the case severity. Total removal and alteration of the fractured implants are among the high-up solutions proposed which consequently demanding another term of healing and rehabilitation processes. Preventive measures are thus needed to improve early diagnosis and preoperative planning of the treatment so as to avoid secondary or revision operation and secure effective bone healing.

This present literature study is limited by the non-systematic approach. Future studies and findings are expected to provide a clearer insight on overloading conditions particularly on how it affects the bone–implant interaction and consequent implant longevity. To date, there are no prospective and randomised trials performed on humans in regard to occlusal overloading because of ethical concerns. Alternatively, preclinical experiments and/or meticulous computational analyses with high-level validation are the way forward to determine adequate thresholds of bone adaptation to loading.

10.5 Conclusion

The effect of overloading on dental implants has widely been debated, but a clear causal association of implant failure to overloading factors has not emerged. There are no solid quantitative justifications found on how mechanics of the overloaded mastication influence the implant failure, especially at the bone–implant interface that making it still debatable. The high ceiling of dental implant survival has clouded the inquiries of why implant failure cases are still reported. To the best of the authors' knowledge, the effect and correlation of the biomechanical overloading factors are not well justified and understood up till now within the implant system assembly in terms of ideal bone adaptation theories. An optimal set of the overloading factors is therefore desired to secure better implant durability.

Acknowledgements The authors would like to acknowledge the support from Fundamental Research Grant Scheme (FRGS) under a grant number of **FRGS/1/2020/TK0/UNIMAP/03/2** from the **Ministry of Higher Education Malaysia**. The authors reported no conflicts of interest related to this study.

References

1. Kunjappu J, Mathew V, Abdul Kader M, Kuruniyan M, Mohamed Ali A, Shamsuddin S (2019) Fracture of dental implants: an overview. Int J Prev Clin Dent Res 6(1):21–23
2. Marcelo CG, Filié Haddad M, Gennari Filho H, Marcelo Ribeiro Villa L, Dos Santos DM, Aldiéris AP (2014) Dental implant fractures-aetiology, treatment and case report. J Clin Diagn Res 8(3):300–304
3. Stoichkov B, Kirov D (2018) Analysis of the causes of dental implant fracture: a retrospective clinical study. Quintessence Int 49(4):279–286
4. Chrcanovic BR, Kisch J, Albrektsson T, Wennerberg A (2018) Factors influencing the fracture of dental implants. Clin Implant Dent Relat Res 20(1):58–67
5. Gargallo Albiol J, Satorres-Nieto M, Puyuelo Capablo JL, Sanchez Garces MA, Pi Urgell J, Gay Escoda C (2008) Endosseous dental implant fractures: an analysis of 21 cases. Med Oral Patol Oral Cir Bucal 13(2):E124–E128
6. Ali B, Ould Chikh EB, Meddah HM, Merdji A, Bachir Bouiadjra BA (2013) Effects of overloading in mastication on the mechanical behaviour of dental implants. Mater Des 47:210–217
7. Lian ZQ, Guan H, Loo YC (2011) Optimum degree of bone-implant contact in bone remodelling induced by dental implant. Proc Eng 14:2972–2979
8. Yousef A, Hadeel A (2019) Occlusal considerations in dental implantology. EC Dent Sci 18(8):1872–1883
9. Chrcanovic BR, Kisch J, Albrektsson T, Wennerberg A (2016) Bruxism and dental implant failures: a multilevel mixed effects parametric survival analysis approach. J Oral Rehabil 43(11):813–823
10. Chrcanovic BR, Kisch J, Albrektsson T, Wennerberg A (2017) Bruxism and dental implant treatment complications: a retrospective comparative study of 98 bruxer patients and a matched group. Clin Oral Implant Res 28(7):e1–e9
11. Quirynen M, Naert I, Van Steenberghe D (1992) Fixture design and overload influence marginal bone loss and future success in the Brånemark® system. Clin Oral Implant Res 3(3):104–111
12. Rangert B, Krogh PH, Langer B, Van Roekel N (1995) Bending overload and implant fracture: a retrospective clinical analysis. Int J Oral Maxillofac Implants 10(3):326–334
13. Mangano FG, Shibli JA, Sammons RL, Iaculli F, Piattelli A, Mangano C (2014) Short (8-mm) locking-taper implants supporting single crowns in posterior region: a prospective clinical study with 1-to 10-years of follow-up. Clin Oral Implant Res 25(8):933–940
14. Chitumalla R, Halini Kumari K, Mohapatra A, Parihar A, Anand K, Katragadda P (2018) Assessment of survival rate of dental implants in patients with bruxism: A 5-year retrospective study 9(6):278–282
15. Pommer B, Bucur L, Zauza K, Tepper G, Hof M, Watzek G (2014) Meta-analysis of oral implant fracture incidence and related determinants. J Oral Implantol 2014
16. Glauser R, Ree A, Lundgren A, Gottlow J, Hammerle CHR, Scharer P (2001) Immediate occlusal loading of brånemark implants applied in various jawbone regions: a prospective, 1-year clinical study. 3(4):204–213
17. Kilic E, Doganay O (2020) Evaluation of stress in tilted implant concept with variable diameters in the atrophic mandible: three-dimensional finite element analysis. J Oral Implantol 46(1):19–26
18. Valera-Jiménez JF, Burgueño-Barris G, Gómez-González S, López-López J, Valmaseda-Castellón E, Fernández-Aguado E (2020) Finite element analysis of narrow dental implants. Dent Mater 36(7):927–935
19. Cicciù M, Cervino G, Milone D, Risitano G (2018) FEM investigation of the stress distribution over mandibular bone due to screwed overdenture positioned on dental implants. Materials (Basel, Switzerland) 11(9):1512
20. Duyck J, Ronold HJ, Van Oosterwyck H, Naert I, Vander Sloten J, Ellingsen JE (2001) The influence of static and dynamic loading on marginal bone reactions around osseointegrated implants: an animal experimental study. Clin Oral Implants Res 12(3):207–218

21. Naert I, Quirynen M, Steenberghe DV, Darius P (1992) A study of 589 consecutive implants supporting complete fixed prostheses. Part II: prosthetic aspects. J Prosthet Dent 68(6):949–956
22. Shackleton JL, Carr L, Slabbert JCG, Becker PJ (1994) Survival of fixed implant-supported prostheses related to cantilever lengths. J Prosthet Dent 71(1):23–26
23. Romanos GE, Gupta B, Gaertner K, Nentwig GH (2014) Distal cantilever in full-arch prostheses and immediate loading: a retrospective clinical study. Int J Oral Maxillofac Implants 29(2):427–431
24. Romeo E, Tomasi C, Finini I, Casentini P, Lops D (2009) Implant-supported fixed cantilever prosthesis in partially edentulous jaws: a cohort prospective study 20(11):1278–1285
25. Sadowsky SJ (2019) Occlusal overload with dental implants: a review. Int J Implant Dent 5(1):29–29
26. Romeo E, Storelli S (2012) Systematic review of the survival rate and the biological, technical, and aesthetic complications of fixed dental prostheses with cantilevers on implants reported in longitudinal studies with a mean of 5 years follow-up. 23(s6):39–49
27. Freitas da Silva EV, Dos Santos DM, Sonego MV, de Luna Gomes JM, Pellizzer EP, Goiato MC (2018) Does the presence of a cantilever influence the survival and success of partial implant-supported dental prostheses? Systematic review and meta-analysis. Int J Oral Maxillofac Implants 33(4):815–823
28. Storelli S, Del Fabbro M, Scanferla M, Palandrani G, Romeo E (2018) Implant-supported cantilevered fixed dental rehabilitations in fully edentulous patients: Systematic review of the literature. Part II 29(S18):275–294
29. Belser UC, Mericske-Stern R, Bernard J-P, Taylor TD (2000) Prosthetic management of the partially dentate patient with fixed implant restorations. Note 11(s1):126–145
30. Zhang X, Mao J, Zhou Y, Ji F, Chen X (2020) Study on statics and fatigue analysis of dental implants in the descending process of alveolar bone level. Proc Inst Mech Eng H 0954411920926080
31. Vidyasagar L, Apse P (2003) Biological response to dental implant loading/overloading. Implant overloading: Empiricism or science. Stomatologija 5
32. O'Mahony A, Bowles Q, Woolsey G, Robinson SJ, Spencer P (2000) Stress distribution in the single-unit osseointegrated dental implant: finite element analyses of axial and off-axial loading. Implant Dent 9(3):207–218
33. Holmes DC, Loftus JT (1997) Influence of bone quality on stress distribution for endosseous implants. J Oral Implantol 23(3):104–111
34. Gupta S, Gupta H, Tandan A (2015) Technical complications of implant-causes and management: a comprehensive review. Natl J Maxillofac Surg 6(1):3–8
35. Schwarz MS (2000) Mechanical complications of dental implants. 11(s10:156–158
36. Hernandez-Rodriguez MAL, Contreras-Hernandez GR, Juarez-Hernandez A, Beltran-Ramirez B, Garcia-Sanchez E (2015) Failure analysis in a dental implant. Eng Fail Anal 57:236–242
37. Farawati FA, Nakaparksin P (2019) What is the optimal material for implant prosthesis? Dent Clin North Am 63(3):515–530
38. Shetty M, Shetty NHG, Jaiman R (2014) Implant abutment connection: biomechanical perspectives. Nitte Univ J Health Sci 4(2)
39. Kitagawa T, Tanimoto Y, Iida T, Murakami H (2020) Effects of material and coefficient of friction on taper joint dental implants. J. Prosthodont Res
40. Scholander S (1999) A retrospective evaluation of 259 single-tooth replacements by the use of Branemark implants. Int J Prosthodont 12(6):483–491
41. Al Jabbari YS, Fournelle R, Ziebert G, Toth J, Iacopino AM (2008) Mechanical behavior and failure analysis of prosthetic retaining screws after long-term use in vivo. Part 1: characterization of adhesive wear and structure of retaining screws 17(3):168–180
42. Najeeb S, Mali M, Yaqin SAU, Zafar MS, Khurshid Z, Alwadaani A, Matinlinna JP (2019) 21-Dental implants materials and surface treatments. Woodhead Publishing
43. Tretto PHW, dos Santos MBF, Spazzin AO, Pereira GKR, Bacchi A (2020) Assessment of stress/strain in dental implants and abutments of alternative materials compared to conventional titanium alloy-3D non-linear finite element analysis. Comput Methods Biomech Biomed Engin 23(8):372–383

44. Huang YS, McGowan T, Lee R, Ivanovski S (2017) 7.23 Dental implants: biomaterial properties influencing osseointegration. Elsevier, Oxford
45. Robau-Porrua A, Pérez-Rodríguez Y, Soris-Rodríguez LM, Pérez-Acosta O, González JE (2020) The effect of diameter, length and elastic modulus of a dental implant on stress and strain levels in peri-implant bone: A 3D finite element analysis. Biomed Mater Eng 30:541–558
46. Reis TAD, Zancopé K, Karam FK, Neves FDD (2019) Biomechanical behavior of extra-narrow implants after fatigue and pull-out tests. J Prosthet Dent 122(1):54.e1–54.e6
47. Ogle OE (2015) Implant surface material, design, and osseointegration. Dent Clin North Am 59(2):505–520
48. Ortega-Oller I, Suárez F, Galindo-Moreno P, Torrecillas-Martínez L, Monje A, Catena A, Wang H-L (2014) The influence of implant diameter on its survival: a meta-analysis based on prospective clinical trials. J Periodontol 85(4):569–580
49. Al-Johany SS, Al Amri MD, Alsaeed S, Alalola B (2017) Dental implant length and diameter: a proposed classification scheme. J Prosthodont 26(3):252–260
50. Lee CT, Chen YW, Starr JR, Chuang SK (2016) Survival analysis of wide dental implant: systematic review and meta-analysis. Clin Oral Implants Res 27(10):1251–1264
51. Graves CV, Harrel SK, Rossmann JA, Kerns D, Gonzalez JA, Kontogiorgos ED, Al-Hashimi I, Abraham C (2016) The role of occlusion in the dental implant and peri-implant condition: a review. Open Dent J 10:594–601
52. Shin SW, Bryant SR, Zarb GA (2004) A retrospective study on the treatment outcome of wide-bodied implants. Int J Prosthodont 17(1):52–58
53. Klein MO, Schiegnitz E, Al-Nawas B (2014) Systematic review on success of narrow-diameter dental implants. Int J Oral Maxillofac Implants 29(Suppl):43–54
54. Zembic A, Johannesen LH, Schou S, Malo P, Reichert T, Farella M, Hammerle CH (2012) Immediately restored one-piece single-tooth implants with reduced diameter: one-year results of a multi-center study. Clin Oral Implants Res 23(1):49–54
55. Vigolo P, Givani A, Majzoub Z, Cordioli G (2004) Clinical evaluation of small-diameter implants in single-tooth and multiple-implant restorations: a 7-year retrospective study. Int J Oral Maxillofac Implants 19(5):703–709
56. Saad M, Assaf A, Gerges E (2016) The use of narrow diameter implants in the molar area. Int J Dent 2016(8)
57. Misch CE (2008) Contemporary implant dentistry. Mosby Elsevier
58. Misch CE, Strong JT, Bidez MW (2015) Chapter 15-Scientific rationale for dental implant design. Mosby, St. Louis
59. Demenko V, Linetskiy I, Linetska L, Yefremov O (2019) Load-carrying capacity of short implants in edentulous posterior maxilla: a finite element study. Med Eng Phys
60. Pirmoradian M, Naeeni HA, Firouzbakht M, Toghraie D, khabaz MK, Darabi R (2020) Finite element analysis and experimental evaluation on stress distribution and sensitivity of dental implants to assess optimum length and thread pitch. Comput Methods Programs Biomed 187:105258
61. Griffin TJ, Cheung WS (2004) The use of short, wide implants in posterior areas with reduced bone height: a retrospective investigation. J Prosthet Dent 92(2):139–144
62. Nielsen HB, Schou S, Isidor F, Christensen AE, Starch-Jensen T (2019) Short implants (≤8mm) compared to standard length implants (>8mm) in conjunction with maxillary sinus floor augmentation: a systematic review and meta-analysis. Int J Oral Maxillofac Surg 48(2):239–249
63. Amine M, Guelzim Y, Benfaida S, Bennani A, Andoh A (2019) Short implants (5–8 mm) vs. long implants in augmented bone and their impact on peri-implant bone in maxilla and/or mandible: systematic review. J Stomatol Oral Maxillofac Surg 120(2):133–142
64. Jung RE, Al-Nawas B, Araujo M, Avila-Ortiz G, Barter S, Brodala N, Chappuis V, Chen B, De Souza A, Almeida RF, Fickl S, Finelle G, Ganeles J, Gholami H, Hammerle C, Jensen S, Jokstad A, Katsuyama H, Kleinheinz J, Kunavisarut C, Mardas N, Monje A, Papaspyridakos P, Payer M, Schiegnitz E, Smeets R, Stefanini M, ten Bruggenkate C, Vazouras K, Weber H-P, Weingart D, Windisch P (2018) Group 1 ITI consensus report: the influence of implant length and design and medications on clinical and patient-reported outcomes. Clin Oral Implant Res 29(S16):69–77

65. Malchiodi L, Giacomazzi E, Cucchi A, Ricciotti G, Caricasulo R, Bertossi D, Gherlone E (2019) Relationship between crestal bone levels and crown-to-implant ratio of ultra-short implants with a microrough surface: a prospective study with 48 months of follow-up. J Oral Implantol 45(1):18–28
66. Meijer HJA, Boven C, Delli K, Raghoebar GM (2018) Is there an effect of crown-to-implant ratio on implant treatment outcomes? A Syst Rev 29(S18):243–252
67. de Dias FJN, Pecorari VGA, Martins CB, Del Fabbro M, Casati MZ (2019) Short implants versus bone augmentation in combination with standard-length implants in posterior atrophic partially edentulous mandibles: systematic review and meta-analysis with the Bayesian approach. Int J Oral Maxillofac Surg 48(1):90–96
68. Caricasulo R, Malchiodi L, Ghensi P, Fantozzi G, Cucchi A (2018) The influence of implant-abutment connection to peri-implant bone loss: a systematic review and meta-analysis. Clin Implant Dent Relat Res 20(4):653–664
69. Pjetursson BE, Tan K, Lang NP, Brägger U, Egger M, Zwahlen M (2004) A systematic review of the survival and complication rates of fixed partial dentures (FPDs) after an observation period of at least 5 years. Clin Oral Implant Res 15(6):625–642
70. Koutouzis T (2019) Implant-abutment connection as contributing factor to peri-implant diseases. Periodontology 2000 81(1):152–166
71. Da Silva EF, Pellizzer EP, Quinelli Mazaro JV, Garcia Júnior IR (2010) Influence of the connector and implant design on the implant–tooth-connected prostheses 12(3):254–262
72. Sasada Y, Cochran DL (2017) Implant-abutment connections: a review of biologic consequences and peri-implantitis implications. Int J Oral Maxillofac Implants 32(6):1296–1307
73. Maeda Y, Satoh T, Sogo M (2006) In vitro differences of stress concentrations for internal and external hex implant–abutment connections: a short communication 33(1):75–78
74. Jeng M-D, Lin Y-S, Lin C-L (2020) dBiomechanical evaluation of the effects of implant neck wall thickness and abutment screw size: a 3D nonlinear finite element analysis. Appl Sci 10(10)
75. Möllersten L, Lockowandt P, Lindén L-Å (1997) Comparison of strength and failure mode of seven implant systems: an in vitro test. J Prosthet Dent 78(6):582–591
76. Alkan A, Bulut E, Arici S, Sato S (2008) Evaluation of treatments in patients with nocturnal bruxism on bite force and occlusal contact area: a preliminary report. Eur J Dent 2(4):276–282
77. Haïat G, Wang H-L, Brunski J (2014) Effects of biomechanical properties of the bone-implant interface on dental implant stability: from in silico approaches to the patient's mouth. Annu Rev Biomed Eng 16(1):187–213
78. Szmukler-Moncler S, Salama H, Reingewirtz Y, Dubruille JH (1998) Timing of loading and effect of micromotion on bone-dental implant interface: review of experimental literature. J Biomed Mater Res 43(2):192–203

Chapter 11
Sensitivity Creep Simulation for the Equipment's Component Experiencing Extreme Elevated Temperature

Hashim Othman, Nurul Hana Kamaruzaman, and Ir. Hambali Chik

Abstract Generally, prolonged exposure of low alloy carbon steel at an elevated temperature beyond the material's creep threshold temperature under load will result in slow material deformation even though it is below the yield stress, which eventually leads to rupture (API Recommended Practice 571: damage mechanisms affecting fixed equipment in refining industry. American Petroleum Institute, pp 32–38, 2011). Thus, it is crucial to ensure the right selection or the use of a suitable barrier by means of internal refractory to ensure the material can perform as per the intended design life. This paper presents the sensitivity creep simulation results using the MPC Project Omega Data (API 579–1/ASME FFS-1. Fitness-For-Service. American Petroleum Institute, pp. 10.1–10B.57, 2016) model for the component exposed to an extremely elevated temperature for a short duration of approximately a day; in the context of creep damage and the remaining life of the affected component. Several site testings were conducted to support the sensitivity creep simulation to ensure the study results are precise and reliable. The microstructure evaluation was also included to assess the extent of the creep damage. The creep remaining life prediction covered several scenarios of exposure durations to the extremely elevated metal temperature. The simulation results are used as an input to the plant owner in order to establish proper mitigation and monitoring program to avoid similar recurrence.

Keywords Low-alloy carbon steel · Refractory spalling · Creep · Elevated temperature · MPC Project Omega Data · Remaining life

---

H. Othman · N. H. Kamaruzaman (✉) · Ir. H. Chik
Group Technical Solutions, Petroliam Nasional Berhad, 50088 Kuala Lumpur, Malaysia
e-mail: hana_kamaruzaman@petronas.com

S. Abdullah et al. (eds.), *Structural Integrity Cases in Mechanical and Civil Engineering*, Structural Integrity 23, https://doi.org/10.1007/978-3-030-85646-5_11

11.1 Introduction

All components material shall be selected based on the service requirement with consideration of mechanical strength and corrosion resistance to ensure it can perform as per intended design life [1]. For a component that operates at elevated temperature, creep damage is a concern if the component metal temperature exceeds its creep threshold temperature [2].

This case study describes a component designed to be operated at 460 °C with internal castable refractory installed to protect the material of low-alloy carbon steel, 1.25Cr–0.5Mo from exposure to process gas at an extremely elevated temperature of 707 °C. However, an incident where the internal castable refractory spalled-off (Fig. 11.1) resulted in the exposure to the elevated temperature of process gas for approximately a day.

During the incident, the maximum component metal temperature was measured using IR Thermography and this data was used as an input for creep simulation. The maximum external surface component metal temperature recorded was approximately 602 °C. The additional technical information of the analysed component is shown in Table 11.1.

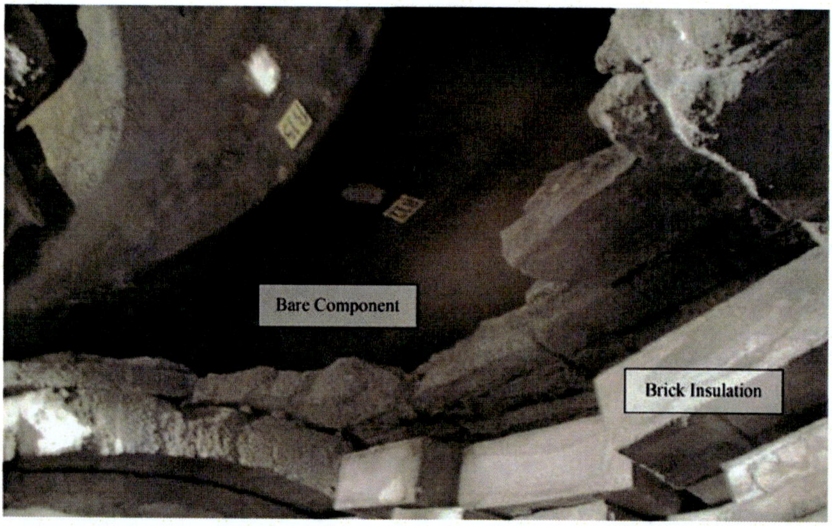

Fig. 11.1 The internal castable refractory spalled-off resulted in the component material had exposed to the process gas

Table 11.1 Additional technical information of the component

Description	Values
Design code	ASME Section VIII Div. 1
Design pressure	40.0 barg
Design temperature	460 °C
Nominal outside diameter	2.4 m
Nominal shell thickness	55.0 mm

11.1.1 Objective

This paper aims to share the sensitivity creep simulation findings performed on the metal component that experienced extremely elevated temperature during operation. The findings focus on the prediction of creep damage and the remaining life assessment of the component.

11.2 Methodology

The sensitivity creep simulation was carried out to predict creep damage and the affected component's remaining life. It requires several analyses with supporting site testing results to ensure the simulation findings are precise and reliable. The following is the summary of the methodology for sensitivity creep simulation and site testing validation.

11.2.1 MPC Project Omega Data Creep Simulation

Creep was identified as the primary damage mechanism for the component. It was exposed to an extremely high temperature beyond the design temperature and creep threshold temperature for 1.25Cr–0.5Mo [2] for approximately a day. MPC Project Omega Data [3] was selected as a method and acceptance criteria for the prediction of creep damage and the remaining life of the affected component.

Prior to the detailed creep simulation, a collection of crucial information was carried out, including operating temperature and pressure trending history to evaluate the equipment's operating cycles and hours. Thickness data review revealed no significant internal thinning on the affected component as it is internally protected by refractory.

A review of the skin metal temperature survey record from IR Thermography (Fig. 11.2) during the incident and a comparison with previous survey records was made to determine the affected component's maximum temperature exposure. The skin metal temperature trending (Fig. 11.3) from survey records indicate that the

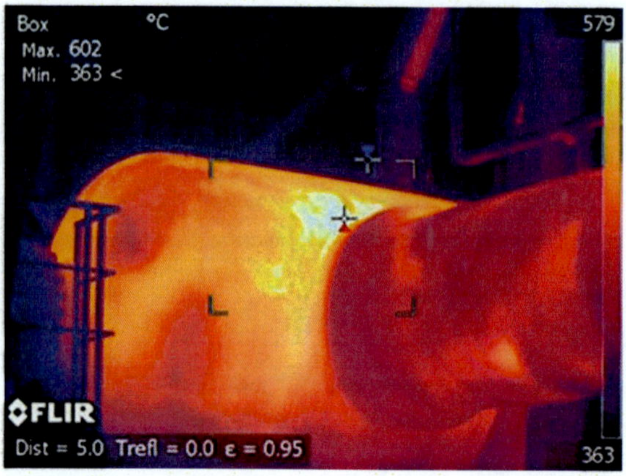

Fig. 11.2 Skin metal temperature survey from IR Thermography during the incident

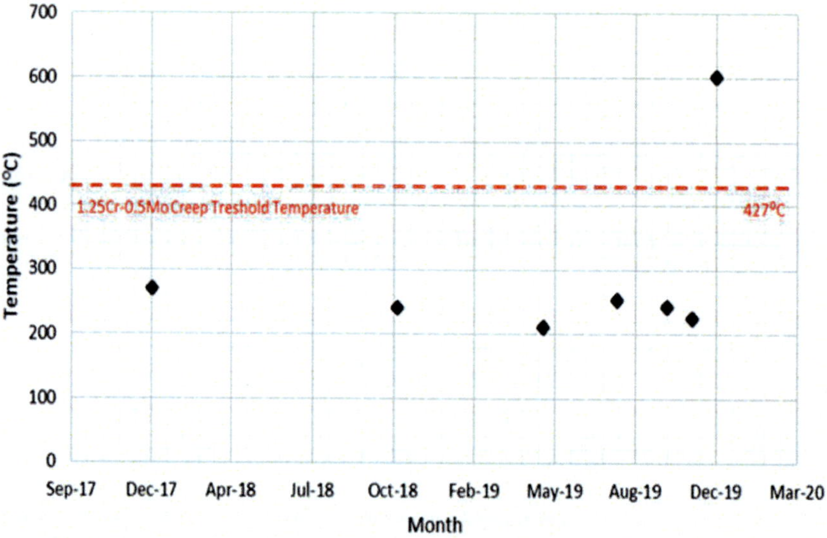

Fig. 11.3 Skin temperature survey records for the affected component

affected component had experienced high metal skin temperature of 602 °C during the incident [4] and was expected to experience creep damage.

Based on skin metal temperature records, the mid-wall metal temperature profile was derived from Fourier Law Equation based on operation data and component information i.e. process temperature data, material thermal conductivity, and metal skin temperature. This data was used as the main input for creep simulation besides operating pressure, operating hours and MPC Project Omega Data.

The component's principal stress "(1)–(3)" for the time increment ${}^{n}t$ are determined based on Elastic Stress Solution Based on Mean Diameter Stress Eq. (11.2).

$$\sigma_1 = \sigma_c = \frac{PD_{\text{mean}}}{2t_{\text{comp}}} \cdot L_f \tag{11.1}$$

$$\sigma_2 = \sigma_m = \frac{PD_{\text{mean}}}{4\left(t_{\text{comp}} - t_{sl}\right)} \cdot L_f \tag{11.2}$$

$$\sigma_3 = \sigma_r = 0.0 \tag{11.3}$$

Load reference stress is calculated as per Eq. (11.4). The primary load reference stress, ${}^{n}\sigma^{P}{}_{\text{ref}}$ is determined based on the value of yield strength, σ_{ys} that is derived at temperature ${}^{n}T$ at time increment ${}^{n}t$ [3].

$${}^{n}\sigma_e = \frac{1}{\sqrt{2}}\left[\left({}^{n}\sigma_1 - {}^{n}\sigma_2\right)^2 + \left({}^{n}\sigma_1 - {}^{n}\sigma_3\right)^2 + \left({}^{n}\sigma_2 - {}^{n}\sigma_3\right)^2\right]^{0.5} \tag{11.4}$$

Effective stress, ${}^{n}\sigma_{e,}$ is calculated using Eq. (11.5) as in input to uniaxial Omega damage parameter "(X)".

$${}^{n}L = \frac{1}{\dot{\varepsilon}_{co\Omega_m}} \tag{11.5}$$

The initial creep strain rate $\dot{\varepsilon}_{co}$ is based on the stress state and temperature. A_0–A_4 are material coefficients for the MPC Project Omega strain-rate-parameter [3].

$$log_{10}\dot{\varepsilon}_{co} = -\left\{\left(A_0 - \Delta_{\Omega}^{sr}\right) + \left[\frac{1}{460 + {}^{n}T}\right]\left[A_1 + A_2 S_1 + A_3 S_1^2 + A_4 S_1^3\right]\right\}$$

Ω_m as the multiaxial Omega damage parameter [3].

$$\Omega_m = \Omega_n^{\delta_{\Omega+1}} + \alpha_{\Omega} \cdot n_{BN}$$

Ω_n as the adjusted uniaxial Omega damage parameter at the specific period [3].

$$\Omega_n = \max[(\Omega - n_{BN}), 3.0]$$

The uniaxial Omega damage parameter, Ω, can be further derived using the following formula where B_0–B_4 are material coefficients for the MPC Project Omega Omega-parameter [3].

$$\log_{10}\Omega = \left(B_o + \Delta_{\Omega}^{cd}\right) + \left[\frac{1}{460 + {}^{n}T}\right]\left[B_1 + B_2 S_1 + B_3 S_1^2 + B_4 S_1^3\right]$$

The damage parameter exponent for MPC Project Omega Life Assessment Model, δ_Ω can be derived from the following formula where the Prager factor, β_Ω is set as 0.33 for the MPC project Omega model [3].

$$\delta_\Omega = \beta_\Omega\left(\frac{{}^n\sigma_1 + {}^n\sigma_2 + {}^n\sigma_3}{{}^n\sigma_e} - 1.0\right)$$

The value of the Bailey Norton coefficient, $n_{BN,}$ is derived using the following formula and utilised in the multiaxial Omega damage parameter formula.

$$n_{BN} = -\left\{\left[\frac{1}{460 + {}^nT}\right]\left[A_2 + 2A_3S_1 + 3A_4S_1^2\right]\right\}$$

S_l is the stress parameter derived from $\log_{10}$ effective stress for a specific period.

$$S_l = \log_{10}\left({}^n\sigma_e\right)$$

The stress level, ${}^nL$, at Eq. (11.5) for the time increment, ${}^nt$, in the component's current operating cycle, mth shall be determined. The accumulated creep damage at Eq. (11.6) is calculated for all points in the mth cycle. The creep damage simulation is terminated when $D_c^{\text{total}} = 1.0$.

$$ {}^nD_C = \sum_{i=0}^{N} \frac{{}^nt}{{}^nL} \tag{11.6}$$

The creep life simulation was conducted under two scenarios to assess the impact of elevated temperature exposure time on the final accumulated creep damage and the remaining life. The two scenarios that was considered are exposed to elevated temperature process gas for 24 h only and continuously exposed to elevated temperature process gas [4]. The total allowable creep damage, D_c^{allow}, is 1.0 (100% of total creep life). However, it is recommended to set the creep damage at 0.8 (80% of total creep life) as a warning limit to allow the asset owner to make necessary preparation and planning prior to the predicted creep rupture failure.

11.2.2 In Situ Surface Replication Evaluation

In situ surface replication testing was selected as one of the methods to validate the creep life simulation findings, specifically from a microstructure evaluation perspective to assess the current degradation of the material.

In situ surface replication testing was performed on several critically identified locations of the component's external and internal surface such as internal refractory spalling and non-spalling area. These testing results were evaluated using

Table 11.2 Microstructure rating chart [5]

Class rating	Microstructure condition
1	Subject to creep loading without creep cavities
2a	Individual isolated creep cavities
2b	Numerous creep cavities, randomly orientation
3a	Numerous creep cavities, specific orientation
3b	Chains of creep cavities; individual grain boundary separation
4	Micro-crack
5	Macro-crack

Microstructure Rating Charts for Evaluating the Microstructure and Creep Damage of High-Temperature Steels for High-Pressure Piping and Boiler Components (Table 11.2).

The findings from this testing were evaluated against the simulated creep damage results.

11.2.3 Hardness Evaluation

Hardness testing was selected as one of the methods to determine the changes in tensile strength of 1.25 Cr–0.5Mo material and changes in other material properties such as toughness and ductility [3].

The hardness testing was conducted at a similar location of in-situ surface replication spots. The testing results were evaluated against the metallurgical requirement for 1.25Cr–0.5Mo. The material measured hardness shall meet or exceed the minimum hardness value of 130 HB [3].

11.3 Results and Discussions

11.3.1 MPC Omega Data Creep Assessment

The results of creep damage and creep remaining life for the affected component based on the two simulation scenarios are shown in Fig. 11.4.

The result shows a substantial increase of cumulative creep damage with a significant reduction of creep remaining life when the affected component being continuously exposed to the extremely elevated temperature of 707 °C. The predicted

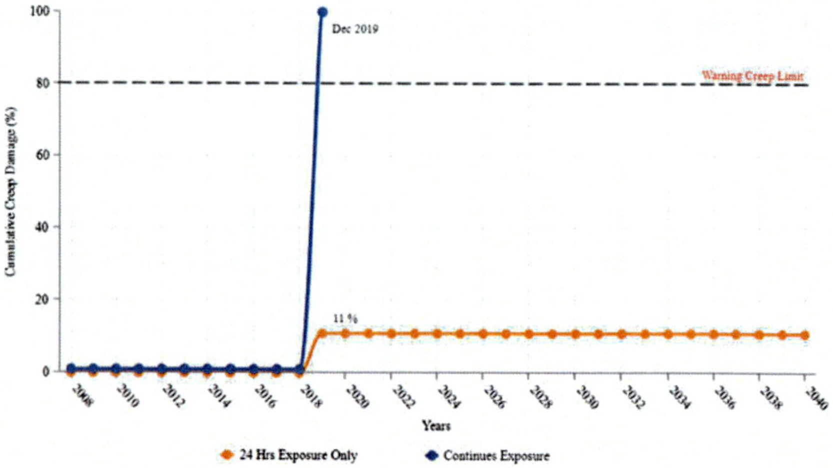

Fig. 11.4 Cumulative creep damage simulation results

Table 11.3 Projected remaining life from creep life simulation

Continuously exposed to elevated temperature process gas	Exposed to elevated temperature process gas for a day only
10 days	>20 years

remaining life indicated that only 10 days required to reach 100% total creep damage (Table 11.3).

However, a day exposure duration to the extremely elevated temperature resulted in 11% of total simulated creep damage with sufficient desired remaining life, i.e. more than 20 years.

11.3.2 In Situ Surface Replication Evaluation

The in situ surface replica result conducted on the internal surface of the affected component, focusing on the refractory spalling area indicated a sign of creep exposure which is depicted by the dissolution of bainite with the presence of ferrite and carbides in Figs. 11.5 and 11.6. However, the material did not experience any significant creep and/or voids after being exposed to an extremely elevated temperature for a day.

A comparison of in situ surface replica results was made with the internal non-spalling area, taken from the external surface (Figs. 11.7 and 11.8) to evaluate the extent of creep damage. The result revealed the presence of ferrite and bainite without dissolution into carbide. This result indicated that the severity of creep damage at this area is less severe as compared to the refractory spalling area.

Fig. 11.5 In situ surface replica results at the refractory spalling area of the internal surface at 200x magnification

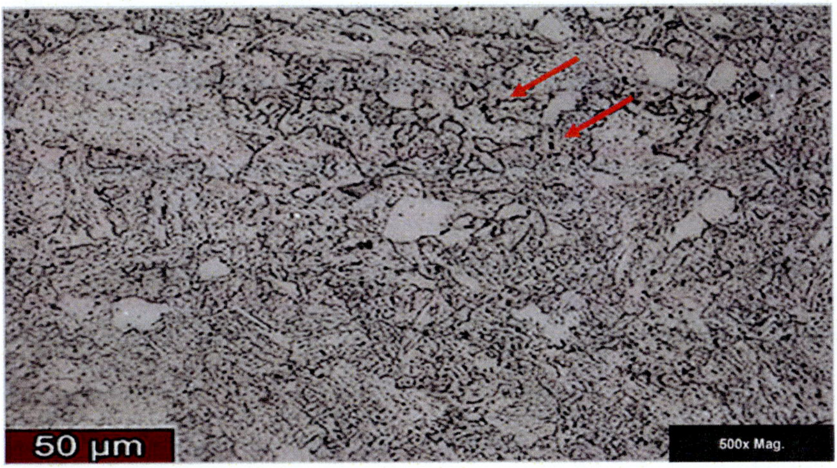

Fig. 11.6 Close-up view of in situ surface replica result at the affected area. The red arrow indicated the presence of carbide at the grain boundaries

Based on the above discussion, the in situ surface replica results for the affected component at refractory spalling and non-spalling area can be categorized as Class Rating 1; the material had subjected to creep loading without creep cavities [4, 5].

As part of the validation process, the result of the MPC Omega Data creep assessment is found to be correlated with the findings from in situ surface replication evaluation which indicated that the material had been subjected to creep damage for approximately 11% without creep cavities.

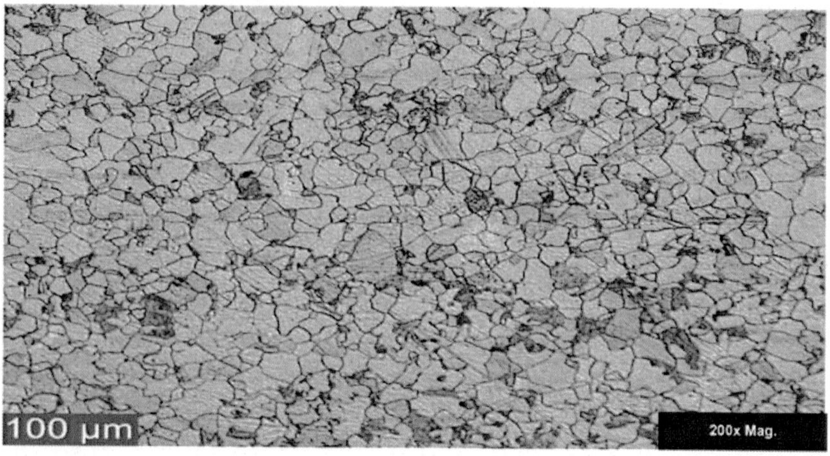

Fig. 11.7 In situ surface replication at baseline area at 200x magnification

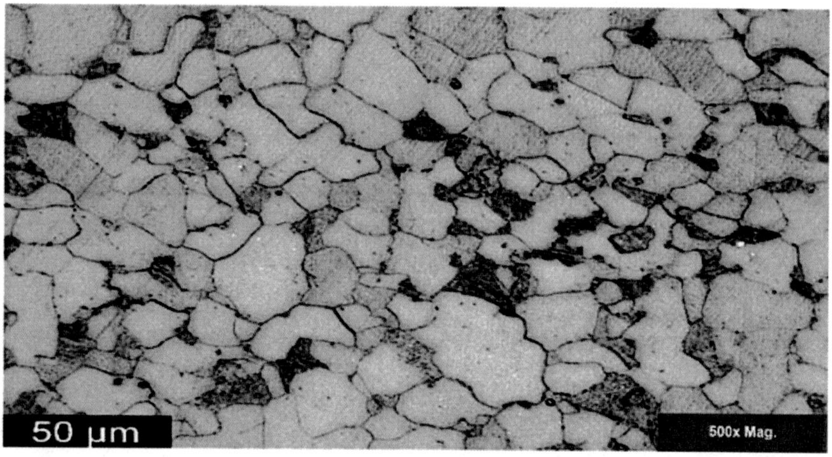

Fig. 11.8 In situ surface replication at baseline area at 500x magnification

11.3.3 Hardness Evaluation

Hardness testing results were carried out on the same spot location of the in situ surface replication area. The hardness readings were analysed to determine the changes in tensile strength of 1.25 Cr–0.5Mo material and other material properties such as toughness and ductility.

The overall results (Table 11.4) showed all hardness values including the refractory spalling area had exceeded the minimum required hardness value of 130 HB for 1.25Cr–0.5Mo material [3].

Table 11.4 Overall hardness evaluation results [6]

Locations	Minimum measured hardness value (HB)	Evaluation status
Refractory spalling area		
Location 1	165	Passed
Location 2	176	Passed
Location 3	169	Passed
Non-spalling area		
Location 1	142	Passed
Location 2	152	Passed
Location 3	150	Passed

Based on the above results, the tensile strength, toughness and ductility properties of 1.25Cr–0.5Mo were not affected during the incident as discussed above.

11.4 Conclusion

Based on the results from MPC Omega Data creep assessment, In situ Replication and Hardness Testing Evaluation, the assessed component is deemed to be fit-for-service and shall be operated at the design operating parameters after a day "hot-spot" incident based on the following:

(a) MPC Omega Data creep assessment result indicated that the calculated accumulation of creep damage is approximately 11% which is far below the warning creep damage limit, i.e. 80%.
(b) The projected remaining life based on current accumulated creep damage is estimated to be more than 20 years.
(c) In situ surface replication results had revealed that the material had experienced no significant creep damage and/or voids after being exposed to an extreme elevated process gas temperature of approximately 707 °C for a day and had been categorized as Class Rating 1 [5].
(d) The overall hardness results on the affected locations still meet the minimum required hardness value of 130 HB [3] and the hardness evaluation results indicated that the tensile strength, toughness and ductility properties of the affected component are less likely to be affected by the incident.

It is recommended to conduct regular skin metal temperature monitoring using IR thermography at the affected component for early detection of any abnormal skin metal temperature.

The MPC Omega Data creep assessment shall be reassessed if the component is subjected to another abnormal high metal skin temperature that exceeds the creep threshold temperature as the current creep life assessment result will be nullified.

References

1. Peter E (2001) Choose materials for high-temperature environment. http://www.aiche.org/cep
2. API Recommended Practice 571 (2011) Damage mechanisms affecting fixed equipment in refining industry. American Petroleum Institute, pp 32–38
3. API 579–1/ASME FFS-1 (2016) Fitness-For-Service. American Petroleum Institute, pp 10.1–10B.57
4. Kamaruzaman NH et al (2020) Fitness for service assessment for reformed gas cooler exchangers and associated transfer lines. PETRONAS
5. VGB-TW 507 (2005) Guideline for the assessment of microstructure and damage development of creep exposed materials for pipes and boiler components. VGB Power Tech Service GmbH, pp 2
6. Brian R (2012) Creep life simulation and assessment of high temperature piping systems and components. In: Proceedings of the ASME 2012 pressure vessels & piping division conference
7. ASME Boiler and Pressure Vessel Code (2015) Section II Part D. American Society of Mechanical Engineers, pp 38–39
8. ASTM E-140 (2002) Standard hardness conversion tables for metals relationship among Brinell hardness, Vickers hardness, Rockwell hardness, Superficial hardness, Knoop hardness, and Scleroscope hardness. American Society for Testing and Materials, pp 4
9. Kadir J et al (2019) Remaining creep life prediction of high temperature pressure primary superheater components of heat recovery steam generator. In: The 4th international conference on the science and engineering of materials 2019
10. Pornsak T (2015) Remaining creep life assessment of service superheat tube boiler. Trans Tech Publications, Switzerland (2015)

Chapter 12
Characterization of Fatigue and Crack Growth on AZ31B Magnesium Alloy

A. H. Isahak, M. F. Abdullah, M. K. Faidzi, Aidy Ali, and M. M. Mubasyir

Abstract This research studied the characteristics of fatigue and Crack Growth (CG) behaviour of AZ31B magnesium alloy at room temperature when the material was used for designing. The experiments were carried out using a fatigue machine, which involved the experimental testing of stress–strain material and fatigue and CG tests. The fatigue test was conducted using a standard ASTM E8 and dog bone samples at 90, 80, 70, 60 and 50% of applied yield stress. Meanwhile, a CG test was done using ASTM E647 standard Compact Tension (CT) and specimen at 90% of yield stress. Both tests were conducted under a sinusoidal cycle by applying the Constant Amplitude Loading (CAL) at a frequency of 10 Hz. The fatigue endurance showed that at 1×10^6, the stress was 48 Mpa using the Basquin equation prediction. The predicted lives were in good agreement with an R^2 of 0.91. The CG behaviour of the specimen, indicating a fatigue CG (a-N) with failure of the structure, and the Paris Law constants were determined.

Keywords Fatigue life · Crack growth · Constant amplitude loading · S–N diagram · Paris Law constants

12.1 Introduction

The endurance and life limit of body armour material have been the two major challenges for researchers in the military industry. Magnesium alloy, considered the best alloy in the twenty-first century, is a significant structural material for the future for its physical qualities, such as lightweight, strong, recyclable and good machinability. Since the 1940s, magnesium alloy has been used for military purposes, from ground vehicles to aircraft parts. Magnesium alloy innovation's past, existing and future mechanical and physical infrastructures have been studied to fix the issues surrounding the US military applications [1]. According to Abdullah et al. [2], throughout this paper, magnesium alloy has the potential for missile and military

A. H. Isahak · M. F. Abdullah (✉) · M. K. Faidzi · A. Ali · M. M. Mubasyir
Faculty of Engineering, Department of Mechanical Engineering, Universiti Pertahanan Nasional Malaysia, Kem Sungai Besi, 57000 Kuala Lumpur, Malaysia
e-mail: m.faizal@upnm.edu.my

S. Abdullah et al. (eds.), *Structural Integrity Cases in Mechanical and Civil Engineering*, Structural Integrity 23, https://doi.org/10.1007/978-3-030-85646-5_12

applications. In addition, the weight of an armoured vehicle is one of the factors that influence its durability, which affects its operational performance.

Since fracture mechanics was first applied to fatigue Crack Growth (CG) exposure in metals in the mid-1960s [3], an enormous measure of data has been gathered in terms of factors, for example, mean stress, high or low frequency, microstructure, environment and temperature [4]. A military operation vehicle is an example of a structure that undergoes fatigue phenomenon, especially when being shot by a high-velocity ammunition bullet [5], which reduces its operational performance. Fatigue failure occurs under cyclic or fluctuating loading. The fatigue properties of magnesium combinations are fundamentally influenced by factors like heat treatment, sur-rounding situations, applied load, and high or low frequency.

In relation to other series of magnesium alloys, the AZ31B is the strongest due to its lower density, ability to withstand large loads and resistance to high temperature. However, the weaknesses of AZ31B must also be considered to avoid any failure that may affect safety. One disadvantage of magnesium alloys is limited formability as it easily rusts and corrodes under high chemical reactivity [6]. The market price of magnesium alloy is also higher compared to aluminium and iron alloys [7].

The expected fatigue life should be possible in two unique manners. The first manner, as indicated by Paris Law [8], links the growth rate of subcritical cracks to various stress intensity factors. On the other hand, the second manner is based on the Wöhler curve [9], where the range of stress applied is determined by the number of cycles to failure. Using self-equations and incomplete fractal modelling, these distinct approaches can be intimately linked, so that the crack and specimen sizes are considered.

This study aims to investigate the characteristics of fatigue and CG behaviour of AZ31B magnesium alloy. The material properties, fatigue durability, and CG of AZ31B magnesium alloy were tested in the Constant Amplitude Loading (CAL) by using a frequency of 10 Hz.

12.2 Materials and Methods

The AZ31B magnesium alloy material was considered in this study to determine the characteristics of fatigue and CG behaviour. This study focused on two key approaches of AZ31B magnesium alloy, which are static and dynamic experiments. Figure 12.1 demonstrates the process flow of the methodology.

The rectangular tension test and compact tension, CT were fabricated carefully at the National Defense University of Malaysia laboratory using EDM-wire cut with coolant, this caused the microstructure of material was not affected by the heat generated by the machine. The samples for tensile, fatigue and crack growth test with dimension were discussed methodology subtopics. The specimens were cut based on the ASTM E8 [10] for tensile and fatigue test, meanwhile ASTM E647 [11] compact tensions, CT for crack growth test.

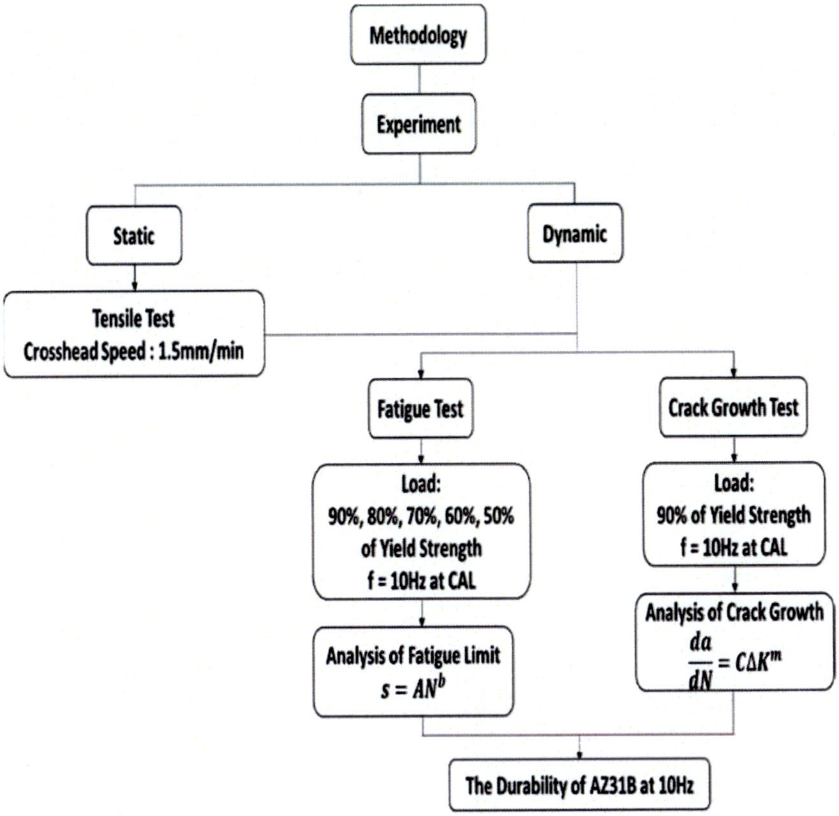

Fig. 12.1 Process flow for the conducted research study

12.2.1 Tensile Tests

The fatigue and CG testing and analysis were characterised by the mechanical properties of AZ31B magnesium alloy. In this study, the test was repeated [12, 13] multiple times using new tensile samples according to the standard method for the tensile test of metallic materials (ASTM E8) to obtain a value close to the actual value. The selected cross-head speed force [14] on the samples was 1.5 mm/min. During the test, the axial load was consistently applied on the two closures of the samples. The strain that was delivered when shear forces were applied on the samples was estimated by an extensometer. For each sample, the dimension is sub-sized rectangular, with a thickness of 6 mm and a gauge length of 25 mm for fatigue and tensile tests, as shown in Fig. 12.2a and b. The ultimate strength, 0.2% of yield strength, fracture strength and the modulus of elasticity are obtained from the resulting stress graph. Figure 12.3 shows the tensile test machine; Universal Testing Machine Instron.

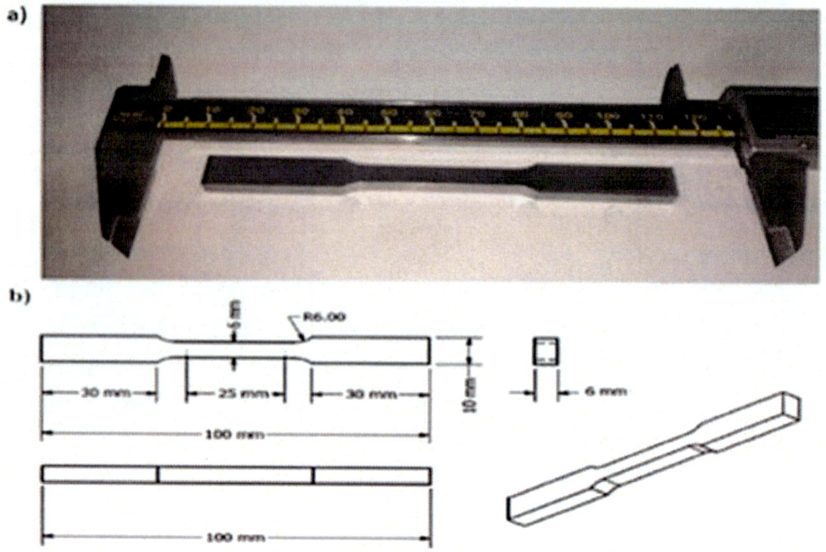

Fig. 12.2 **a** Tensile and fatigue AZ31B magnesium alloy specimens, and **b** the dimensions (mm)

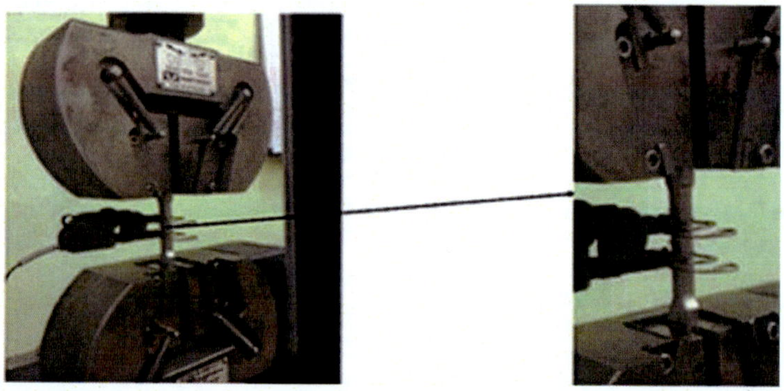

Fig. 12.3 Tensile test machine

12.2.2 Fatigue Tests

The experiment continues with axial CAL to determine the S–N curve. As shown in Fig. 12.4, the standard fatigue test is conducted according to applicable ASTM E466 [15] at the National Defence University of Malaysia using a fatigue test machine; 100 kN servo-hydraulic, with a constant load cycle of 90% of yield strength of AZ31B magnesium alloy. The specimen needs to be placed straight, at 90 degrees, by using a magnetic angle finder to reduce possible errors during process fracture. The machine

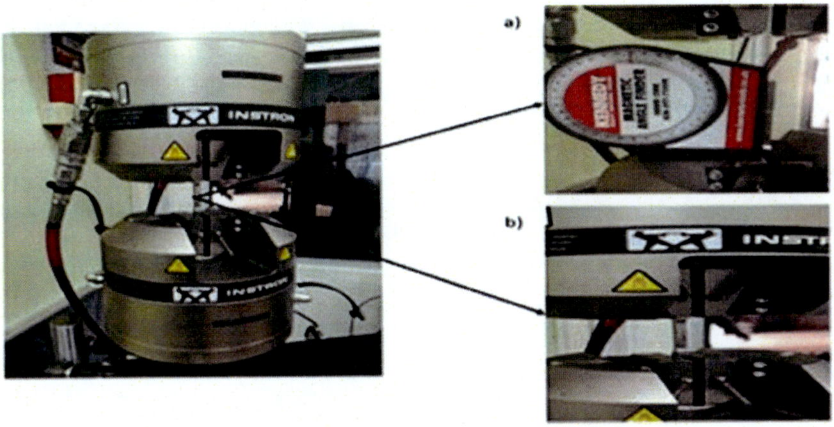

Fig. 12.4 The 100 kN servo-hydraulic fatigue test machine

is set to stop automatically when the specimen reaches 1 million cycles or fails. The waveform of the fatigue test is sinusoidal, and all fatigue tests are performed under CAL using a frequency of 10 Hz at room temperature. The test is carried out twice for each load cycle [16]. It is then repeated using a loading of 80, 70, 60 and 50% of yield strength or until the cycle exceeds 106 load cycles to obtain a uniform load for the experimental data [17]. Next, the S–N curve is generated from the results of the fatigue test. The S–N curve is the result of stress versus the number of cycles needed to fail.

12.2.3 Fatigue Analysis

The strain-based technique is a general method that can be applied to fatigue systems at both high and low cycles. In order to have a long-life application, the plastic strain has to be inconsistent, and the complete strain-life condition decreases to the Basquin equation, which is also used for the stress-life (S–N) method. The conditions related to full suitability of strain (εa) and life (N) are as follows [18]:

$$\frac{\Delta\varepsilon}{2} = \varepsilon_{\alpha} = \frac{\Delta\varepsilon_e}{2} + \frac{\Delta\varepsilon_p}{2} \tag{12.1}$$

$$\frac{\Delta\varepsilon}{2} = \frac{\sigma_f'}{E}(2N_f)^b + \varepsilon_f' + (2N_f)^c \tag{12.2}$$

The fatigue tests were conducted until the specimens failed completely. The results found the S–N curve performed under CAL using a frequency of 10 Hz at room temperature. The fatigue limit or endurance was the stress level below an infinite number of loading cycles applied to a test material without causing fatigue failure. The Basquin equation was used to adequately fit the obtained data as expressed in Eq. (12.3).

$$S = AN^b \tag{12.3}$$

where S is the applied stress, 'A' and 'b' are the constant and the curve exponent, respectively, and N is the number of cycles.

12.2.4 Crack Growth Test

Compact Tension (CT) specimens were prepared, and the geometry and dimension are shown in Fig. 12.5. The width and the thickness of the specimens were 50 and 10 mm, respectively. The initial crack length, a_o, was 12.5 mm, and the ratio of initial

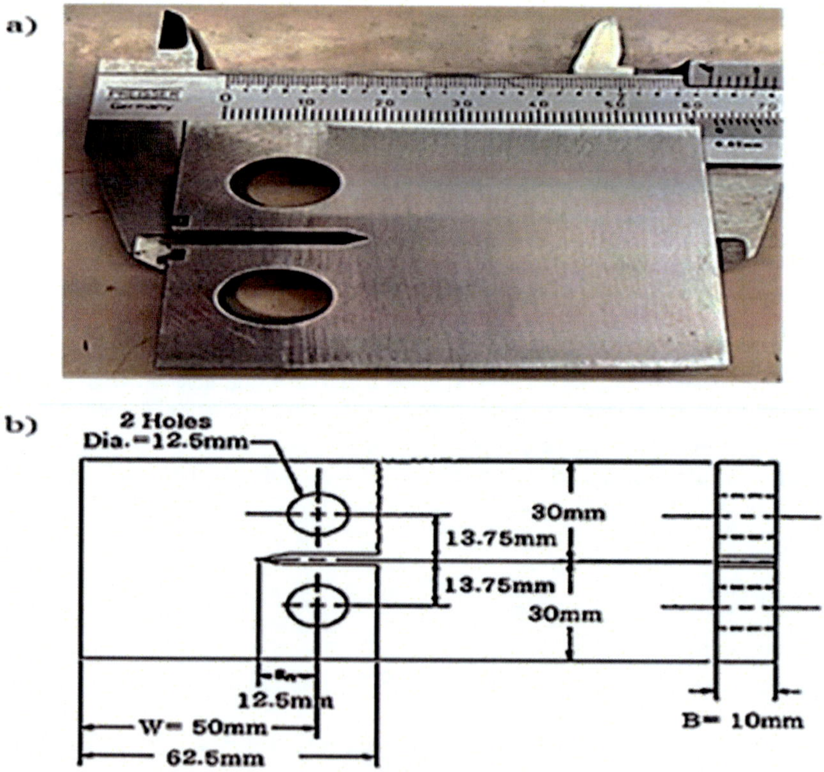

Fig. 12.5 a CT of AZ31B magnesium alloy specimen, and b the dimension (mm)

crack length, a, to specimen width, W, was (*a/W*). After fatigue tests were carried out, the maximum load of AZ31B magnesium alloy was chosen in this study for the CG test with 90% of yield strength. The CG tests were conducted on the 100 kN servo-hydraulic testing machine system at room temperature. The experiment employed a frequency of 10 Hz with a sinusoidal loading pattern. The load ratio, R = Pmin/Pmax, was held at R = 0.1, where the applied loading was at maximum strength. The technique was employed to measure crack length, *a*, at a certain number of cycles, *N*. For each specimen, a notch length, an, of 10 mm and pre-cracked 2.5 mm were set up before fatigue CG test was conducted according to the ASTM standard: E647-11. Three samples were prepared to obtain an average value. The CG was monitored and measured at every 2 mm increment [19]. The crack length reported was the projection of the actual crack in the horizontal direction perpendicular to the loading direction. In the experimental relationship between the crack length and the number of loading cycles, a parabolic curve was used to best fit a set of five successive data points. The midpoint derivative of the fitting curve was defined as the growth rate of experimental cracks.

12.2.5 Crack Growth Analysis

By plotting the log–log *da*/*dN* versus ΔK, the foundation reaction of the CG was obtained under CAL. The plot created a sigmoidal form, which can be divided into three main regions [20]. Region I indicated the threshold value close to the threshold, ΔKth, below which Failure Crack Growth (FCG) was not observed. Below ΔKth, fatigue cracks were characterized as non-propagating cracks. This threshold occurred at FCG rates in the order of 1×10^{-10} m/cycle or less. Region II essentially indicated a linear relationship between *da*/*dN* and ΔK, which corresponded to the following formula:

$$\frac{da}{dN} = C\Delta K^m \tag{12.4}$$

Equation (12.4) is known as the Paris equation, where m is the slope of the line, and C is the coefficient found by extending the straight line to $\Delta K = 1$ MP m. The Paris equation does not consider the stress ratio, R. Since the CG rate in a material depends on the stress ratio, and if the Paris equation is to be used for a non-zero stress ratio, then a suitable C value has to be used for the desired stress ratio, R. In region III, FCG rates increased as they approached instability, and little FCG life was involved. The CG results were represented as the stress intensity factor range, ΔK. The following formulae were used to calculate the stress intensity factor, *K*, for rectangular CT specimens, respectively. According to E.S. Ameh [21], the geometry and Eq. (12.5) are legitimate for $a/W \geq 0.2$.

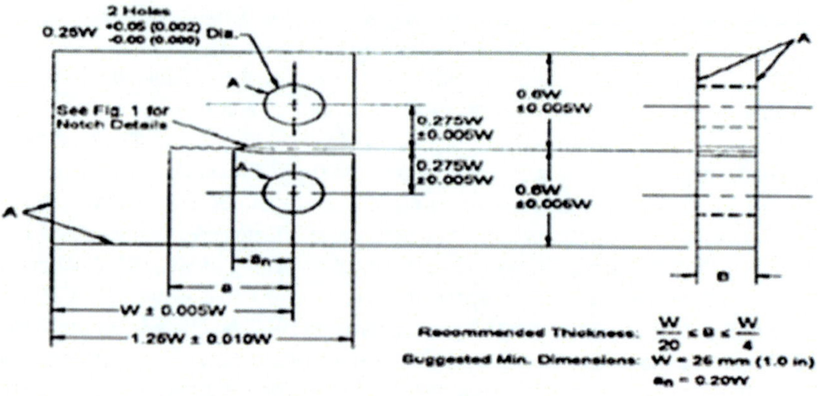

Fig. 12.6 American standard test materials (ASTM E647, 11)

$$\Delta K = \frac{\Delta P}{B\sqrt{W}(1-\alpha)^{\frac{3}{2}}(0.886 + 4.64\alpha - 13.32\alpha^2 + 14.72\alpha^3 - 5.6\alpha^4)} \tag{12.5}$$

where W is the specimen width, B is the specimen thickness, P is the applied load, and in $\alpha = a/W$, a is the crack length measured from the applied load to the crack tip as shown in Fig. 12.6.

Region III is often ignored in practical engineering as it does not affect the total crack propagation life. The behaviour of transition from Region II to Region III depends on the stress ratio, stress intensity factor and yield strength of the material.

12.3 Results and Discussion

12.3.1 Tensile Test

The stress–strain curve resulting from the samples is shown in Fig. 12.7. The test is repeated twice using a tensile sample according to ASTM E8 standard test method. The best data of stress–strain readings can be obtained from the stress–strain curve, such as 0.2% of yield strength, ultimate tensile strength, fracture strength, Young's modulus and elongation percentage of the sample, as shown in Table 12.1. From Table 12.1, the 0.2% of yield strength and ultimate tensile strength of AZ31B magnesium alloy are 140 and 237.5 MPa, respectively. In addition, the percentage elongation and strength at break are 18.2% and 72.3 MPa, whereas Young's modulus is 43.7 GPa. According to the material properties of ASTM, AZ31B has yield strength and ultimate tensile strength of 145 and 275 MPa, respectively, and Young's modulus of 45 Gpa. By comparing the material properties obtained during the experiment with the standard values, a maximum percentage of <20% errors is obtained.

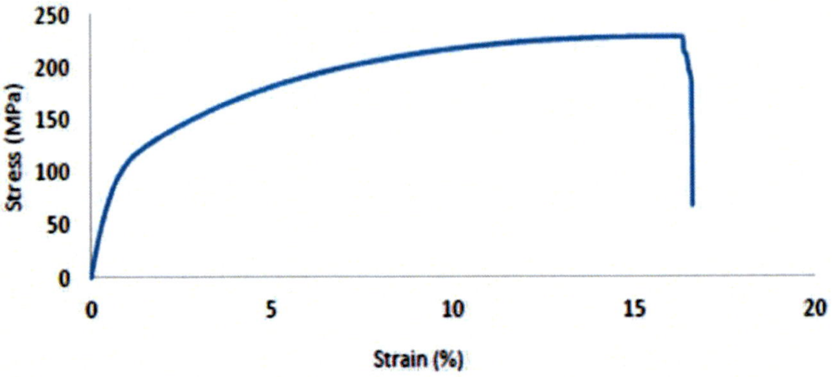

Fig. 12.7 Sample tensile test of stress–strain curves of AZ31B magnesium alloy

Table 12.1 Tensile test results for the AZ31B magnesium alloy

Mechanical properties	Experiment	Standard	Error, %
Yield strength	140.0 MPa	145 MPa	3.0
Modulus of elasticity	43.7 Gpa	45 GPa	2.8
Ultimate tensile strength	237.5 MPa	275 MPa	13.6
Strength at break	72.3 MPa	–	–
Elongation at break	18.0%	15%	20

12.3.2 Fatigue Test

This experiment involved two repetitive tests for each load in cyclic loads of 90, 80, 70, 60 and 50% of yield strength of AZ31B magnesium alloy with 140 MPa or up to 106 fatigue life cycles. The experimental fatigue data of axial fatigue testing with a cycle frequency of 10 Hz were obtained on a standard dog bone specimen, ASTM E8. The S–N curve and data results are presented in Fig. 12.8 and Table 12.2, respectively. Based on the results obtained from the experiment, the applied stress versus fatigue life is plotted on a semi-log scale to view the significance of fatigue strength at higher stress, compared to lower stress on the specimen until it ruptures.

According to the data in Table 12.2, fatigue tests are conducted until 50% of yield strength, but they do not complete the 106 cycles and the material is expected to experience failure at a smaller load. The sample completes 106 cycles at a load of 30% of yield strength predicted using the Basquin equation. The sample completes 106 cycles late at a load of 30% of yield strength because the fatigue test uses 10 Hz and the vibrate load is too fast. In terms of the fatigue theory pattern, the fatigue life increases with decreasing stress range.

The relationship between stress, $\Delta\sigma$, and the number of cycles to failure, N_f, is shown in the form of the S–N curve in Fig. 12.8. The fatigue limit value is predicted at 48 MPa, which is equal to 106 cycles. The load versus life cycle curve constructed

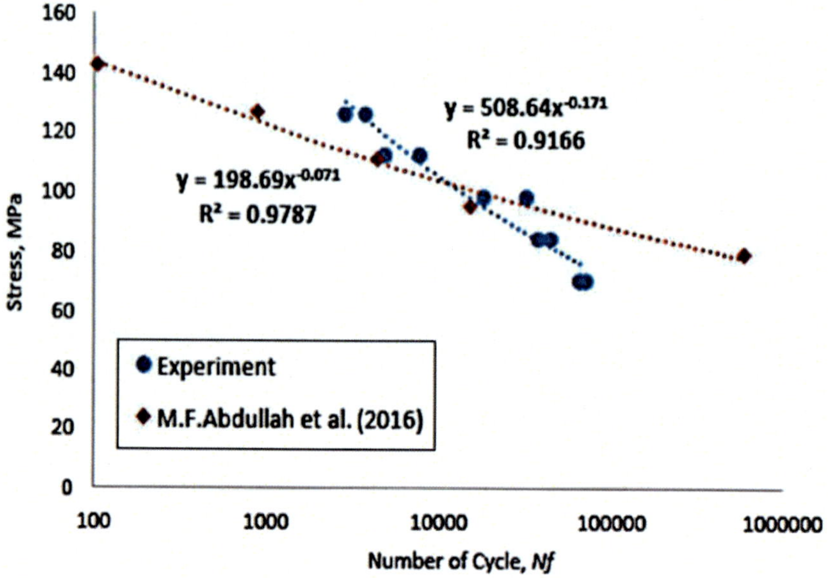

Fig. 12.8 Trend of S–N curve for AZ31B magnesium alloy

Table 12.2 Fatigue test results of the AZ31B magnesium alloy

Load applied	No. of samples	Area (m^2)	Stress (MPa)	Load (kN)	No. of cycles
0.9YS	1	3.6×10^{-5}	126	4.536	2868
	2	3.6×10^{-5}	126	4.536	3767
0.8YS	3	3.6×10^{-5}	112	4.033	4881
	4	3.6×10^{-5}	112	4.033	7750
0.7YS	5	3.6×10^{-5}	98	3.529	18,213
	6	3.6×10^{-5}	98	3.529	32,124
0.6YS	7	3.6×10^{-5}	84	3.025	37,762
	8	3.6×10^{-5}	84	3.025	44,451
0.5YS	9	3.6×10^{-5}	70	2.521	65,916
	10	3.6×10^{-5}	70	2.521	71,176

from the data obtained is compared which shown that the pattern is approximately the same as the data from a previous study by Faizal [22]. Based on the graph in Fig. 12.8, although similar in shape, there is a big difference in the shape of the specimen used with different test frequencies. This difference occurs due to distinct sample preparation processes, where the control temperatures are different. The surrounding conditions and the machine used also contribute to the differences in the cycles. The value of R2 for each graph exceeds 0.85, which shows that the value obtained for the cycle is acceptable based on the statistical analysis. This value is in the same range as the values generally observed for other magnesium materials. Therefore, the Basquin equation of AZ31B magnesium alloy, based on the least square regression line, can be rewritten as follows:

$$S = 508.64N^{-0.171} \tag{12.6}$$

12.3.3 Crack Growth Test

The test performed for fatigue CG at 90% of yield stress results in values of *da/dN* from the slope of the curve (a-N) for every point as shown in Table 12.3 and Fig. 12.9.

Table 12.3 Results of 3 CG length samples using a frequency of 10 Hz

Crack length (mm)	No. cycles 1	No. cycles 2	No. cycles 3	Average no. cycles
14	3920	4050	3950	3973
16	6500	6400	6470	6457
18	9600	9900	9100	9533
20	11,100	11,800	11,380	11,427
22	12,500	12,450	12,500	12,483
24	13,600	13,700	13,600	13,633
26	13,950	13,990	13,950	13,963
28	14,300	14,400	14,600	14,433
30	14,490	14,480	14,700	14,557
31	14,550	14,580	14,770	14,633
32	14,600	14,650	14,800	14,683

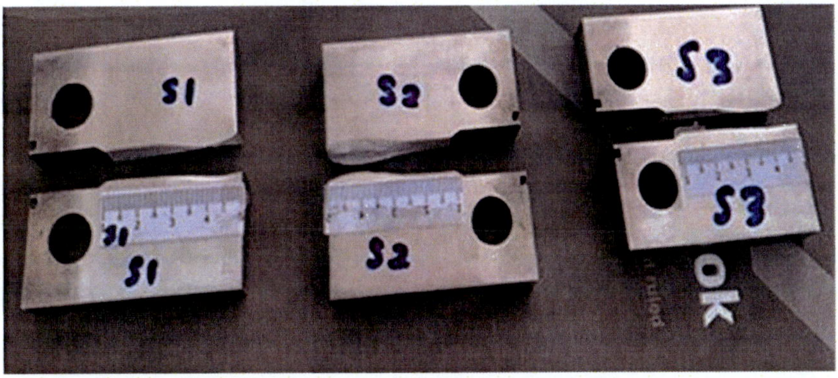

Fig. 12.9 Three samples of AZ31B CT with respective crack lengths

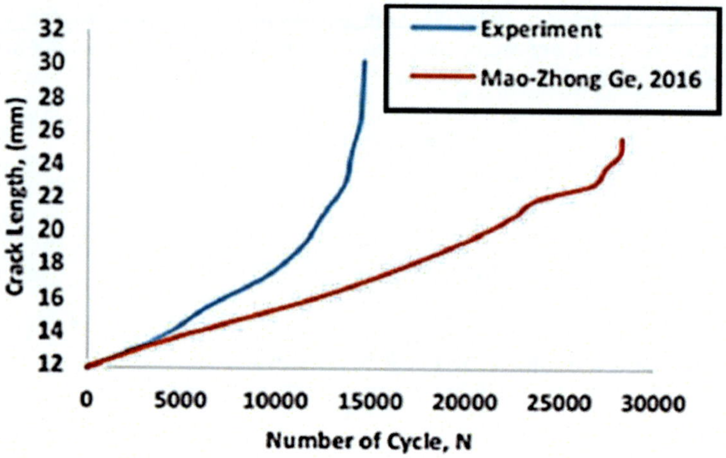

Fig. 12.10 Trend of a-N curve for AZ31B magnesium alloy at 10 Hz

Figure 12.10 shows a remarkable reaction during the early period and stable stage of crack length or fatigue CG with a successive increase in the number of cycles, and a similar reaction happens in the other three samples. Nevertheless, after crack length reaches 32 mm, this reaction attributes an uneven distribution of CG, thus, it can be concluded that when the stress and crack length reach a range of 30 mm to 32 mm, it can make crack length grows faster.

The stress intensity factor range, ΔK, was calculated based on the crack length, the dimension, and the applied load of the AZ31B magnesium alloy specimen in accord-ance to the ASTM E647 standard by using Eq. (12.5). The experimental data for the alloy were fitted to a CG relationship. Figure 12.11 shows the fatigue CG rate versus the stress intensity factor range. The horizontal axis represents the stress

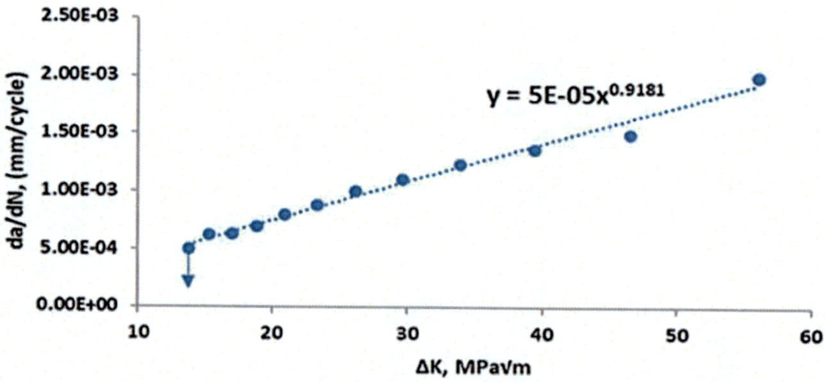

Fig. 12.11 Experiment CG rate curves da/dN versus ΔK for CT using 10 Hz

Table 12.4 Paris law constant, C and m, for the AZ31B magnesium alloy using a frequency of 10 Hz

Types	Load, N	Ratio, R	C	m
AZ31B	4082	0.1	5×10^{-5}	0.9181
AZ31B [23]	3000	0.5	2.6762×10^{-5}	2.4042

intensity factor range and the vertical axis that represents the fatigue CG rate. Region I in the figure indicates the threshold value of the stress intensity factor range. From the results, it is found that the threshold stress intensity factor range obtained is 13.8 MPam1/2. At the top right side of the CG curve, the ΔK becomes very high, approaching the critical stress intensity factor range value of KIC. The final ΔK value measured during the test is 58 MPam1/2. The Paris power-law parameters to draw a fitting curve, C and m, are 5×10^{-5} and 0.9181, respectively. The equation line in Fig. 12.11 represents the fitting curve as shown in Table 12.4 and Eq. (12.7).

$$\frac{da}{dN} = 5 \times 10^{-5} \times \Delta K^{0.9181} \tag{12.7}$$

To describe the relationship of constant, C and m, mathematical formulations of Paris were established as in the study by Zhao et al. [24]. C and m describe the mate-rial parameters as shown in Table 12.4, where the comparison value is almost similar to the previous study, but slightly different, which also depends on the loading applied on the AZ31B samples. Therefore, the larger the curve of (a-N) on samples, the larger the parameters on C and m.

12.4 Conclusion

By the tensile tests carried out, it was observed that the yield strength of AZ31B was 140.0 MPa, the ultimate stress strength was 237.5 MPa, and the strength at break was 72.3 MPa, producing an error of <20% compared to the ASTM value. The fatigue test also showed that the capacity of AZ31B to reach more than 106 cycles was only under a cyclic loading of 30% of the yield stress of the alloy. This means that the AZ31B magnesium alloy does not experience failure if it is subjected to cyclic loading of 48 MPa and below. For the CG test, the threshold stress intensity factor range was 12 MPam1/2, and the final ΔK value measured during the test was 58 MPam1/2. From this study, it can be concluded that AZ31B has the capacity and strength for fatigue life. The focus of this study was to investigate the characteristics of fatigue and CG behaviour of AZ31B magnesium alloy at room temperature by using a frequency of 10 Hz. AZ31B magnesium alloy has been found to be a better material for its lightweight property for military vehicle operations in designs. Further investigation, such as on the effect on fatigue CG by combining the AZ31B magnesium alloy with high strength material, is recommended to be done.

References

1. Mathaudhu SN, Nyberg EA (2014) Magnesium alloys in U.S. military applications: past, current and future solutions, Essent. Readings Magnes Technol 9781118858:71–76
2. Abdullah MF, Abdullah S, Omar MZ, Sohaimi RM (2014) A study of ballistic resistance on magnesium alloy, AZ31B. Aust J Basic Appl Sci 8(19):223–226
3. Meher A, Mahapatra M, Samal P, Vundavilli P, Madavan S (2019) Synthesis, microstructure and mechanical properties of magnesium matrix composites fabricated by stir casting. Mater Today Proc 18:4034–4041
4. Rahman NA, Abdullah MF, Abdullah S (2017) Characterising ballistic limits of light-weight laminated-structure as a protective panel for armoured vehicle 4(3):20–34
5. Chen X, Bu J, Fan X, Lu J, Xu L (2017) Effect of loading frequency and stress level on low cycle fatigue behavior of plain concrete in direct tension. Constr Build Mater 133:367–375
6. Hirsch J, Al-Samman T (2013) Superior light metals by texture engineering: optimized Aluminium and Magnesium alloys for automotive applications. Acta Mater 61:818–843
7. Powell BR, Quinn J, Miller W, Allison J, Hines J, Beals R (2010) Magnesium replacement of aluminum cast components in a production V6 engine to effect cost-effective mass reduction, slaid, general motors, Ford, Chrysler, USA (2010)
8. Paris P, Erdogan F (1963) A critical analysis of Crack Propagation Laws. J Basic Eng 85:528–534
9. Wöhler A (1870) About the strength tests with iron and steel, magazine for construction 20:73–106
10. ASTM International (2016) ASTM E8/E8M-16a, standard test methods for tension testing of metallic materials. ASTM International
11. ASTM International (2011) ASTM E647–11, standard test method for measurement of fatigue crack growth rates. ASTM International
12. Aldeeb T, Abduelmula M (2018) Fatgue strength of S275 mild steel under cyclic loading. World Acad Sci Eng Technol Int J Mater Metall Eng 12(10):564–570
13. Xin R, Wang B, Zhou Z, Huang G, Liu Q (2010) Effects of strain rate and temperature on microstructure and texture for AZ31 during uniaxial compression. Trans Nonferrous Met Soc, China, pp 594–598
14. Jamil WNM, Aripin MA, Sajuri Z, Abdullah S, Omar MZ, Abdullah MF, Zamri WFH (2016) Mechanical properties and microstructures of steel panels for laminated composites in armoured vehicles, 13(3):3742–3753
15. ASTM International (2015) ASTM E466, Standard practice for conducting force controlled constant amplitude axial fatigue tests of metallic materials. ASTM International
16. Mohammad, Khairul Azhar, et al. Fatigue Life for Type 316L Stainless Steel under Cyclic Loading. Advanced Materials Research. Vol. 701. Trans Tech Publications, (2013).
17. Isahak AH et al (2019) Fatigue crack growth behaviour of sandwiched metal panel of aluminium and mild steel under constant amplitude loading, 362–66
18. Cui W (2002) A state-of-the-art review on fatigue life prediction methods for metal structures. J Mar Sci Technol 7(1):43–56
19. Zhou JZ, Huang S, Zuo LD, Meng XK, Sheng J, Tian Q, Han YH, Zhu WL (2014) Effects of laser peening on residual stresses and fatigue crack growth properties of Tie6Ale4V titanium alloy. Opt Laser Eng 52
20. Forman RG (1972) Study of fatigue crack initiation from flaws using fracture mechanics theory. Eng Fract Mech 4(2):333–345
21. Ameh ES, Onyekpe BO (2016) Influence of high strength steel microstructure on fatigue crack growth rate. Am J Eng Res (AJER) 84(7):79–84

22. Abdullah MF et al (2016) Fatigue life of the magnesium alloy AZ31B under specific spectrum loading. Mater Test 58(3):200–205
23. Ge MZ, Xiang JY (2016) Effect of laser shock peening on microstructure and fatigue crack growth rate of AZ31B magnesium alloy. J Alloys Comp 680:544–552
24. Zhao MD, Fan X, Fang QZ, Wang TJ (2015) Experimental investigation of the fatigue of closed-cell aluminum alloy foam. Mater Lett 160:68–71

Chapter 13
The Effects of Mean Strain on the Fatigue Life of the SAE 1541 Carbon Steel Based on the Strain-Life Approach

Teuku Edisah Putra, Husaini, and Rauzatul Akmal

Abstract This study aims to determine the effects of the mean strain on the SAE 1541 carbon steel fatigue life based on the strain-life approach. The strain data were developed with negative, zero, and positive mean values consisting of constant and variable amplitude loadings in the range of 200–2000 $\mu\varepsilon$. The strain data has two variations: tensile-compress and compress-tensile modes. For variable amplitude loadings, the strain data also has two types: of low–high and high-low cycles. The results show that the Coffin–Manson model did not provide a different fatigue life for negative, zero, or positive mean strains. This was because the model assumes that strain data has a zero mean value. Based on the Morrow and SWT models, however, strain data with a negative mean value provided a longer fatigue life than strain data with zero and positive mean values. The results also show that constant and variable amplitude loadings gave a different fatigue life, but tensile-compress and compress-tensile modes gave a similar result. A small difference only happened between low–high and high-low cycles. It concluded that the mean value does not affect the fatigue life of strain data significantly based on the strain-life approach. It can be concluded that the strain-life approach does not consider the cycle sequence effect in its analysis.

Keywords Failure · Stress · Amplitude

13.1 Introduction

Fatigue is one of the phenomena of failures in a material caused by constant or variable amplitude loadings (CALs or VALs). This is a process of changing a component which is continually experiencing a stress or strain that is fluctuating that will cause cracks in the component. Fatigue failure occurs when the stress or strain produces permanent damage on the component, even if the stress or strain is under the yield strength of the material [1, 2].

T. E. Putra (✉) · Husaini · R. Akmal
Department of Mechanical Engineering, Universitas Syiah Kuala, Darussalam, 23111 Banda Aceh, Indonesia
e-mail: edi@unsyiah.ac.id

S. Abdullah et al. (eds.), *Structural Integrity Cases in Mechanical and Civil Engineering*, Structural Integrity 23, https://doi.org/10.1007/978-3-030-85646-5_13

According to Zakaria et al. [2], it is estimated that 70–90% failures of a component are caused by fatigue, so researchers admit that the problem of fatigue is a serious matter. This failure results in large losses that could result in fatalities. Bhat and Patibandla [3] reported that more than 300 fatal accidents have occurred globally in the past 86 years, resulting in almost 2000 deaths, which is an alarming incidence caused by fatigue. The above statistics show that a model for fatigue life assessments is still far from perfect. This is because phenomenon of fatigue is very complex since it is influenced by various factors.

The main factor in analyzing fatigue is mean value, which is more significant at low strain or high-cycle fatigue [4–6]. Therefore, failure parameters caused by fatigue are mostly developed by considering the mean strain or stress. Akmal et al. [7] developed strain-life curves in order to identify the fatigue life characteristics using strain data. However, the study only involved tensile-compress modes and low–high cycles. Thus, the purpose of this work is to identify the influence of the mean strain to the fatigue life considering tensile-compress and compress-tensile modes with low–high and high-low cycles. This study applied the strain-life model since the approach is widely used for predicting the fatigue life.

13.2 Methodology

The strain-life approach is based on the relationship between fatigue life and strain. The Coffin–Manson model [8, 9] combining the elastic and plastic strain to get a correlation between fatigue life and strain is the basis of the strain-life approach, which can be expressed as:

$$\varepsilon = \frac{\sigma'_f}{E}(2N_f)^b + \varepsilon'_f(2N_f)^c \tag{13.1}$$

where ε is the strain amplitude, σ'_f is the fatigue strength coefficient, E is the material modulus of elasticity, N_f is the number of cycles, b is the fatigue strength exponent, ε'_f is the fatigue ductile coefficient, and c is the fatigue ductile exponent.

The Morrow model [10] states that the mean stress can be calculated by adjusting the elastic strain-life curve with the mean stress, which is expressed as:

$$\varepsilon = \frac{\sigma'_f - \sigma_{\text{mean}}}{E}(2N_f)^b + \varepsilon'_f(2N_f)^c \tag{13.2}$$

where σ_{mean} is the mean stress. Another strain-life model that has been suggested by Smith et al. [11] is the SWT model. This model is based on the fatigue test data obtained at a different mean value, which is defined as:

$$\sigma_{max}\varepsilon = \frac{\sigma_f'^2}{E}(2N_f)^{2b} + \sigma_f'\varepsilon_f'(2N_f)^{b+c} \tag{13.3}$$

where σ_{max} is the maximum stress amplitude.

Fatigue damage is determined using the Palmgren–Miner rule [12, 13] based on the recurrent loading. The rule is used to calculate the cumulative fatigue damage of the block loading, which is expressed as:

$$D = \sum\left(\frac{n_i}{N_f}\right) \tag{13.4}$$

If the component suffered 100% damage and caused fatigue fracture, the Palmgren–Miner equation can be rewritten as follows:

$$\sum\left(\frac{n_i}{N_f}\right) = 1 \tag{13.5}$$

where D is the total damage fatigue and n_i is the number of cycles in a block. Fatigue damage ranges from zero to one, with zero indicating no damage and one showing a failure.

In this study, the strain data were developed with the negative, zero, and positive mean values, consisting of CALs and VALs. For CALs, the strain data consisted of tensile-compress and compress-tensile modes. For VALs, besides divided into tensile-compress and compress-tensile modes, they were also divided into low–high and high-low cycles. Each strain data consisted of ten ranges, which are 200, 400, 600, 800, 1000, 1200, 1400, 1600, 1800 and 2000 με. For more details, the developed strain data are illustrated in Fig. 13.1, with the total being 180 strain data. Figures 13.2, 13.3 and 13.4, respectively, show the examples of the developed strain data with the negative, zero, and positive mean values with an amplitude range of 200 με. The material used in the simulation was the SAE 1541 carbon steel, often used in the manufacture of gears, shafts, and other engine components. The mechanical properties of this material are stated in Table 13.1.

13.3 Results and Discussion

The fatigue life of the developed 180 strain data were predicted using the strain-life approach, i.e., based on the Coffin–Manson, Morrow, and SWT models. Fatigue life is the number of cycles of stress or strain received by a component prior to failure. It can be expressed as reversals of blocks for the occurrence of a failure. Reduction of a fatigue life can be caused by high stress or strain amplitudes on the component. The obtained fatigue life were presented in a strain-life curve, as shown in Figs. 13.5, 13.6 and 13.7.

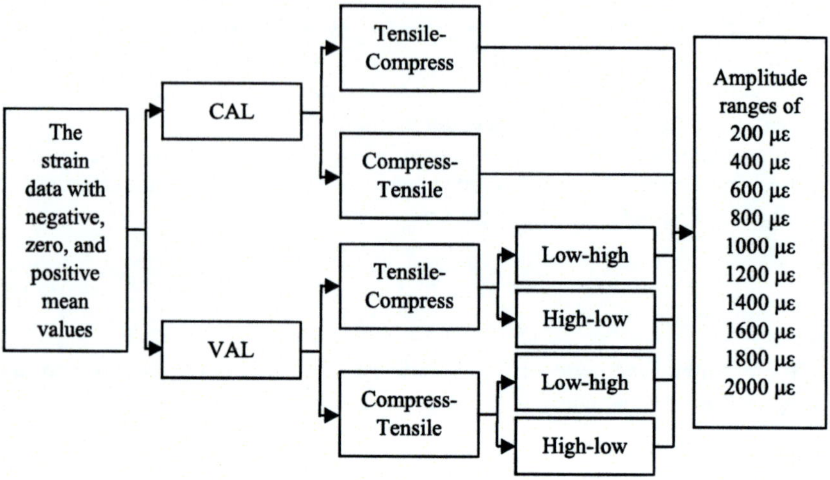

Fig. 13.1 Details of the strain data developed

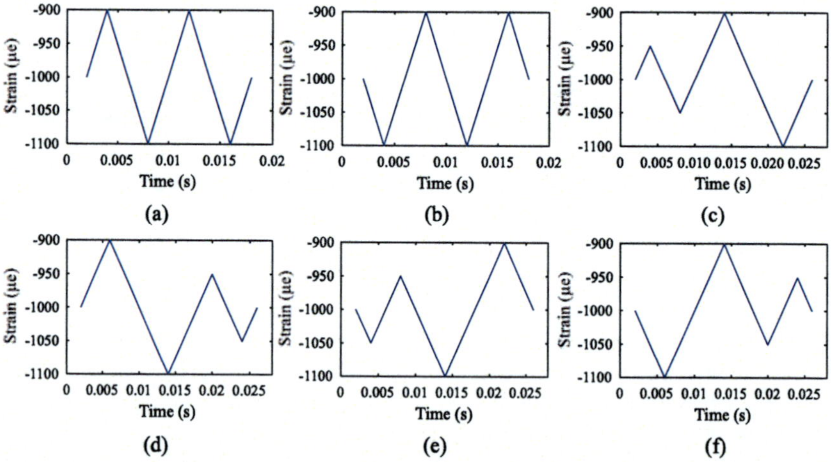

Fig. 13.2 Examples of the developed negative mean value strain data: **a** CAL with tensile-compress mode, **b** CAL with compress-tensile mode, **c** VAL with tensile-compress mode and low–high cycle, **d** VAL with tensile-compress mode and high-low cycle, **e** VAL with compress-tensile mode and low–high cycle, **f** VAL with compress-tensile mode and high-low cycle

Figure 13.5 shows the strain-life curves based on the Coffin–Manson model. The fatigue life trends were identical for the negative, zero, and positive mean strains. This means that no matter if the mean value is negative, zero, or positive, the model provides similar fatigue life. This condition is true not only for CALs, but also for VALs. This is because the Coffin–Manson's theory assumes zero as the mean value of a strain data [15–17].

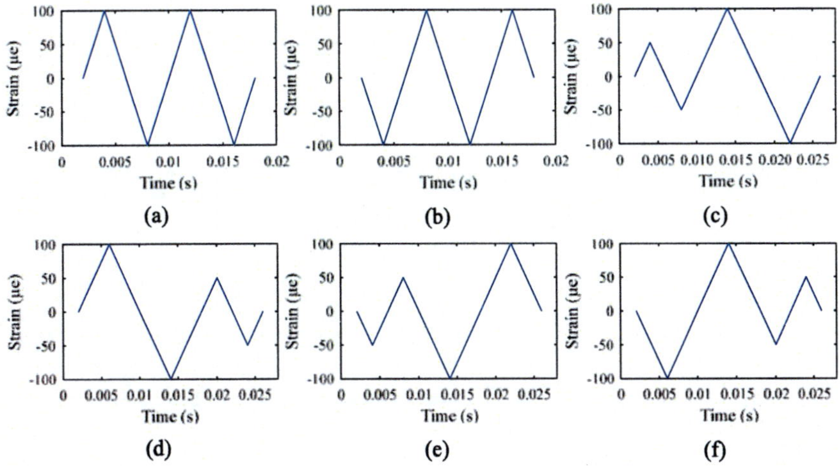

Fig. 13.3 Examples of the developed zero mean value strain data: **a** CAL with tensile-compress mode, **b** CAL with compress-tensile mode, **c** VAL with tensile-compress mode and low–high cycle, **d** VAL with tensile-compress mode and high-low cycle, **e** VAL with compress-tensile mode and low–high cycle, **f** VAL with compress-tensile mode and high-low cycle

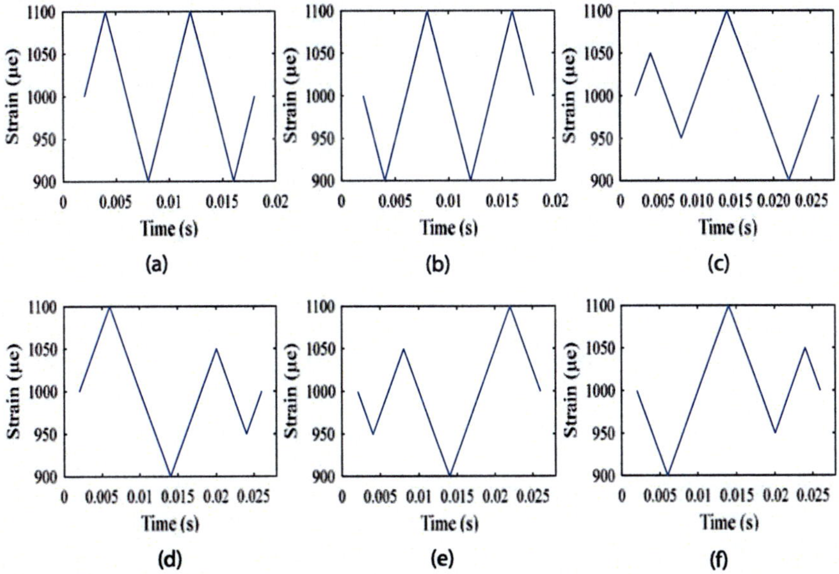

Fig. 13.4 Examples of the developed positive mean value strain data: **a** CAL with tensile-compress mode, **b** CAL with compress-tensile mode, **c** VAL with tensile-compress mode and low–high cycle, **d** VAL with tensile-compress mode and high-low cycle, **e** VAL with compress-tensile mode and low–high cycle, **f** VAL with compress-tensile mode and high-low cycle

Table 13.1 Mechanical properties of the SAE 1541 carbon steel [14]

Mechanical properties	Values
Ultimate tensile strength, S_u [MPa]	1200
Material modulus of elasticity, E [GPa]	207
Fatigue strength coefficient, σ'_f [MPa]	2980
Fatigue strength exponent, b	−0.15
Fatigue ductility exponent, c	−0.61
Fatigue ductility coefficient, ε'_f	0.68

The results also show that the Coffin–Manson model gave the same fatigue life for CALs with tensile-compress (Fig. 13.5a) and compress-tensile (Fig. 13.5b) modes. The longest fatigue life was 2.25E5 reversals of blocks due to the strain amplitude range of 200 μɛ, while 138.5 reversals of blocks was the shortest fatigue life caused by a higher amplitude range, i.e., 2000 μɛ. Also, no difference was found for VALs. The fatigue lives in the tensile-compress mode with low–high cycle (Fig. 13.5c) and high-low cycle (Fig. 13.5d) were similar. The results were even similar to the fatigue life for the compress-tensile mode with low–high cycle (Fig. 13.5e) and high-low cycle (Fig. 13.5f). For all four figures (Fig. 13.5c–f), the longest fatigue life was 4.43E5 reversals of blocks with the amplitude range of 200 μɛ, and the highest amplitude range, i.e., 2000 μɛ, exhausted the fatigue life of 228.7 reversals of blocks. It means that the Coffin–Manson model only considers strain amplitude in predicting the fatigue life. No matter if the strain data is tensile-compress or compress-tensile modes and if the strain data is low–high or high-low cycles, the fatigue lives were similar. This is due to the lack of consideration of cycle sequence effect in the Palmgren–Miner rule associated with the strain-life approach [2, 18, 19].

The fatigue lives obtained from the Morrow model are presented in strain-life curves in Fig. 13.6. The fatigue lives for CALs were shorter than the fatigue lives for VALs. These results were similar to the results from Stanzl-Tschegg et al. [20]. This is because a VAL consists of low and high amplitudes, where low amplitudes contribute to a minimum fatigue damage, extending the fatigue life [15, 17, 19, 21–24].

In contrast to the Coffin–Manson model, the Morrow model gave a different fatigue life for the negative, zero, and positive mean values. The negative mean values provided the longest fatigue life, followed by zero and positive mean values. These results were similar to the results from Zakaria et al. [2], Khan et al. [25], and Bruchhausen et al. [26]. This is because the model was developed by considering the mean strain [17, 27–29], where the mean value of a strain data can be not only zero, but it can also be negative or positive.

However, there was no difference in fatigue life for tensile-compress or compress-tensile modes, as shown in Fig. 13.6a, b. The fatigue life for the negative mean value was 5.75E5 reversals of blocks at an amplitude range of 200 μɛ and 141.3 reversals of blocks at an amplitude range of 2000 μɛ. The zero mean values gave a fatigue life of 2.25E5 reversals of blocks at an amplitude range of 200 μɛ and 138.5 reversals of blocks at the amplitude range of 2000 μɛ. Finally, the positive mean values at an

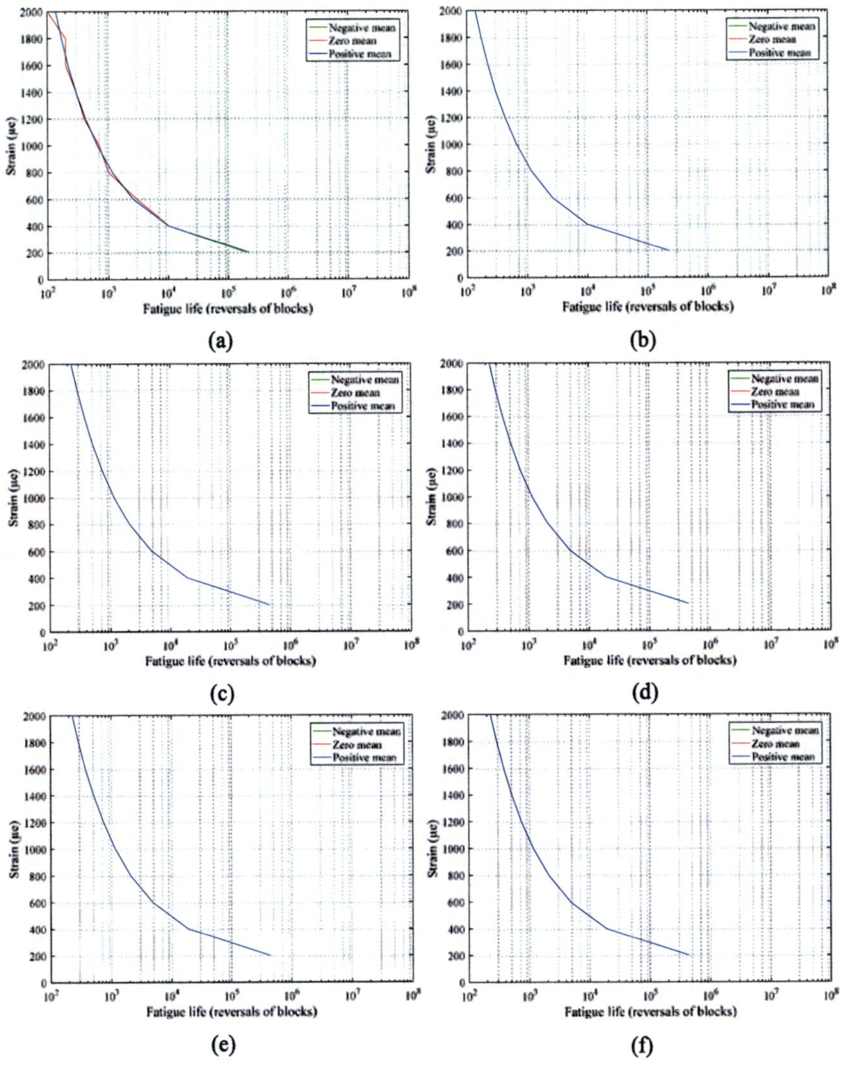

Fig. 13.5 Strain-life curves based on the Coffin–Manson model: **a** CALs with tensile-compress mode, **b** CALs with compress-tensile mode, **c** VALs with tensile-compress mode and low–high cycle, **d** VALs with tensile-compress mode and high-low cycle, **e** VALs with compress-tensile mode and low–high cycle, **f** VALs with compress-tensile mode and high-low cycle

amplitude range of 200 provided a fatigue life of 8.58E4 reversals of blocks and at an amplitude range of 2000 με, the fatigue life was 135.8 reversals of blocks.

Tensile-compress modes with low–high and high-low cycles are shown in Fig. 13.6c, d, respectively, the negative mean value gave a fatigue life of 1.14E6 reversals of blocks at the amplitude range of 200 με and at the amplitude range of

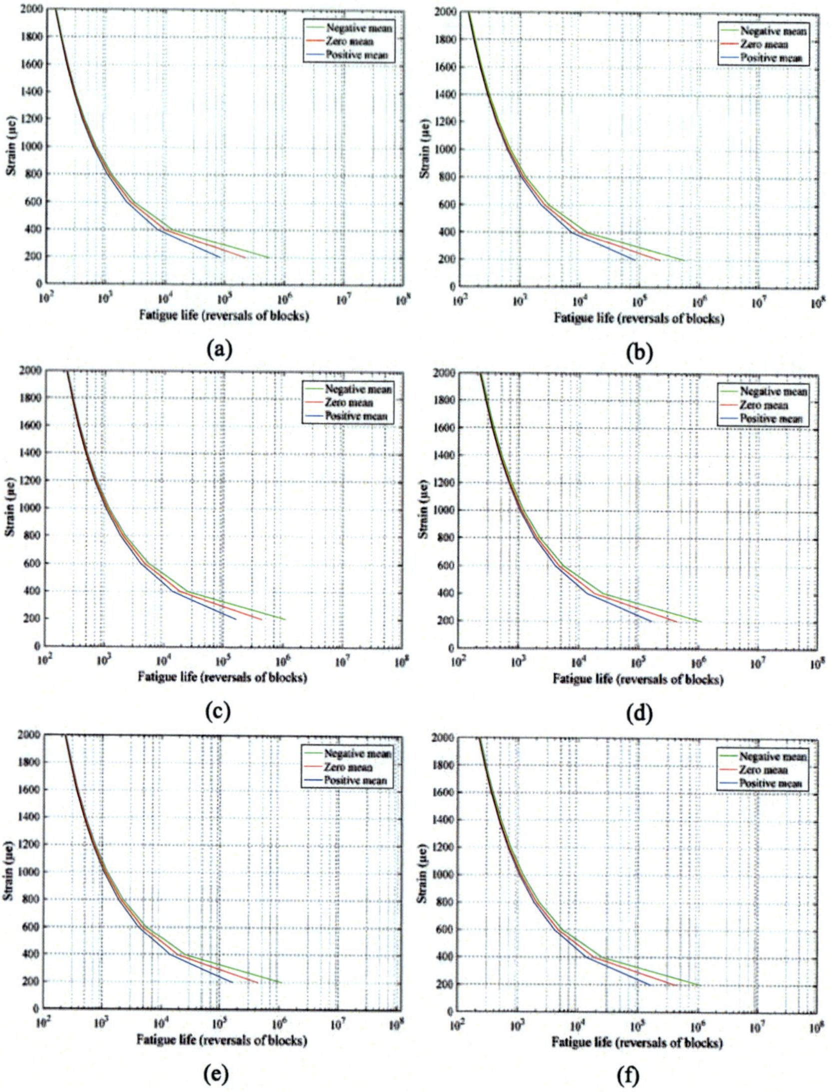

Fig. 13.6 Strain-life curves based on the Morrow model: **a** CALs with tensile-compress mode, **b** CALs with compress-tensile mode, **c** VALs with tensile-compress mode and low–high cycle, **d** VALs with tensile-compress mode and high-low cycle, **e** VALs with compress-tensile mode and low–high cycle, **f** VALs with compress-tensile mode and high-low cycle

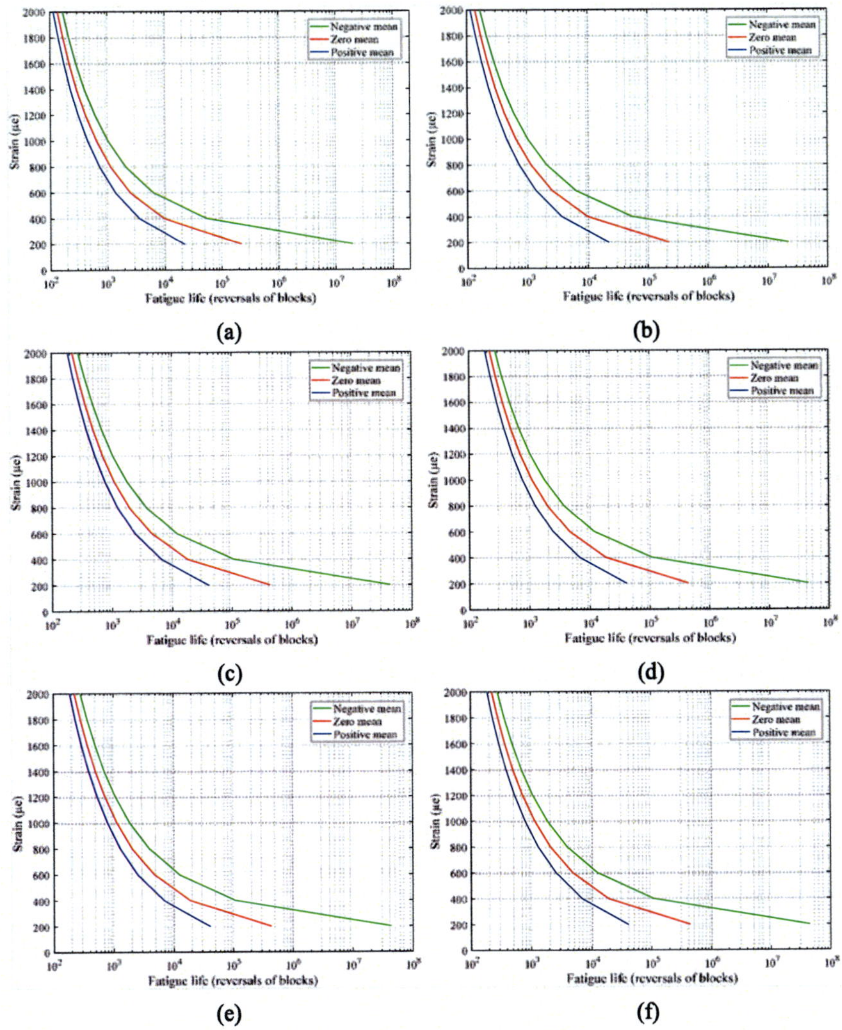

Fig. 13.7 Strain-life curves based on the SWT model: **a** CALs with tensile-compress mode, **b** CALs with compress-tensile mode, **c** VALs with tensile-compress mode and low–high cycle, **d** VALs with tensile-compress mode and high-low cycle, **e** VALs with compress-tensile mode and low–high cycle, **f** VALs with compress-tensile mode and high-low cycle

2000 $\mu\varepsilon$, the fatigue life was 233.7 reversals of blocks. The zero mean value produced a fatigue life of 4.43E5 reversals of blocks at an amplitude range of 200 $\mu\varepsilon$ and 228.5 reversals of blocks at an amplitude range of 2000 $\mu\varepsilon$. While for the positive mean value, the fatigue life at an amplitude range of 200 $\mu\varepsilon$ was 1.68E5 reversals of blocks and at the amplitude range of 2000 $\mu\varepsilon$, the fatigue life was 223.3 reversals of blocks.

A higher fatigue life of one reversal of block was found for compress-tensile mode in Fig. 13.6e, f for the amplitude range of 200 με.

The SWT model gave an identical fatigue life as the Morrow model, as presented in Fig. 13.7, where the negative mean values provided the longest fatigue life, followed by the zero and positive mean values, but with a more significant difference. This is because the equation is based on the assumption that the strain amplitude is different from the mean value, which remains constant [30]. Figure 13.7a, b show the strain-life curves for CALs with tensile-compress and compress-tensile modes, respectively. The negative mean value gave the fatigue life of 2.21E7 reversals of blocks at an amplitude range of 200 με and at the amplitude of 2000 με, the fatigue life was 169 reversals of blocks. For the zero mean value, the fatigue life was 2.25E5 reversals of blocks and 138.5 reversals of blocks at the amplitude range of 200 με and 2000 με, respectively. For the positive mean value, the fatigue life was 2.25E4 reversals of blocks and 116 reversals of blocks.

For tensile-compress modes with low–high (Fig. 13.7c) and high-low (Fig. 13.7d) cycles, the negative mean value gave the fatigue life of 4.42E7 reversals of blocks at the amplitude range of 200 με and the amplitude range of 2000 με gave the fatigue life of 279.3 reversals of blocks. The zero mean value produced the fatigue life of 4.43E5 reversals of blocks at an amplitude range of 200 με and 226.8 reversals of blocks at an amplitude range of 2000 με. While for the positive mean value, the fatigue life at an amplitude range of 200 με was 4.13E4 reversals of blocks and at the amplitude range of 2000 με, the fatigue life was 188.5 reversals of blocks. Opposite of the Morrow model, the SWT model gave a higher fatigue life about five reversals of blocks for compress-tensile mode in Figs. 13.7e, f at a higher amplitude range, namely, 2000 με.

The strain-life approach is accurate in predicting fatigue life by considering plastic build-up in the local area. This approach typically uses ductile materials at low fatigue cycles [6, 31], because it is cost-efficient and does not require extensive crack observation. In addition, this approach can also be used in high fatigue cycles comprising some plasticity that takes into account plastic strain [30] and as a substitute for the stress-life approach [32]. However, a disadvantage is associated with the strain-life approach, which is in line with the Palmgren–Miner linear cumulative fatigue damage rule assuming the sequence changes that occur in non-uniform cycles do not affect fatigue life. Cycle sequence cannot be neglected when predicting fatigue life for VAL because it contributes substantially to this process. Therefore, the accuracy of the strain-life approach is doubtful. The VAL contains a cycle mix of low and high amplitudes. The higher amplitude cycle affects the increase in the effective stress for the subsequence low amplitude cycle, thereby increasing its crack growth rate. Even, cycles below the fatigue limit lead to significant failure [33].

From this research, it can be concluded that each strain-life model gave inconsistent results, depending on the mean strain value. Unfortunately, a consistent model is still being developed. It proves that it is difficult to combine fatigue life assessment for general use due to the complexity of fatigue phenomena. Meanwhile, the strain-life approach is still regarded as an important approach for predicting the fatigue life of a component [6]. The result analyses based on this approach are acceptable at this time [34].

13.4 Conclusions

This study investigates the characteristics of the fatigue life of the SAE 1541 carbon steel based on the strain-life approach modifying the mean value of the strain data. From the results, it was found that the Coffin–Manson model gave a similar fatigue life without considering the mean value of the strain data. This condition happens to CALs and VALs. The Coffin–Manson model also does not consider if the strain data is tensile-compress or compress-tensile modes. Even, low–high or high-low cycles do not influence the fatigue life as well. Different results were found based on the Morrow and SWT models. Both models show that CALs decrease the fatigue life. Both models also agree that the positive mean value harms the fatigue life, meanwhile the negative mean value prolongs the fatigue life. However, the SWT model gave more significant differences. Both models did not give a difference in fatigue life for tensile-compress mode with low–high and high-low cycles. A small difference was found in fatigue life for the compress-tensile mode with low–high and high-low cycles. Increasing the fatigue life of one reversal of block at the amplitude range of 200 $\mu\varepsilon$ was produced by the Morrow model, and about five reversals of blocks at a higher amplitude range, which is 2000 $\mu\varepsilon$, was found by the SWT model.

Acknowledgements The authors would like to express their gratitude to Universitas Syiah Kuala for financial support for this research through Grant No. 10/UN11.2.1/PT.01.03/DPRM/2021.

References

1. Ye XW, Su YH, Han JP (2014) A state-of-the-art review on fatigue life assessment of steel bridges. Math Prob Eng 1–13 (2014)
2. Zakaria KA, Abdullah S, Ghazali MJ (2016) A review of the loading sequence effects on the fatigue life behaviour of metallic materials. J Eng Sci Technol Rev 9:189–200
3. Bhat S, Patibandla R (2011) Metal fatigue and basic theoretical models: a review. Alloy steel-properties and use. Chapter in Book. Intech, Rijeka
4. Klimkeit B, Nadot Y, Castagnet S, Nadot-Martin C, Dumas C, Bergamo S, Sonsino CM, Büter A (2011) Multiaxial fatigue life assessment for reinforced polymers. Int J Fatigue 33:766–780
5. Kovacs S, Beck T, Singheiser L (2013) Influence of mean stresses on fatigue life and damage of a turbine blade steel in the VHCF-regime. Int J Fatigue 49:90–99
6. Ghafoori E, Motavalli M, Nussbaumer A, Herwig A, Prinz GS, Fontana M (2015) Determination of minimum CFRP pre-stress levels for fatigue crack prevention in retrofitted metallic beams. Eng Struct 84:29–41
7. Akmal R, Husaini, Putra TE (2021) The characteristics of the fatigue life of the SAE 1541 carbon steel based on the strain-life approach. Key Eng Mater 872:1662-9795
8. Coffin LF Jr (1954) A study of the effects of cyclic thermal stresses on a ductile metal. Trans ASME 76:931–950

9. Manson SS (1965) Fatigue: a complex subject—some simple approximation. Exp Mech 5:193–226
10. Morrow J (1968) Fatigue design handbook. Society of Automotive Engineers, Warrendale
11. Smith KN, Watson P, Topper TH (1970) A stress-strain function for the fatigue of materials. J Mater JMLSA 5:767–778
12. Palmgren A (1924) Die Lebensdauer von Kugellagern. Zeitschrift VDI 68:339–341
13. Miner MA (1945) Cumulative damage in fatigue. J Appl Mech 67:A159–A164
14. nCode: GlyphWorks. nCode International, Ltd., Sheffield (2018)
15. Putra TE, Abdullah S, Nuawi MZ (2011) Data scattering of fatigue damaging segmentation in order to develop the fatigue damaging and Morlet wavelet coefficient relationship. Int Rev Mech Eng 5(4):605–610
16. Nopiah ZM, Abdullah S, Baharin MN, Putra TE, Sahadan SN, Willis KO (2011) Comparative study on data editing techniques for fatigue time series signals. Adv Mater Res 146–147:1681–1684
17. Putra TE, Abdullah S, Schramm D, Nuawi MZ, Bruckmann T (2017) Reducing cyclic testing time for components of automotive suspension system utilising the wavelet transform and the Fuzzy C-Means. Mech Syst Signal Process 90:1–14
18. Cai Y, Zhao Y, Ma X, Yang Z, Ding Y (2018) An extended model for fatigue life prediction and acceleration considering load frequency effect. IEEE 6:21064–21074
19. Zhang J, Fu X, Lin J, Liu Z, Liu N, Wu B (2018) Study on damage accumulation and life prediction with loads below fatigue limit based on a modified nonlinear model. Materials 11
20. Stanzl-Tschegg SE, Meischel M, Arcari A, Iyyer N, Apetre N, Phan N (2016) Combined cycle fatigue of 7075 aluminum alloy-fracture surface characterization and short crack propagation. Int J Fatigue 91:352–362
21. Abdullah S, Putra TE, Nuawi MZ, Nopiah ZM, Arifin A, Abdullah L (2010) The Morlet wavelet analysis for fatigue feature clustering. WSEAS Trans Math 5(9):345–354
22. Abdullah S, Putra TE, Nuawi MZ, Nopiah ZM, Arifin A, Abdullah L (2010) Extracting fatigue damage features using STFT and CWT. WSEAS Transactions on Signal Processing 3(6):91–100
23. Husaini, Putra TE, Ali N (2018) The Morlet wavelet transform for reducing fatigue testing time of an automotive suspension signal. In: AIP conference proceedings 1983
24. Zhu S-P, Yue P, Yu Z-Y, Wang Q (2017) A combined high and low cycle fatigue model for life prediction of turbine blades. Materials 10
25. Khan SMA, Benyahia F, Bouiadjra BB, Albedah A (2014) Analysis and repair of crack growth emanating from V-notch under stepped variable fatigue loading. Procedia Eng 74:151–156
26. Bruchhausen M, Fischer B, Ruiz A, González S, Hähner P, Soller S (2015) Impact of hydrogen on the high cycle fatigue behaviour of inconel 718 in asymmetric push-pull mode at room temperature. Int J Fatigue 70:137–145
27. Husaini, Putra TE, Ali N (2018) Fatigue feature clustering of modified automotive strain signals for saving testing time. Int J Autom Mech Eng 15(2):5251–5272
28. Putra TE, Husaini, Machmud MN (2020) Predicting the fatigue life of an automotive coil spring considering road surface roughness. Eng Fail Anal 116:104722
29. Putra, T.E., Abdullah, S., Schramm, D.: Effect of cycle amplitude removal of fatigue strain loadings associated to signal energy characteristics. Engineering Failure Analysis 116, 104723 (2020).
30. Ince A, Glinka G (2011) A modification of Morrow and Smith-Watson-Topper mean stress correction models. Fatigue Fract Eng Mater Struct 34:854–867
31. Nagode M (2014) Continuous damage parameter calculation under thermo-mechanical random loading. MethodsX 1:81–89

32. Fajdiga G, Sraml M (2009) Fatigue crack initiation and propagation under cyclic contact loading. Eng Fract Mech 76:1320–1335
33. Carvalho ALM, Martin JP, Voorlwad HJC (2010) Fatigue damage accumulation in aluminum 7050–T7451 alloy subjected to block programs loading under step-down sequence. Procedia Eng 2:2037–2043
34. Taheri S, Vincent L, Le-roux J-C (2013) A new model for fatigue damage accumulation of austenitic stainless steel under variable amplitude loading. Procedia Engineering 66:575–586

Chapter 14
Mechanical Characteristics of Fly Ash-Based Geopolymer Bricks

Mahfodzah Md Padzi, Nor Sarah Zahir, and Mohd Ridzuan Mohd Ali

Abstract The manufacturing of OPC falls as a second cause after the automobile as the main carbon dioxide source that polluted the atmosphere. Nowadays, sustainable development of industrial waste is crucial in building construction. Hence, seeking an alternative option to the current most costly, resource-consuming OPC is inevitable. The introduction of geopolymer has an impact on the environment and is widely used as a replacement for OPC. This research investigated the mechanical performance of fly ash-based geopolymer bricks. OPC was partially replaced by some amount of fly ash with various percentages ranging from 0 to 30%. The bricks were made of OPC, fly ash, sand and water. The test specimens were cubic and cylindrical in shape, which were prepared for the compressive and splitting test, respectively. Both tests were performed in accordance with ASTM C39 and ASTM C496. Cubic specimens were tested for compressive strength at a pace rate of 0.9 kN/mm and cylinder specimens were tested for splitting test at a pace rate of 1 kN/min. The added fly ash has a significant effect on the bricks' mechanical behaviour. A replacement of OPC with 20% of fly ash gave the highest value compared to 10 and 30% fly ash for both compressive and splitting test. This indicates that the added fly ash as a minor replacement to OPC can improve the characteristics of the bricks, thus can be effectively utilized by the structural industry.

Keywords Geopolymer · Fly ash · Ordinary Portland Cement (OPC) · Compressive strength · Splitting tensile strength first section

M. M. Padzi (✉) · N. S. Zahir
Department of Mechanical Engineering, Universiti Kuala Lumpur Malaysia France Institute, Seksyen 14 Jalan Teras Jernang, 43650 Bandar Baru Bangi, Selangor, Malaysia
e-mail: mahfodzah@unikl.edu.my

M. R. M. Ali
Department of Civil Engineering, Universiti Teknologi MARA, Shah Alam3 Jalan Ilmu 1/1, 40450 Shah Alam, Sleangor, Malaysia

S. Abdullah et al. (eds.), *Structural Integrity Cases in Mechanical and Civil Engineering*, Structural Integrity 23, https://doi.org/10.1007/978-3-030-85646-5_14

14.1 Introduction

The introduction of geopolymer give an impact to environment and is widely used as a replacement for ordinary Portland cement. Geopolymer is a mixture of cement and ash. By using geopolymer fly ash based some factors can be reduced such as greenhouse gas emission and environmentally friendly construction. Geopolymer is a replacement of cementitious material that contain source material which are rich in silica and alumina. The application of broader include fire resistance coating, toxic and waste encapsulation, cement and concrete etc. Geopolymer bricks can produce ceramics bricks by using polycondensation process [1].

The geopolymer application also used also include in coating application which resulted to thermal insulation but the preparation of paraffin-based is needed [2]. Also, it is proven as a non-combustible structural material and appropriate for construction on building that involve with high degree of temperature which acts as fire resistance but consume less cost [3]. Geopolymer bricks is an example of friendly material from fly ash. The application of geopolymer become more popular because the amount of waste being disposed into the landfill is increased globally. Basically, fly ash is fused and float out of the furnace. As it rises, it cools and solidifies into spherical glass particles which called as fly ash. The fly ash is collected with bag filters or electrostatic precipitator. A geopolymer bricks is a mixture of ash and cement. It is introduced to sustain the good environment. The production of ordinary Portland cement caused higher carbon dioxide because the process focussed on heating and combustion. In contrast, geopolymer brick only produce with an estimation of less than 40% emission. Moreover, it is recommended to utilize because it can gain ultimate strength within 24 h [4]. Even though geopolymer bricks is a game changing in construction industry, its mechanical behaviour properties still need to be evaluated. The ratio of the ash and cement should be correct in order to gain high strength [5]. This is because to maintain the good condition of long-term mechanical properties.

The geopolymer bricks can reduce up to 80% of carbon dioxide gases that caused by the ordinary cement. Although geopolymer solve cement production problem, the properties of geopolymer bricks are vary due to different percentage and has no standard percentage mixture. Based on previous study, a few types of ash has been studied such as pond ash and crusher sand ash. The optimum level of strength of pond ash is found to be at 40% [6]. However, the optimum level of strength of crusher sand ash is at 20%. Therefore, this research is focussed on mechanical behaviour of geopolymer bricks at different percentage to investigate the optimum amount of fly ash will affect the bricks mechanical performance.

14.2 Methodology

The methodology involves specimen preparation and experimental testing. Samples were prepared with four different composition of fly ash which are 0, 10, 20, and 30%. There will be three specimens for each composition. Compressive and splitting test were performed according to ASTM C39 and ASTM C496, respectively. The data were recorded and observed for further analysis. Figure 14.1 shows an overview of the research work.

14.2.1 Fly Ash Chemical Analysis

A handheld XRF spectrometer was used to analyze the element of fly ash. It released 50 kV X-ray tube that is characteristic of a specific element which is registered by the detector in the XRF meter. The fly ash was placed in a tube inside the handheld analyser. The x-ray signal was then released from the front end of the analyser.

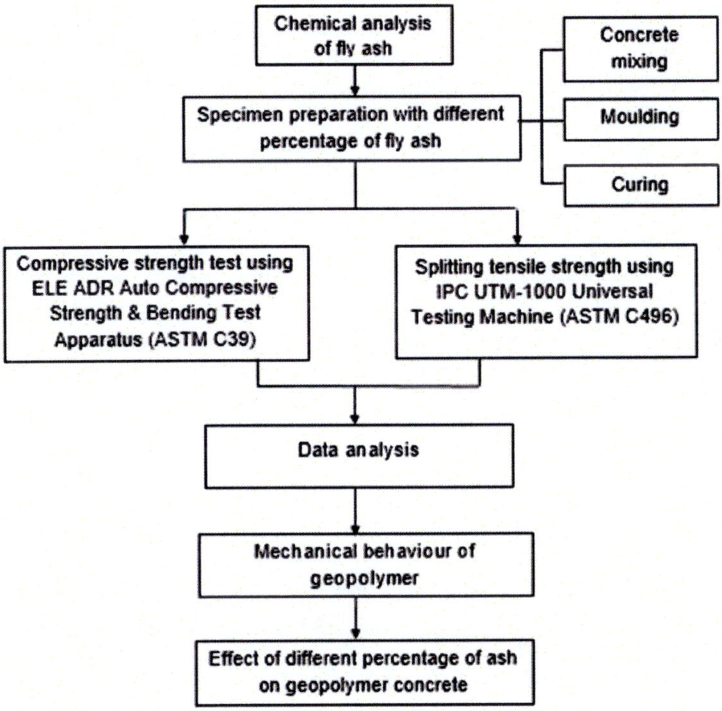

Fig. 14.1 Research flowchart

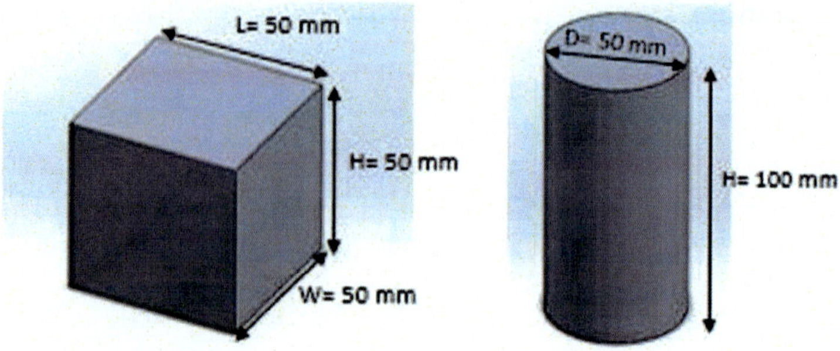

Fig. 14.2 Sample dimension (**a**) cube and (**b**) cylinder

14.2.2 Specimen Preparation

The test specimens were cubic bricks of 50 × 50 mm and cylindrical bricks of 50 mm (d) × 100 mm (h). Cube bricks and cylinder bricks were prepared according to the dimension followed. Figure 14.2 shows sample dimension for both cube and cylinder geometry.

Mixing Process. Fly ash, OPC and sand are first mixed and the mixture blended on a watertight non-absorbent platform until change to uniform colour. The binder mixture at a cement-to-fly ash ratio of 90:10% (by weight) when making the geopolymer bricks. The mixture was blended for 10 min after the water was added. Then the process repeated with different percentage of fly ash, 10, 20, and 30% respectively. The w–c of water is 0.38.

Moulding Process. The mixture was then inserted into the half the mould first. Then a tamping rod was used stroke the mixture. The strokes were 25 times for each specimen. This is because to allow the layer of the mixture to close any holes left and to release any big bubbles of air that might have been trapped. After that, the specimen was left under the room temperature for 24 h before demolding it.

Curing Process. The curing process will be 28 days and all the specimens were put into the curing pool shown in Fig. 14.3 at the temperature of 23 °C.

14.2.3 Mechanical Testing

Compressive strength was conducted at pace rate of 0.9 kN/s while splitting tensile strength was conducted at pace rate of 1 kN/min.

Compressive Test. ELE ADR Auto Compressive Strength & Bending Test Apparatus was used for compressive strength test. The capacity of this machine was 3000 kN. The pace rate for load was 0.9 kN/s which was automatically set for cubic specimen for the size of 50 mm. The load was continuously applied until the specimen failed. The result of compressive strength will be appeared on the screen of the

Fig. 14.3 Curing pool

machine. The test will be running for two times for each specimen to get average value. Figure 14.4a, b show a ELE ADR Auto Compressive Strength & Bending Test Apparatus, respectively.

Splitting Test. The splitting tensile machine that has been used was iIPC iUTM-1000 Universal Testing Machine. The capacity of this machine was 1000 kN. The pace rate for load was 1 kN/min the cylinder specimen. The load was continuously iapplied until the specimen failed. The result of splitting tensile will be displayed ion the computer which was connected to the machine. The computer was set up

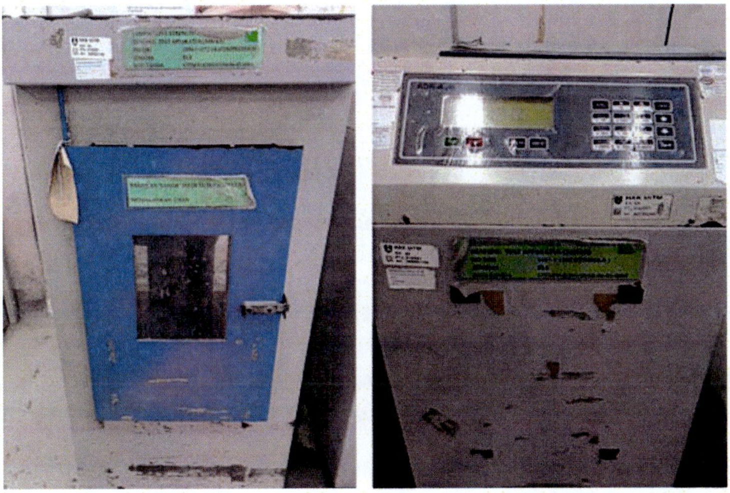

(a) (a)

Fig. 14.4 Compressive test equipment, (**a**) auto compressive machine and (**b**) bending test machine

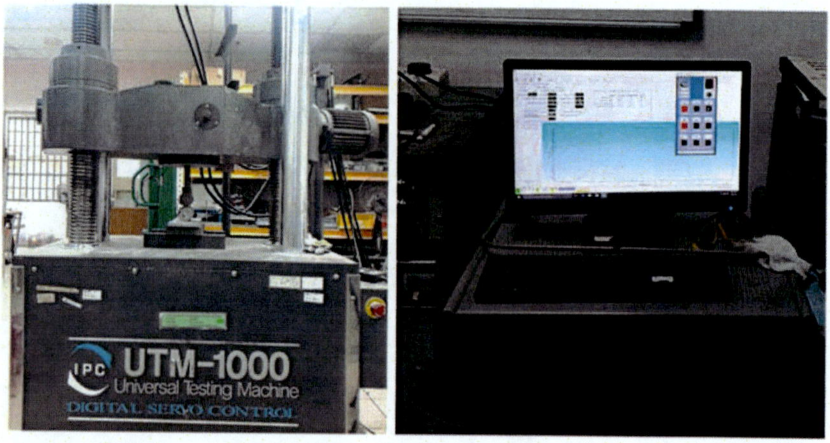

Fig. 14.5 Splitting Test, (**a**) IPC UTM-1000 Universal Testing Machine and (**b**) Computer to monitor and record the splitting tensile testing

the iproperties of ithe specimen such as dimension and monitor the splitting tensile testing. The test will ibe running ifor itwo itimes ifor ieach ispecimen to get average value. Figure 14.5a shows the IPC UTM-1000 Universal Testing Machine that has been used for the testing while Fig.14.5b shows the computer that monitor the splitting tensile test and record the result.

14.3 Results and Discussions

14.3.1 Fly Ash Chemical Analysis

The composition of fly ash was assessed and presented in Table 14.1. The primary compounds that have been determined in fly ash were oxides of silicon which consisted 67.709%, aluminium oxide contained 14.889%. In a previous research [7], they stated that the fly ash can actually be classified either Class F or Class C based on the its composition. Class F has low calcium fly ash which is less than 10% while Class C has more 20% of calcium. Hence, the fly ash used for this experiment was categorized as Class F type because the calcium is only 4.639%. Although, the previous research [8] assessed that the presence of silicon dioxide, aluminium oxide and calcium oxide contributed to high strength. Graph in Fig. 14.6 shows the composition of Fly Ash and OPC.

This indicates that the high reaction of silica content requires the formation of high formation of aluminosilicates which resulting in high mechanical strength of the resulting material. The fracture of the specimen is one of the fundamental mechanical behaviours due to essential role in the crack growth of the brick's structural elements.

Table 14.1 Xrf analysis of fly ash

Composition	Percentage (%)
Silicon dioxide, SiO2	67.709
Aluminium oxide, Al2O3	14.889
Iron (III) Oxide, Fe2O3	8.176
Calcium carbonate, CaCO3	4.639
Potassium oxide, K2O	1.862
Magnesite, MgCO3	< LOD
Titanium dioxide, TiO2	1.195
Phosphorus pentoxide, P2O5	0.465
Sulphur trioxide, SO3	0.276
Manganese oxide, MnO	0.195

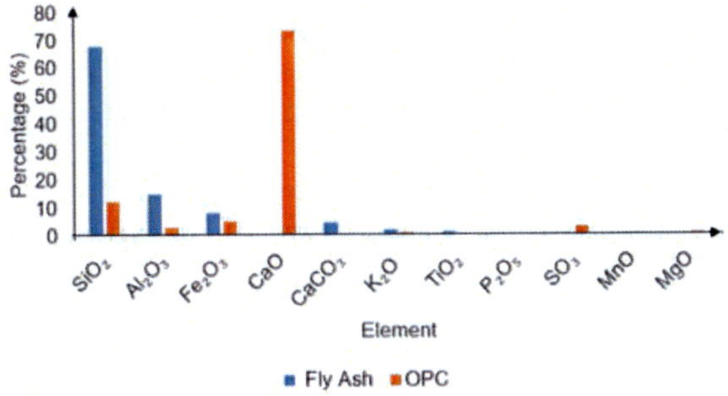

Fig. 14.6 Comparison percentage between composition of fly ash and OPC

14.3.2 Fracture Analysis

For compressive strength fracture, the crack was initiated from the edge. Then the specimen was considered failed when the new crack propagate and reached at another edge. The crack on the specimen appeared as in diagonal cracking. The specimen was split into two where there was no crack propagation even after the load was continuously applied until the specimen fracture. Besides that, the surface of the specimen was basically rough and flat fracture as the failure occur along the middle of vertical diameter. Figure 14.7 shows the fractography of the specimen.

14.3.3 Compressive Strength

The optimum percentage of fly ash for this study was 20% replacement of fly ash. In addition, when the ash was replaced the OPC by 20%, the results achieved was the highest, 43.355 N/mm^2 compare to 10 and 30% of fly ash which was 40.38 and

Fig. 14.7 Fractography surface of specimen: (a) Cube and (b) Cylinder

41.885 N/mm^2. Table 14.2 shows compressive strength at different percentage of fly ash. Figure 14.8 shows the comparison of compressive strength between this research and previous research by Awodiji and Onwuka [9]. Percentage error was calculated to compare the compressive strength values. The error obtained was in the range of 40–60%. However, the trend shown in Fig. 14.8 is similar where the highest value obtained at 20% added ash and OPC. Current compressive strength are higher than previous research due to fly ash were added to the mixture. This explained that added fly ash gives a significant effect to the compressive strength.

Table 14.2 Compressive strength at different fly ash

Specimen	Fly ash (%)	Average compressive strength (N/mm^2)	Previous average compressive strength (N/mm^2)	Percentage error (%)
$Cube_0$	0	44.655	18.22	59.8
$Cube_{10}$	10	40.380	23.85	40.9
$Cube_{20}$	20	43.355	25.33	41.5
$Cube_{30}$	30	41.885	20.59	50.8

Fig. 14.8 Comparison of compressive strength between this research and previous research

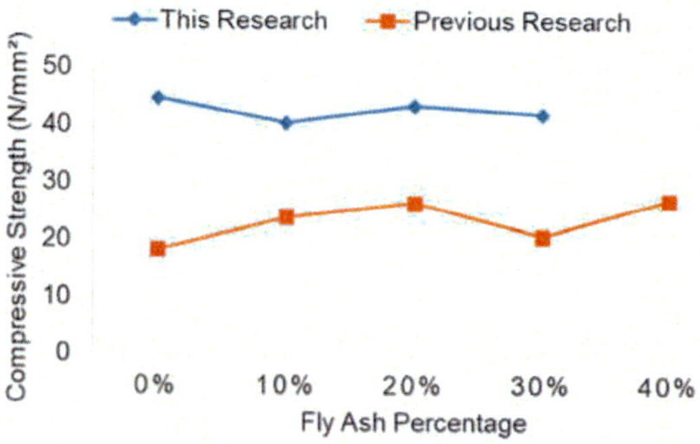

Table 14.3 Splitting tensile strength results

Specimen	Fly Ash (%)	Average splitting tensile strength (N/mm^2)	Previous Average splitting tensile strength (N/mm^2)	Percentage error (%)
Cylinder0	0	14.666	2.100	85.6
Cylinder10	10	13.152	2.905	77.9
Cylinder20	20	12.361	2.700	78.1
Cylinder30	30	11.652	2.585	77.8

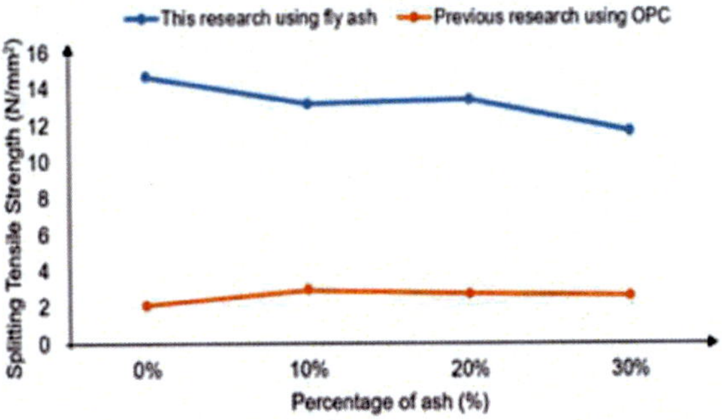

Fig. 14.9 Comparison of splitting tensile strength between this research and previous research

14.3.4 Splitting Tensile Strength

Table 14.3 shows the tensile strength at different percentage of fly ash comparing with previous research by Olujide, 2017 [10] while Fig. 14.9 shows the trend of splitting strength values at different percentage ash and OPC. The strength was the highest when no fly ash was added which is 14.666 N/mm^2. The values keep decreasing with added fly ash from 10 to 30%. Splitting strength for 10, 20, and 30% fly ash were 13.152, 12.361, and 11.652 N/mm^2, respectively. Percentage error was obtained in the range of 77–86% which is considered high. This is due to fly ash was added replacing the OPC. Splitting strength for fly ash specimens were high compared to previous research which were used OPC. This shows fly ash give a significant effect to splitting strength.

14.4 Conclusion

Both testing results show that the percentage replacement of fly ash was optimum when added 20% with the concrete mixture. The compressive strength data was validated with the previous studies' results which use OPC in the cement mixture. Both results (current and previous) show the same fluctuate trend of compressive strength. The compressive strength was higher at 20% fly ash which is 43.355 N/mm^2. For splitting tensile strength, the results for this project can be wrapped up to has higher strength compare to other bricks that have no fly ash added. Previous studies used different technique which can affect the splitting tensile strength. For this research, geopolymer bricks achieved higher splitting tensile strength at 20% compare to 10 and 30% which is 13.361 N/mm^2.

References

1. Ferone C, Colangelo F, Cioffi R, Montagnaro F, Santoro L (2011) Mechanical performances of weathered coal fly ash based geopolymer bricks. Proc Eng 21:745–752. https://doi.org/10.1016/j.proeng.2011.11.2073S
2. Zhang C, Zhu JK, Sin O, Mok PKT (1999) A novel ultrathin elevated channel low-temperature poly-Si TFT. IEEE Electron Device Lett. 20:569–571
3. Kumar SG, Aleem MIA, Dinesh S (2016). Application of geopolymer concrete. Int Res J Eng Technol 96–99
4. Lv X, Guo P, Liu H, Cui L, Cui X (2018) Preparation of paraffin-based phase-change microcapsules and application in geopolymer coating. J Coat Technol Res 15(4):867–874. https://doi.org/10.1007/s11998-018-0071-6
5. Lyon RE, Sorathia U, Balaguru PN, Foden A, Davidovits J, Davidovics M (1996) Fire response of geopolymer structural composites. In: Proceedings of the First International Conference on Fiber Composites in Infrastructure (ICCI '96). pp 972–981
6. Nuruddin MF, Malkawi AB, Fauzi A, Mohammed BS, Almattarneh HM (2016) Geopolymer concrete for structural use: recent findings and limitations. IOP Conf Ser Mater Sci Eng 133(1). https://doi.org/10.1088/1757-899X/133/1/012021
7. Simon S, Hemamathi A, Jenishtalouis J (2019) Strength assessment on flyash based geopolymer concrete. Int J Innov Technol Explor Eng 9(1):3956–3960. https://doi.org/10.35940/ijitee.A5034.119119
8. Tanikella P, Olek J (2017) Updating physical and chemical characteristics of fly ash for use in concrete. 112. https://doi.org/10.5703/1288284315213
9. Awodiji CTG, Awodiji OO, Onwuka DO (2016) Re-Investigation of the compressive strength of ordinary portland cement concrete and lime concrete. Int J Geol Agric Environ Sci 4
10. Olujide O (2008) Ph.D, Concrete as a promising structural material. Int Res J Eng Sci Technol (IREJEST) 5(1)

Chapter 15
Effect of Bio Lubricants on the Dynamic Performance of Rotor Bearing System: A Mathematical Model

Muhammad Imran Sadiq, Wan Aizon Wan Ghopa, Mohd Zaki Nuawi, Mohammad Rasidi Mohammad Rasani, and Tajammal Imran

Abstract High-performance machinery operates on fluid-film bearings due to the operating requirements and conditions such as high speed of the shaft. Such extreme operating conditions demand a carefully designed rotor bearing system with suitable lubricant. The lubricant tribological, rheological, and dynamical properties play pivotal role in having the smooth operation of the rotor bearing system. The rising temperature of the lubricant during operation causes reduction in its viscosity and affects the operating bearing clearance. Radial bearing clearance and oil temperature are very critical in rotor bearing system as they may cause flow induced instability like oil whirl and oil whip. These flow induced instabilities that are known to be detrimental for the smooth operation of the rotating machines. Thus, the main objective of current work is to review and summarize the methodologies that are implemented to improve and enhance the dynamic performance of the fluid film journal bearing using the bio-lubricants. The mathematical model was also developed in order to simulate the coefficient of friction and the friction torque of lubricant. It was simulated based on the data available in literature. It is found out that the results are comparable to the available literature as discussed in this paper. It has given the glimpse of potential of the bio-based lubricants. Furthermore, the bio lubricants will further be tested to evaluate their damping characteristics.

Keywords Rotor bearing system · Fluid-film bearing · Bio lubricants · Rotor-dynamics · Viscosity · Mathematical model

---

M. I. Sadiq · W. A. W. Ghopa (✉) · M. Z. Nuawi · M. R. M. Rasani
Faculty of Engineering and Built Environment, University Kebangsaan Malaysia, 43600 UKM Bangi, Selangor, Malaysia
e-mail: waizon@ukm.edu.my

T. Imran
Mechanical Engineering Department, University of Hafr Albatin, Hafar Al Batin 39524, Saudi Arabia

S. Abdullah et al. (eds.), *Structural Integrity Cases in Mechanical and Civil Engineering*, Structural Integrity 23, https://doi.org/10.1007/978-3-030-85646-5_15

15.1 Introduction

Bearings are one of the most critical components in any rotating machinery. Early investigators did not recognize the importance of the bearing properties, so many analytical rotor models (e.g., Jeffcott rotor) had rigid supports at the bearing locations (infinite stiffness, no moment restraint). Thus, an understanding of how bearing is characterized for the best performance of the machine is required and is reported by [1].

Most high-performance commercial machinery which operates on fluid-film bearings experiences rising temperature of the lubricant, which induces thermal expansion and affects the operating bearing clearance [2].

The overall power loss of the machine can be reduced by using an oil with better physical and dynamical properties. For example, on the one hand, the optimized dynamic viscosity curve can be used for the formulation of oil, with suitable additives [3]. On the other hand, bio-based lubricants can also be used as potential lubricants due to their better lubricant properties and environment friendliness as reported by [4].

Similarly it is reported that frictional characteristics of the lubricants are significantly influenced by the clearance [5]. It was also reported that additives in the lubricant also contributed to achieve improved journal bearing performance [6].

So, it can be said that a better tribological performance can be achieved by selecting a bio-based lubricant blended with nano particles to overcome the issues of oil whirl and oil whip. It will also be beneficial for better future design of the rotor-bearing systems.

There are many experimental investigations conducted on the performance of bio lubricant in a journal bearing [7]. However, the area of bio lubricant is less reported from a theoretical perspective. Therefore, in this work a review is made on the efforts which have reported to mathematically predict the static and dynamic performance of the bio lubricants in a journal bearing.

Likewise, it is essential to experimentally simulate the lubricant properties. It is also very important to consider repeatability of test conditions and parameters [8]. Easy acquisition of measurement data and keeping in mind provisions for future, amendments should also be considered [9]. Keeping these requirements in mind, a Journal Bearing Test Rig is also developed to experimentally investigate the lubricant properties.

Similarly, less work is reported in terms of mathematical or experimental evaluation of bio lubricants for flow induced instabilities and damping characteristics. Therefore, in this work, a review is done on the efforts which have been reported to investigate the static and dynamic performance of the bio-lubricants mathematically in a journal bearing. Also, the mathematical equations are further used to see the trend of reported bio lubricant properties and further match the results with Raimondi and Boyd type curves used for journal bearing design.

15.2 Rotor Model

The self-excited instability or otherwise known as oil whirl and oil whip problem of rotor-system is a subject of an active area of research since many years. Oil whirl and oil whip are sub-synchronous instability phenomena related to wide speed range of rotor system that is caused by the dynamic forces of the fluid film bearing and was first identified by [10 and 11]. Oil whirl occurs near half the rotational critical speed (ω) of the rotors (or journal), and oil whip occurs when the oil whirl coincides with the first flexural natural frequency of the rotor, hence, near twice this natural frequency or critical speed (ω).

Figure 15.1 illustrates a two degree of freedom horizontal isotropic Jeffcott rotor is normally considered to study the dynamic response of a rotor bearing system. Further, it is common to model the rotor system as a Jeffcott rotor to study its dynamical response [12–20]. Such model is simple in its presentation yet covers most aspects of rotor system response such as shaft whirl due to unbalance weight [13–17] and self-excited vibrations such as oil whirl and oil whip [16–20]. This model is also capable of studying frequency response, Campbell diagrams, orbital motion, and consequently the instability features such as oil whirl and oil whip of rotor system.

Mathematically speaking, to study rotor system instabilities is a thorough problem and has attracted many researchers [12–23]. A more specific definition of rigid and flexible rotor is given by [24].

It is important to study the performance of journal bearing by considering Reynolds equation of oil film lubrication. Fundamental understanding on the performance of journal bearing has been the subject of many investigations and thus most reported investigations are referenced here [10–24].

In this section, the performance of journal bearing is investigated mathematically by considering three bio lubricants. This is because the oil whirl and oil whip instabilities are associated mainly with the performance of lubricant in the journal bearing [25]. The nomenclature of the journal bearing is shown in Fig. 15.2.

Simple calculations are made in the current work for static performance and then compared with the published theoretical and experimental work that is reported for the performance of lubricants in a journal bearing.

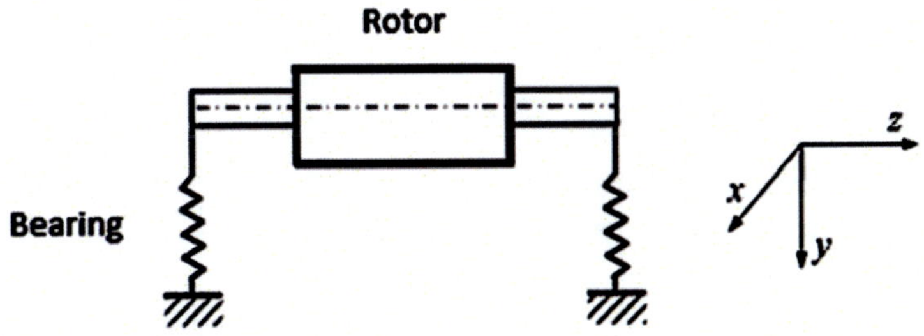

Fig. 15.1 Jeffcott Rotor, a simple rotor system mounted on bearings and support

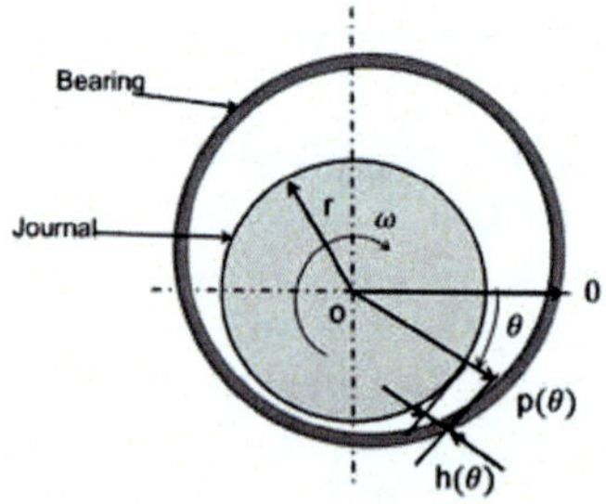

Fig. 15.2 Equilibrium position of journal bearing

It is a common practice [26] to derive the Reynold's Eq. (15.1) in two dimensional Cartesian coordinates where Navier–Stokes equations are providing the basis of this derivation. To achieve an exact solution for simple 1-D oil flow condition, Eq. (15.1) is then simplified to one-dimension form (2). This is achieved by making assumptions of constant viscosity, incompressibility, and full oil flow conditions. It is done by considering three different types of boundary conditions.

$$\frac{\partial}{\partial x}\left(\frac{h^3}{\eta}\frac{\partial p}{\partial x}\right)+\frac{\partial}{\partial z}\left(\frac{h^3}{\eta}\frac{\partial p}{\partial z}\right)=6U\frac{\partial h}{\partial x}+12\frac{\partial h}{\partial t} \tag{15.1}$$

After simplifications and replacing some terms in Eq. (15.1), Reynold's equation can be written as

$$\frac{\partial}{\partial x}\left(\frac{h^3}{\eta}\frac{\partial p}{\partial x}\right)+\frac{\partial}{\partial z}\left(\frac{h^3}{\eta}\frac{\partial p}{\partial z}\right)=6.R.\omega.\frac{\partial h}{\partial x} \tag{15.2}$$

Equation (15.2) is then reduced for short bearing. The short bearing condition refers to the ratio of L/D < < 1. Equation (15.2) can be reduced to 1-D,

$$\frac{\partial}{\partial z}\left(h^3\frac{\partial p}{\partial z}\right)=6\eta.\omega.\frac{dh}{d\theta} \tag{15.3}$$

Integrating Eq. (15.3) with respect to "z" takes the following form in terms of pressure distribution for short bearing with the boundary conditions of p = 0 at ±L/2,

$$p=\frac{3\eta\omega}{C^2}\left(\frac{L^2}{4}-z^2\right)\frac{\varepsilon\sin\theta}{(1+\varepsilon\cos\theta)^3} for 0<\theta<\pi \tag{15.4}$$

The performance calculations of the lubricant are done for static conditions.

The load capacity can be found by solving the pressure distribution Eq. (15.4). Manipulating the integrals as reported in [20, 21] will give the resultant load 'W' as

$$W = \frac{\eta R\omega L^3}{4C^2} \cdot \frac{\varepsilon}{\left(1-\varepsilon^2\right)^2} \cdot \left[\pi^2\left(1-\varepsilon^2\right) + 16\varepsilon^2\right]^{\frac{1}{2}} \tag{15.5}$$

And the attitude angle is

$$\tan\phi = \frac{W_y}{W_x} = \frac{\pi}{4} \cdot \frac{\left(1-\varepsilon^2\right)^{\frac{1}{2}}}{\varepsilon} \tag{15.6}$$

Equation (15.5) provides a unique relationship between eccentricity ratio and bearing load. The values of eccentricity ratio are normally found by iteration at known values of applied load W.

For practical reasons, it is common practice to relate the eccentricity ratio to that of Sommerfeld number. So, it is important to define the Sommerfeld number 'S' in terms of eccentricity ratio. The Sommerfeld number is

$$S = \left(\frac{D}{L}\right)^2 \frac{\left(1-\varepsilon^2\right)^2}{\pi.\varepsilon.\left[\pi^2\left(1-\varepsilon^2\right) + 16\varepsilon^2\right]^{\frac{1}{2}}} \tag{15.7}$$

The solution of Eq. (15.7) will give the unique values of eccentricity ratio against each value of S. Similarly, the lubricant friction force (viscous), the coefficient of friction, and the friction torque can be estimated as

$$F_f = \frac{\eta L R^2 \omega}{C} \cdot \frac{2\pi}{\left(1-\varepsilon^2\right)^{1/2}} \tag{15.8}$$

$$f = \frac{F_f}{W} \tag{15.9}$$

$$T_f = \frac{\eta L R^3 \omega}{C} \cdot \frac{2\pi}{\left(1-\varepsilon^2\right)^{1/2}} \tag{15.10}$$

15.3 Design of Journal Bearing Test Bench

A test rig is an apparatus used to experimentally simulate the rotor-dynamic response and flow induced instabilities. It is very important to design a test rig to simulate the operating conditions [27] and to be able to record the measurement data with repeatability and accuracy [28]. It is likely that if different operators are using the same equipment the results might vary depending upon the level of experience and sensitivity to equipment. Standard operating procedure may be prepared beforehand to avoid any such problems [29]. It is essential to develop a system which can easily be tested for different lubricants with additives. Furthermore, to keep provision for sensors at axial and radial locations for vibration analysis is also needed for testing and analysis purpose. The test rig needs to be flexible in order to deal with shafts of large diameters and high speeds. Moreover, there should be provision for modernization and further development of test rig [30].

The test rig model can be seen in Fig. 15.3. The oil is supplied to the bearing through the oil inlet port which is at the top of the journal bearing. It can contain up to 350 ml of oil. A DC motor equipped with speed control unit which can operate up to 3000 rpm (50 Hz) is directly attached to the journal. There is tachometer near the shaft to record the speed of the shaft. The load can be applied to the journal bearing through a lever connected to the journal-bearing set up. There is a provision for temperature measurement. The main control unit can set the desired rpm, record the actual rpm of the shaft, and record the temperature of oil. The specifications of the journal bearing test rig are given below in Table 15.1. Vibration data is measured through a data acquisition system known as Rionote FFT Analyzer. The details of the vibration analyzer are given in Table 15.2.

Figure 15.3 presents the journal bearing test bench model in a simplified way.

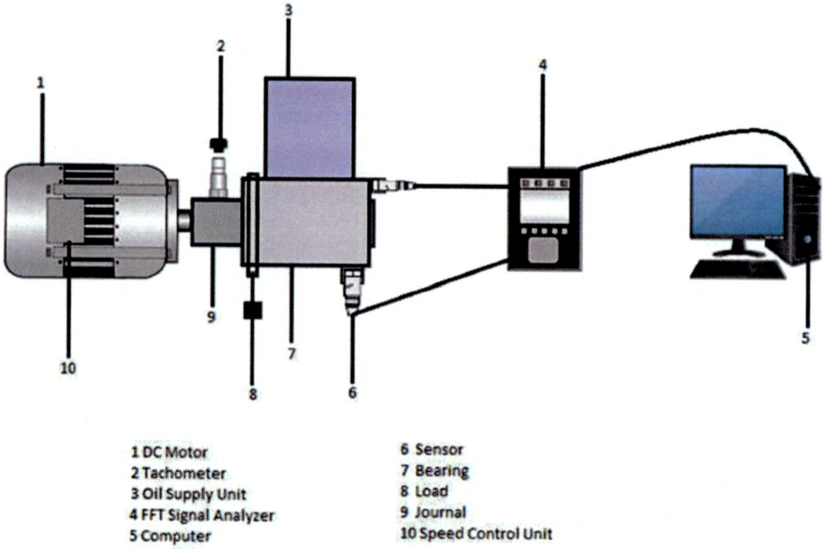

Fig. 15.3 Schematic diagram of journal bearing test bench

Table 15.1 Specifications of the journal bearing test bench

No	Description	Specification
1	L, bearing length	45 mm
2	d, inner diameter for plain bearing	30.04 mm
3	Φ, shaft diameter	30 mm
4	W, Weight of journal shaft	8.2 N
5	C°, total clearance	0.04 mm
6	C, radial clearance	0.02 mm
7	L/D ratio	1.5
8	Operating speed	60 ~ 3000 rpm

Table 15.2 Specifications of Analyzer

No. of recording channels	1 to 4 channels
Frequency range	1 to 20 kHz
Recorded data	Wave format
Maximum input voltage	± 13 V
Trigger source	Wave form, time, external, rotation speed

15.4 Results and Discussions

Rheology of bio lubricant is important to study their performance and application. Study of various rheological models showed that the bio lubricant viscosity is greatly dependent on temperature as noted by many researches such as [31]. It is also reported that some of the bio lubricants such as palm oil and sunflower oil can behave as Newtonian and Non-Newtonian. This was noted by comparing the values of flow behavior index as estimated through Ostwald de-Waele (n) and Herschel–Bulkley (nH) models. The performance analysis of bio lubricant is made in the current study by considering the short bearing theory. It is also assumed that bio lubricant behaves as Newtonian fluid. Three bio lubricants are selected for the static performance and properties as reported by [32]. It is assumed that the operating temperature of the lubricant is 40 °C. Table 15.3 is referring the properties of the selected bio lubricants [32].

The operating conditions for static performance are given in Table 15.4. The results are then presented in usual way that is Oil film thickness 'h' and Pressure distribution 'p'. Eccentricity ratio '$\varepsilon = e/C$', against Sommerfeld number 'S' (or bearing duty number), Attitude angle 'ϕ', Load carrying capacity, Friction force, and Friction coefficient are also plotted for three bio lubricants against eccentricity ratio ε.

Figure 15.4 is representing a typical pressure distribution for the Soybean oil at the center of the short journal bearing ($-L/2 < z < +L/2$) as a function of angular variation between the centers of journal and bearing. Further, Fig. 15.5 represents the oil film

Table 15.3 Physical properties of selected bio-lubricants referred from [32]

Properties	Soybean oil	Rapeseed oil	Palm oil
Viscosity @ 40 °C (mm2/sec)	29.0	34.8	39.6
Density (kg/m3)	913	917	918
Absolute Viscosity (Pa-s)	0.02648	0.03919	0.03635
Flash point (°C)	328	323	267
Pour point (°C)	−10.0	−15.0	–
Viscosity Index	246	218	–

Table 15.4 The operating conditions selected for calculations

Operational parameter	Range
Rotation of the journal, N [rpm]	600
Diameter of the journal, D, [m]	0.2
Inside Diameter of the bearing, DB, [m]	0.2004
Eccentricity ratio, = εe/C, [-]	0→1
Length of the journal, L, [m]	0.1
Clearance between the journal and bearing, C, [m]	2 × 10–4

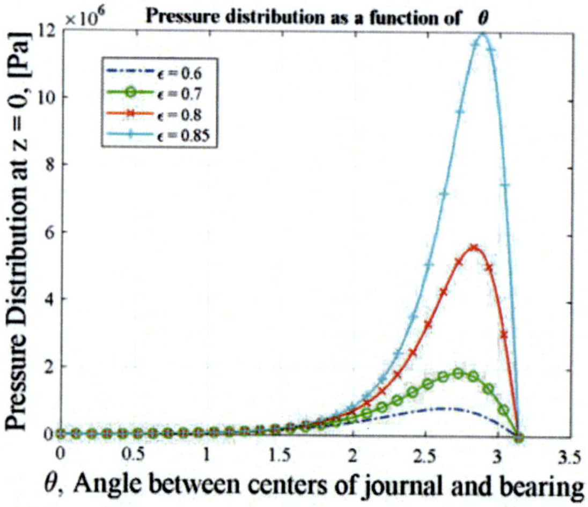

Fig. 15.4 Pressure distribution as a function of angle θ for Soybean oil with different eccentricity ratios

distribution as a function of circumferential direction. The values of the magnitude of pressure and oil film thickness in Figs. 15.4 and 15.5 are in accordance with the published work in the literature [12–19]. The results are compared for different eccentricity ratios. The pressure amplitude is higher when the eccentricity ratio is approaching to maximum value (i-e; → ε1), and the oil film thickness is at minimum at this value of 'ε'. It is important to note that the higher value of eccentricity 'e'

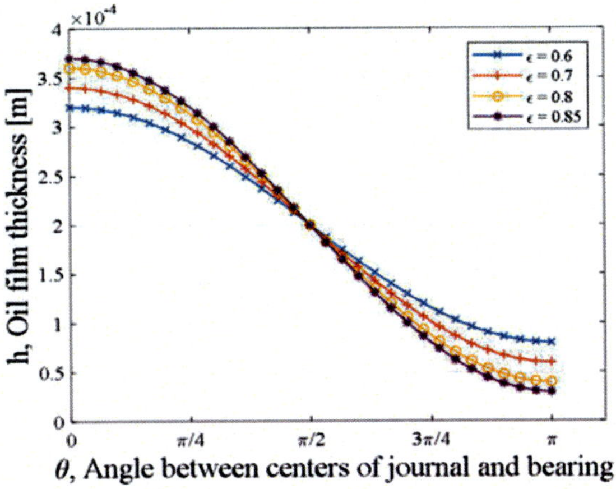

Fig. 15.5 Oil film thickness as a function of angle θ for Soybean oil with different eccentricity ratios

means less value of clearance 'C'. This consequently results in lesser oil trapped in between journal and bearing surfaces and hence results in maximum pressure. The effect of viscosity is also noticeable at the same eccentricity ratio (ε=0.6) for three different bio lubricants. This is shown in Fig. 15.6. More viscous bio lubricant will have higher pressure distribution for same eccentricity ratio,'ε'.

The journal bearing performance parameters depend upon the eccentricity ratio 'ε'. It is a common practice to estimate the unique values of ε through iteration process. A quick analytical method is proposed by [33] to do the simple design calculation of the short journal bearing to avoid long manual iteration process based on the work of Raimondi and Boyd as in [34]. Similarly, a simple expression to calculate eccentricity ratio ε or Sommerfeld number, S, is proposed by [9]. However, in the current work, the eccentricity ratio ε values are assumed first to be varied from 0 to 1 and then the calculations are made for the performance parameters of

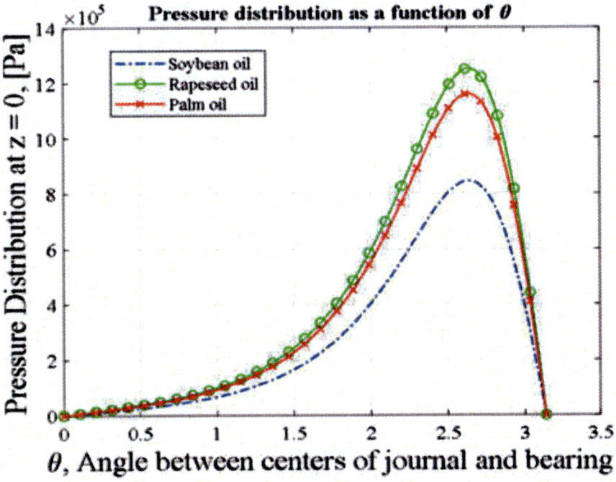

Fig. 15.6 Pressure distribution as a function of angle θ for three bio lubricants with = ε0.6

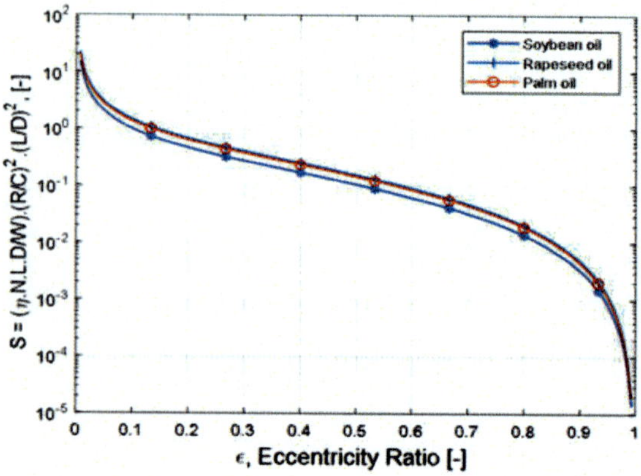

Fig. 15.7 Sommerfeld number S as a function of eccentricity ratio ε for three bio lubricants

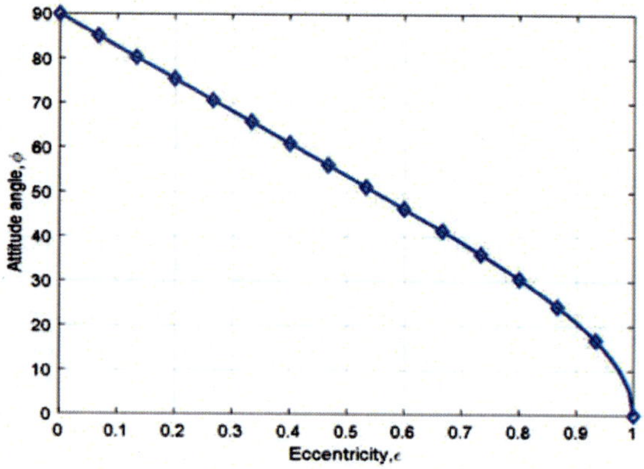

Fig. 15.8 Attitude angle ϕ as a function of Eccentricity ratio ε for three bio lubricants

short journal bearing for three bio lubricants. The results of bearing parameters are presented in Figs. 15.7, 15.8, and 15.9.

Bearing performance curves are also presented and compared to Raimondi and Boyd type curves [34]. Dimensionless film thickness and attitude angle are plotted in Fig. 15.10. The form of results is matching with Raimondi and Boyd type curves.

15.5 Conclusion

Bio lubricants have gained good reputation as potential replacement of mineral based lubricants having exhibited good tribological properties. This work is a step toward exploring more opportunities in the area of bio lubricants. Still some features of bio lubricants are unexplored like damping characteristics and whirling patterns under varying loading and operating conditions. The mathematical model in this work is

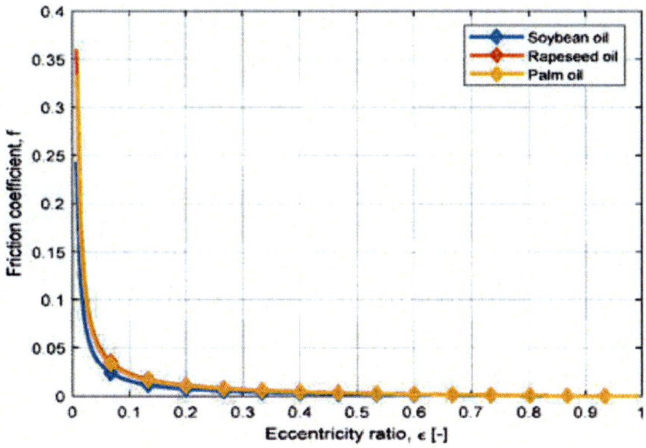

Fig. 15.9 Coefficient of friction as a function of Eccentricity ratio ε for three bio lubricants

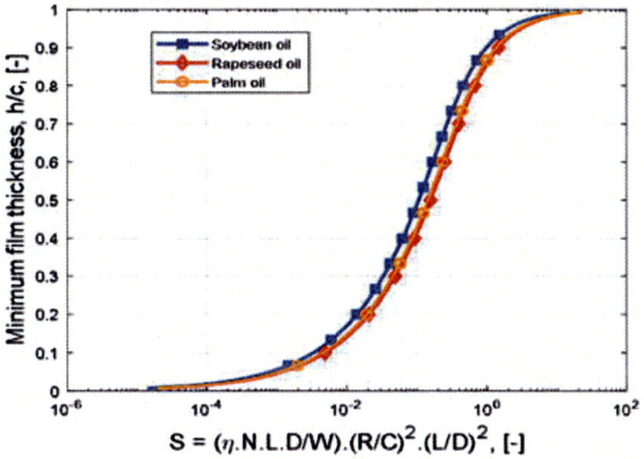

Fig. 15.10 Dimensionless minimum film thickness h/c as a function of Sommerfeld number for three bio lubricants

simulated based on the data available in literature. It is found out that the results are comparable to the available literature as is discussed in results and discussions. It has given the glimpse of potential of the bio-based lubricants. It will be further explored and justified when these lubricants will be experimentally tested. The prime objective of this work is to simulate oil flow parameters mathematically and experimentally in order to come up with future recommendations for bio-based lubricants. The environment friendly nature of these bio-based lubricants is another major factor which has attracted many researchers to work on these potential lubricants.

Acknowledgements We are grateful to Universiti Kebangsaan Malaysia for their Research Grant No. GUP-2018-078. We are also grateful to Higher Education Commission (HEC) of Pakistan for funding the student.

References

1. Cao H, Niu L (2018) Mechanical model development of rolling bearing-rotor systems: a review. MechSyst Signal Process 102:37–58
2. Tong X, Palazzolo A, Suh J (2017) A review of the rotordynamic thermally induced synchronous instability (morton) effect. Appl Mech Rev 69. https://doi.org/10.1115/1.4037216
3. Chatterton S, Pennacchi P, Vania A (2018) Optimized tribo-design of lubricants for power loss reduction in journal bearings used in process industry. 1. https://doi.org/10.1007/978-3-319-99262-4_32
4. Syahir AZ, Zulkifli NWM, Masjuki HH, Kalam MA, Alabdulkarem A, Gulzar M, Khuong LS, Harith MH (2017) A review on bio-based lubricants and their applications. J Clean Prod 168:997–1016
5. Cho IS, Jung JY (2014) The friction characteristics of the journal bearing in a reciprocating compressor for refrigeration and air conditioning systems. J Mech Sci Technol 28, 1473–1480
6. Fetisov A, Kornaev A, Tokmakov N (2018) Friction and vibration in journal bearings operating with nanofluids: experimental results. ICIE
7. Baskar S, Sriram G, Arumugam S (2017) Tribological analysis of a hydrodynamic journal bearing under the influence of synthetic and biolubricants. Tribol Trans 60(3):428–436
8. Marey N, Hegazy E-S, Aly A (2018) Design and setup for a journal bearing universal test rig. Port-Said Eng Res J 22:101–106
9. Jang JY, Khonsari M (2004) Design of bearings on the basis of thermohydrodynamic analysis. Proc Inst Mech Eng Part J Eng Tribol 218:355–363. https://doi.org/10.1243/1350650042128012
10. Newkirk BL (1924) Shaft whipping. Gen Electr Rev 27:169–178
11. Newkirk BL, Taylor HD (1925) Shaft whipping due to oil action in journal bearing. Gen Electr Rev 28:559–568
12. Genta G (2009) Elementary rotordynamics. Jeffcott Rotor. https://doi.org/10.1007/978-0-387-79580-5_23
13. Tiwari R (2017) Rotor systems, analysis and identification
14. Childs D (1993) Turbomachinery rotordynamics, phenomena, modeling, and analysis
15. Vance JM, Zeidan FY, Murphy BG (2010) Machinery vibration and rotordynamics. Willey ISBN: 978–0–471–46213–2
16. Agnieszka M (2005) Rotordynamics. ISBN-13: 978–0824723996
17. Bently DE, Hatch CT (2002) Fundamentals of rotating machinery diagnostics. Bently Pressurized Bearing Company
18. Rao J (2011) History of rotating machinery. Dynamics. https://doi.org/10.1007/978-94-007-1165-5
19. Hooke C (2009) Fluid film lubrication, theory and design, Andras Z Szeri. TribolInt 32:345–347
20. Ding W (2012) Self-excited vibration, theory, paradigms, and research methods. Tsinghua University Press, Beijing and Springer-Verlag Berlin Heidelberg. ISBN 978–7–302–24296–3
21. Surovec R, Bocko J, Šarloši J (2014) Lateral rotor vibration analysis model. Am J Mech Eng 2:282–285. https://doi.org/10.12691/ajme-2-7-23
22. Javorova J, Sovilj B, Sovilj-Nikic I (2009) On the derivation of dynamic force coefficients in fluid film bearings. Mach Des 1:283–288
23. Kramer E (1993) Dynamics of rotors and foundations. Springer-Verlag Berlin Heidelberg GmbH. ISBN 978–3–662–02800–1
24. Neto RR, Bogh DL, Flammia M (2008) Some experiences on rigid and flexible rotors in induction motors driving critical equipment in petroleum and chemical plants. IEEE Trans Ind Appl 44(3):923–931
25. Berry EJ (2005) Oil whirl and whip instabilities - within journal bearings. Machinery Lubrication, online magazine published by NORIA, Technical Associates of Charlotte
26. Lee C-W (1993) Vibration analysis of rotors. https://doi.org/10.1007/978-94-015-8173-8
27. Wale D, Mba, D (2007) A journal bearing test rig with reduced uncertainty: some design considerations. J Test Eval 35. https://doi.org/10.1520/JTE100116

28. Mazdrakova A, Javorova J, Radulescu A, Mirev A, Rakanov Y (2016) hydrodynamic journal bearing test rig with elastic deformations of contact surfaces capabilities. In: Proceedings of the 15th International Scientific Conference. RE & IT - 2016, SMOLYAN—BULGARIA
29. Ciulli E, Forte P, Libraschi M, Nuti M (2018) Set-up of a novel test plant for high power turbomachinery tilting pad journal bearings. Tribol Int 127. https://doi.org/10.1016/j.triboint.2018.06.014
30. Li Q, Zhang S, Wang Y, Xu W, Wang Z (2018) A dynamic response test rig of a full-scale rotor–journal bearing system. Proc Inst Mech Eng Part J J Eng Tribol 135065011881718. https://doi.org/10.1177/1350650118817183
31. Nik WB, Ani FN, Masjuki HH, Giap SE (2005) Rheology of bio-edible oils according to several rheological models and its potential as hydraulic fluid. Ind Crops Prod 22:249–255. https://doi.org/10.1016/j.indcrop.2005.01.005
32. Ali SI, Reshi M (2018) Rheological review on potential of bio-lubricants. Int J Chem Stud 6(3):1135 1138
33. Naffin RK, Chang L (2010) An analytical model for the basic design calculations of journal bearings. J Tribol 132(2):1–7
34. Budynas RG, Nisbett, JK (2015) Shigley's: mechanical engineering design, 10th edn. McGraw-Hill Inc

Chapter 16
Experimental Investigation of Acoustic Emission Characteristics at Different Type of Sensors for Granite

Noorsuhada Md Nor, Khairul Afinawati Hashim, Shahrum Abdullah, and Nur Fazlinieza Khairosam

Abstract The use of suitable sensors on determination of the acoustic emission characteristics for identification of the integrity of granite is significantly important. It is because the use of suitable frequency of sensors reflect to the data captured during testing especially on granite. Hence, this paper presents the acoustic emission characteristics of granite using two different types of sensors at different source location. Two sensors of VS75–V and VS150–M were used. Both sensors were fixed on top of the granite which set up longitudinally. The acoustic emission test was carried out by using the pencil lead fracture on the selected position. The pencil lead fracture method was carried out on six source locations of 0 mm, 20 mm, 40 mm, 60 mm, 80 mm and 100 mm of sensor position. For each source location, three replicates of pencil lead fractures were carried out. Two type of acoustic emission characteristics namely energy and peak frequency were analysed. It is found that the energy for collected from the VS150–M is lower than that of VS75–V. Meanwhile, the peak frequency for VS150–M is higher than VS75–V. It can be inferred that the sensor plays an important role on investigation of the acoustic emission characteristics of granite. This study is significantly benefit for identification of suitable sensor for collecting genuine data for investigation the granite integrity.

Keywords Granite · Rock · Sensor · Acoustic emission · Source location

16.1 Introduction

Acoustic emission is generally used to determine the integrity of a structure or a sample. It is defined as a phenomenon of elastic waves propagation in a material that occurs when the internal sample of a material undergoes permanent changes.

N. M. Nor (✉) · K. A. Hashim · N. F. Khairosam
School of Civil Engineering, College of Engineering, Universiti Teknologi MARA, Cawangan Pulau Pinang, Malaysia
e-mail: ida_nsn@uitm.edu.my

S. Abdullah
Faculty of Engineering and Built Environment, Department of Mechanical and Manufacturing Engineering, Universiti Kebangsaan, Malaysia, Malaysia

S. Abdullah et al. (eds.), *Structural Integrity Cases in Mechanical and Civil Engineering*, Structural Integrity 23, https://doi.org/10.1007/978-3-030-85646-5_16

The elastic waves were generated by the rapid release of internal energy in the material itself [1]. The acoustic emission is a one of non-destructive techniques which normally used for the structural health monitoring [2]. It is widely used for detection of crack occurrence in a structure. Sophisticated investigation on the use of acoustic emission for the damage classification of reinforced concrete beam under various cyclic loadings has been carried out by Md Nor et al. [3]. From the acoustic emission data analysis, the damage classifications of the beam namely tensile crack and shear crack was identified. Damage assessment of the pre-cracked reinforced concrete beam repaired with epoxy injection has been investigated by Mat Saliah et al. [4]. They found that the acoustic emission becomes the useful tool for identification of the crack in the repaired beam. The acoustic emission has been also used for identification of damage on various types of structures such as beam-column joint [5] and beam [3, 6]. However, the use of acoustic emission for identification of the integrity of rock is still limited.

The use of acoustic emission technique for investigation of sand stone fracture has been carried out by Mayr et al. [7]. Recently, Wang et al. [8] have been used to study the mechanical properties, crack dynamic process, failure modes, and damage-fracture evolution on a prefabricated rock so called as rock-like sample under uniaxial compression. They found that the acoustic emission results are in good agreement with the observed microscopic failure of rock. Zhang et al. [9] have investigated the effectiveness and the accuracy of acoustic emission monitoring on coal resource mining process. They found that the acoustic emission is positively correlated with the stress level under uniaxial loading. They also found the correlation between acoustic emission parameters and the damage evolution process of the coal and rock sample. Acoustic emission has been also used to investigate the fracture of granite under high temperature [10], fatigue condition [11], under heating, and cooling cycles [12]. Even though recently the acoustic emission has also been under high intention to be used for investigation of fracture on rock such as granite, unfortunately, investigation on the use of suitable sensor on investigating the granite rock is still limited.

Suitable sensor for investigating of such material is significantly important in order to obtain a reliable and genuine data. For instance, Guo and Wong [13] have used a small size of sensor, PICO which has a good bandwidth and very high sensitivity to signals, and generally has resonant frequency of 250 kHz with operating frequency range of 250–750 kHz on investigation of granite behaviour. From the experimental, the average energy rate of up to 10^5/s at failure was obtained. Kallimogiannis et al. [14] have used sensor with frequency in the range of 200 kHz–1 MHz on investigating the fracture mechanism of the granite rock. The energy rate up to 6×10^9 (9.31×10^{-22} J.S^{-1}) has been found at failure. The acoustic emission sensor with resonance frequency of 150 kHz with frequency response of 0–400 kHz has been applied by Zhou et al. [15] on investigation of crack progression in granite rock under uniaxial loading. They found that the average frequency in the initial stage of cracking was up to 240 kHz with the energy of more than 6000 aJ at failure. From the review it is found that different sensors are producing different results. Most of the researchers [13–15] were focusing on the use of specific sensor only on the investigation of the granite behaviour, damage, and so forth. However, investigation of the use of various application of sensors in order to be obtained a reliable data with suitable frequency range is still limited.

It is because different range of sensor frequency would have different purpose of application. For instance, the sensor with frequency response over the range of 150–750 kHz is ideal for applications of large structures like storage tanks and pipelines which requiring high bandwidth for frequency analysis of the AE signals for noise discrimination and source identification. Sensor with low frequency response of the 30–120 kHz is suitable for monitoring of large objects made of highly attenuating material. Hence, investigation on the acoustic emission characteristics at various type of sensors with different frequency range is significantly vital for collecting a reliable data especially on rock granite. The pencil lead fracture technique would be used for this purpose.

16.2 Methodology

For determination of the acoustic emission characteristics at two different sensors, the process flow of this study is presented in Fig. 16.1. It started with the preparation of the samples. A granite was used for the investigation of the effect of the sensor

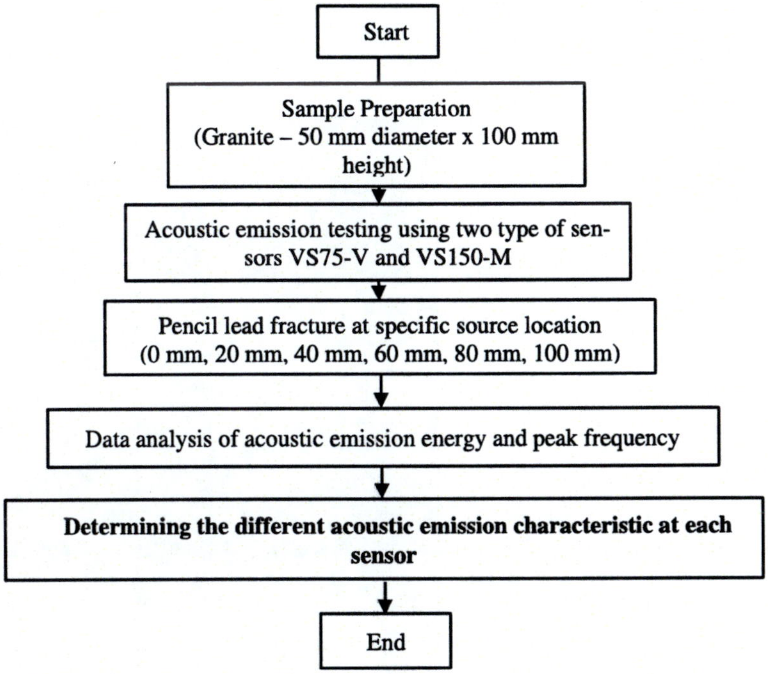

Fig. 16.1 Process flow for determining the acoustic emission characteristics at different sensors

types corresponds to the acoustic emission characteristics. In doing so, the pencil lead fracture was used on selected source location. A total of six source locations were identified of 0, 20, 40, 60, 80 and 100 mm. The data for each source location for each lead broke were analysed. Hence, the acoustic emission characteristics for each sensor for granite were then determined.

16.2.1 Sample Preparation and Position of the Acoustic Emission Source Location

In order to investigate the acoustic emission characteristics of the rock, a granite was used, which is a group of igneous rock. The granite was graded I/II with the density of 2.78 g/cm^3, and the weight of the sample was 545 g. The rock was collected from a soil investigation in a company in Bangi, Selangor, and Malaysia. The granite was cored in a size of 50 mm diameter and 100 mm height as shown in Fig. 16.2. The preparation of the granite was based on the procedure stated I the International Society for Rock Mechanic (ISRM [16]).

Six point of acoustic emission source locations were identified as shown in Fig. 16.2. The 0 mm was positioned at top followed the 20 mm, 40 mm, 60 mm, 80 mm, and 100 mm. The source location of 100 mm was positioned at the bottom part of the granite sample. The source locations were arranged in linear structure.

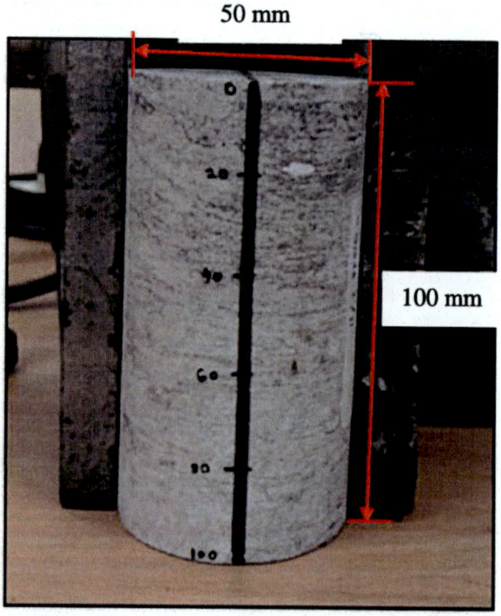

Fig. 16.2 Sample of the rock—granite

The equidistance was selected in order to investigate the trend of wave energy as the pencil lead broke at the selected location and as the wave attenuated.

16.2.2 Test Setup for the Pencil Lead Fracture and the Analyses

Figure 16.3 shows the schematic diagram of the setup of the sensor and location of the acoustic emission source. Pencil lead fracture method was used to generate the acoustic emission energy at the selected position. The same method as Md Nor et al. [17] that applied on a beam for investigation of wave velocity of reinforced concrete beam was applied to this study. A magnetic pencil with a Nielson shoe (Teflon shoe) was used to break a 0.5 mm diameter 2 H lead to generate a simulated acoustic wave against the surface of the granite, which is in accordance with the ASTM E976 [18]. Figure 16.4 shows the handling of the pencil lead which slanting down 30° to the plane of the granite sample. This method was applied throughout the test. For each location, three replicates of pencil lead breaks were carried out. Then, the results were averaged.

Figure 16.5 shows the test setup of the pencil lead fracture in conjunction with acoustic emission test. Two sensors as shown in Fig. 16.6 were used namely VS75–V and VS150–M with the frequency peak of 75 kHz and 150 kHz, respectively. Description of the sensors was tabulated in Table 1. Both sensors have different frequency range of 30–120 kHz and 100–450 kHz for VS75–V and VS150–M, respectively. For each test, one sensor was used and located at top of the sample as illustrated in Fig. 16.3. The sensor was connected to sensor BNC of the pre-amplifier with a gain of 34 dB using a cable. Then, the other cable connected from the + 28VDC output to the ASIP-2 board as shown in Fig. 16.5. From the ASIP-2 board,

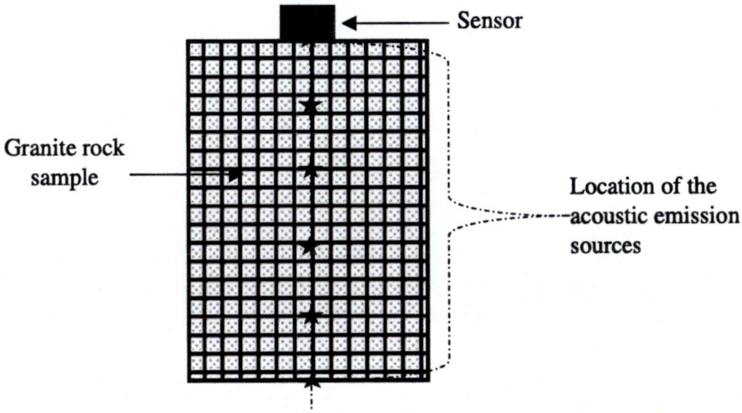

Fig. 16.3 Schematic diagram of the test setup

Fig. 16.4 Handling method of the pencil lead on surface of the granite

Fig. 16.5 Test setup for the pencil lead fracture in conjunction with acoustic emission test

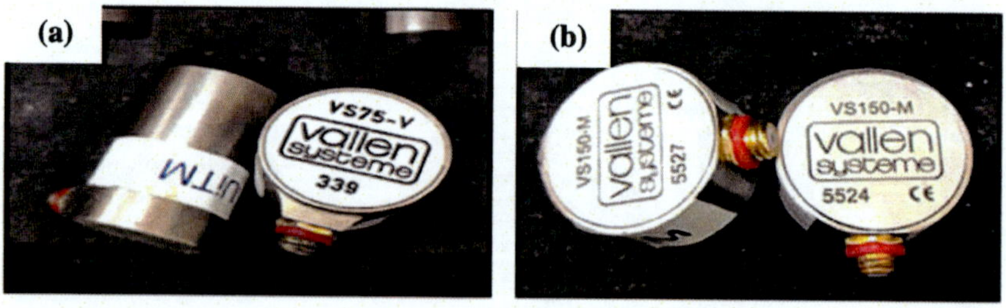

Fig. 16.6 Sensors for the pencil lead fracture **a** VS75-V and **b** VS 150-M

it was connected to a computer for acoustic emission data acquisition and displayed on the acoustic emission visual.

From the data acquisition, the energy and frequency were stored and analysed. The relationship between energy, frequency and location of acoustic emission source was plotted. Then the correlation between energy, frequency and source location was identified and discussed for both sensors.

16.3 Results and Discussion

16.3.1 Acoustic Emission Energy

Table 16.1 presents the energy for each source location for sensors VS75–V and VS150–M. From the table, it is clearly seen that as the source location increases, the energy reduces for both sensors. For instance, sensor VS75 – V, at source location 0 mm, 20 mm, 40 mm, 60 mm, 80 mm and 100 mm, the energy was 10.03×10^6 eu, 7.78×10^6 eu, 6.37×10^6 eu, 6.24×10^6 eu, 5.27×10^6 eu and 4.91×10^6 eu, respectively. Meanwhile for sensor VS150–M, the energy was 2.99×10^6 eu, 2.65×10^6 eu, 2.46×10^6 eu, 1.88×10^6 eu, 1.77×10^6 eu and 1.64×10^6 eu for source location 0 mm, 20 mm, 40 mm, 60 mm, 80 mm and 100 mm, respectively. It is due to as the acoustic emission source location farther to the sensor, and it captured lower energy due to signal loss as the wave travel to the sensor, which known as wave attenuation. At the same time, it is due to the effect of time arrival of wave from the source location to the sensor. Md Nor et al. [17] stated that as the lead breaks on surface of the sample, a wave propagated through a solid medium which carries certain amount of energy. Related to real application, as stated by Dong et al. [19], when a crack is created or propagated, the elastic energy stored in the rock suddenly releases. It generates the elastic stress waves and transmits from the interior of the rock to the boundary, where it is observed as an acoustic emission signal [20].

From the energy captured by two different sensors as presented in Table 16.1 found that the sensor VS75–V captured high energy compared to VS150–M. It can be clearly seen through the percentage difference in Fig. 16.7, where source location of 40 mm has very low energy difference between the two sensors with the percentage difference of 61.4%. However, the percentage difference for source location of 20, 80 and 100 mm are almost similar with the percentage of 66%. Source location of 0 and 60 mm has the percentage of approximately 70%. The energy pattern of the

Table 16.1 Type of sensors for the pencil lead fracture

Type of sensor	Frequency range (kHz)	Frequency peak (kHz)	Size (diameter x height) (mm)	Weight (g)
VS75-V	30–120	75	20.3 × 37	63
VS150-M	100–450	150	20.3 × 14.3	24

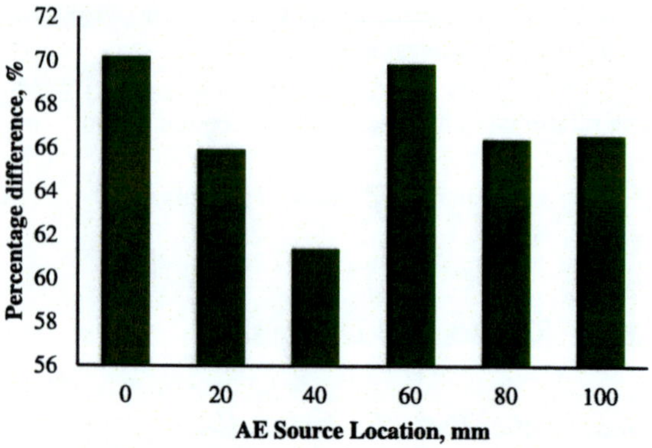

Fig. 16.7 Percentage difference of the energy between sensors VS75–V and VS150–M

granite rock is also can be illustrated through Fig. 16.8 for each source location. The difference of energy between the two sensors is due to the suitability of the frequency range of the sensors. For instance, the VS75–V is a passive piezoelectric sensor which the frequency response is characterized by a peak at 75 kHz where it exhibits a resonance. As stated in Vallen Systeme [21], the sensor VS75–V is categorised as a low-frequency sensor and make this sensor suitable to be used for monitoring large objects which have a high attenuation effect. Meanwhile, the VS150–M is also a passive piezoelectric sensor which has a peak frequency response of 150 kHz. The sensor is recommended to be used for monitoring the integrity of metallic pressure vessels [21]. Hence, based on investigation using the pencil lead fracture, it is found that the sensor VS75–V represents the outstanding sensor rather than the VS150–M since it can captured high energy. The sensor VS75–V is also suitable sensor to be used for monitoring of granite rock.

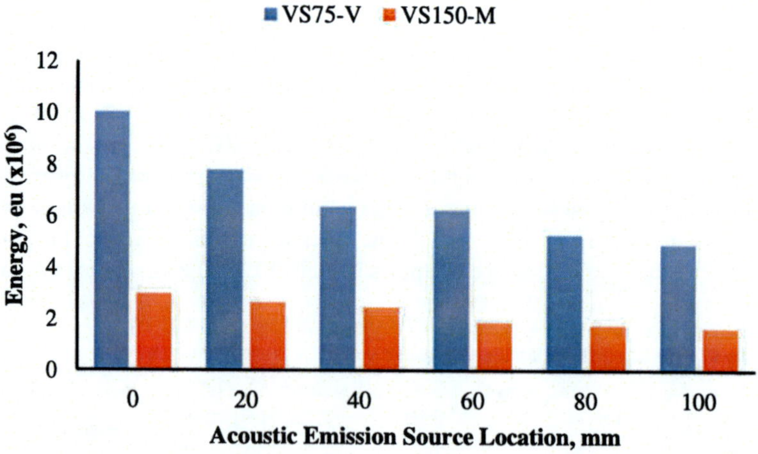

Fig. 16.8 Energy distribution with respect to acoustic emission source location for each sensor

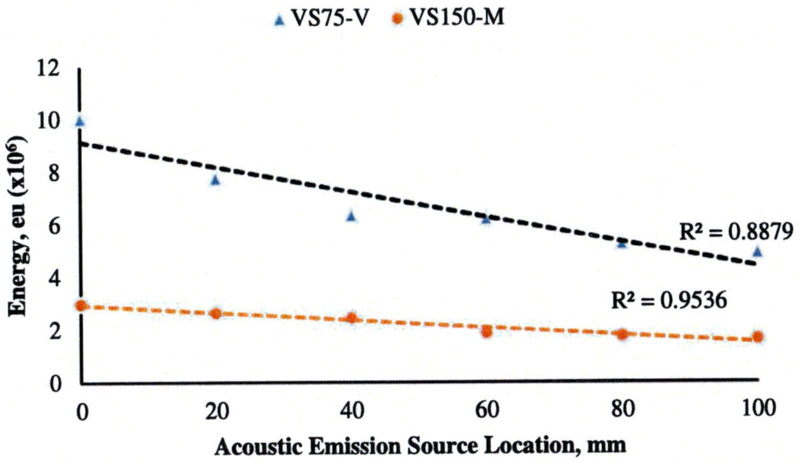

Fig. 16.9 Correlation between energy and the acoustic emission source location

Figure 16.9 shows the energy pattern with respect to the source locations. It is found that the sensor VS150–M represents a good correlation compared to sensor VS75–V. The R^2 for VS150–M is 0.9536 and for VS75–V is 0.8879. From this R^2 indicates that sensor VS150–M has good correlation between the energy and the source distance. The good correlation is only on the aspect of the energy and the location of the acoustic emission source. However, on the aspect of suitability on the application of the convenience sensor for monitoring of granite, this correlation is not a suitable indication.

16.3.2 Peak Frequency at Each Source Location

Although the energy for VS75–V is higher than that of VS150–M, the peak frequency captured by sensor VS75–V is lower than that of VS150–M as presented in Table 16.2. The peak frequency for VS75–V in average is 57.84 kHz, 44.01 kHz, 42.31 kHz, 38.81 kHz, 38.37 kHz and 37.12 kHz for acoustic emission distance of 0 mm, 20 mm, 40 mm, 60 mm, 80 mm and 100 mm, respectively. As the pencil breaks the lead, the sensors captured the peak frequency for sensor VS150–M in average is 64.78 kHz, 63.54 kHz, 63.44 kHz, 62.56 kHz, 62.38 kHz and 60.56 kHz at acoustic emission source of 0 mm, 20 mm, 40 mm, 60 mm, 80 mm and 100 mm, respectively. It can be affirmed that, as the sensor produces the high energy, it would capture the low frequency. It can be presented thorough the frequency pattern in Fig. 16.10 that sensor VS150–M produces high frequency compared to sensor VS75–V.

At the same time, the peak frequency of the granite is reduced as the location of the pencil lead break increases as illustrated in Fig. 16.10. It is because the peak frequency is related to the counts of hits due to the break of the pencil lead to the

Table 16.2 Energy for each source location for both type of sensors

Source location (mm)	VS75 - V				VS150 - M			
	Energy × 10^6 (eu)				Energy × 10^6 (eu)			
	1	2	3	Average	1	2	3	Average
0	10.2	10.3	9.6	10.03	3.08	3.14	2.76	2.99
20	8.12	7.8	7.41	7.78	3.06	2.69	2.21	2.65
40	7.02	6.1	5.99	6.37	2.53	2.58	2.28	2.46
60	8.61	5.1	5.01	6.24	1.91	1.9	1.84	1.88
80	5.7	5.2	4.92	5.27	1.7	1.78	1.84	1.77
100	4.76	5.1	4.86	4.91	1.66	1.64	1.63	1.64

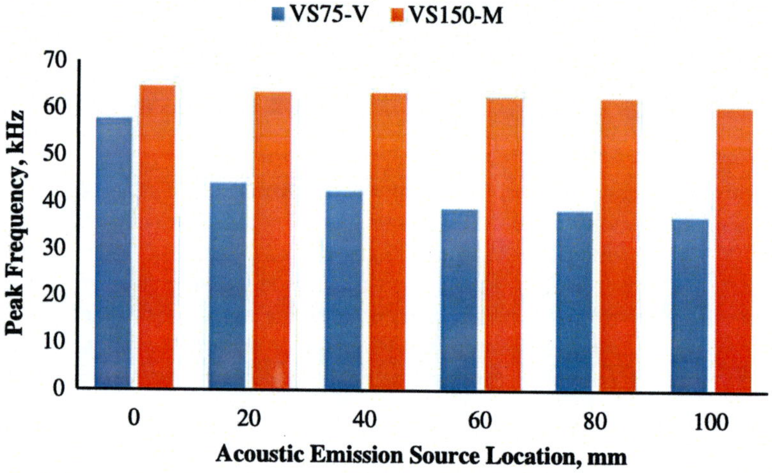

Fig. 16.10 Frequency trend with respect to acoustic emission source location for each sensor

Table 16.3 Peak frequency for each source location for both type of sensors

Source location (mm)	VS75 - V				VS150 - M			
	Peak frequency (kHz)				Peak frequency (kHz)			
	1	2	3	Average	1	2	3	Average
0	85.92	48.25	39.36	57.84	66.49	64.73	63.11	64.78
20	46.34	43.8	41.89	44.01	65.16	63.02	62.44	63.54
40	51.47	38.35	37.12	42.31	64.59	63.73	62.01	63.44
60	36.11	41.72	38.6	38.81	63.91	61.6	62.17	62.56
80	36.92	40.82	37.38	38.37	63.85	61.38	61.9	62.38
100	38.11	36.8	36.44	37.12	61.18	60.17	60.32	60.56

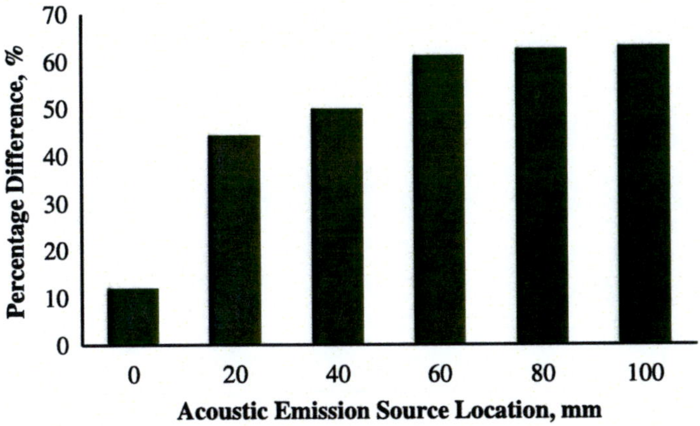

Fig. 16.11 Percentage difference of the peak frequency for sensors VS75–V and VS150–M

surface of the granite rock and the duration of the hits. As the peak frequency is dominant frequency, it becomes the important wavelength in the signal lost. The time of arrival of the wave signal to the sensor also plays a significant indication on the capturing of the frequency at the sensor [17].

The trend of the percentage difference of the frequency with respect to the source location is presented in Fig. 16.11 for both sensors. Location at 0 mm produces a lower percentage difference of 12% compared to other locations. A bigger difference can be seen for source location of 100 mm with the value of 64%. The percentage difference of the frequency for source location of 20 mm, 40 mm, 60 mm and 80 mm is 44%, 50%, 61% and 62%, respectively.

Figure 16.12 represents the correlation between peak frequency and the acoustic emission source locations. From this graph, it is found that the sensor VS150–M represents a good correlation as its R^2 is 0.9174. The same good correlation was also found for acoustic emission energy with respect to the location of the pencil lead breaks. In this context, the sensor VS75–V presents a low correlation with the R^2 of 0.7442.

16.4 Conclusions

Based on the investigation of the acoustic emission characteristic of granite for different type of sensors of VS75–V and VS150–M, several conclusions were addressed.

(i) It can be inferred that the energy for both sensors of VS75–V and VS150–M is reduced as the location of the pencil lead breaks increased. The higher energy was found in the location of 0 mm with the value of 10.03×10^6 eu and 2.99×10^6 eu for sensor VS75–V and VS150–M, respectively. The lower energy

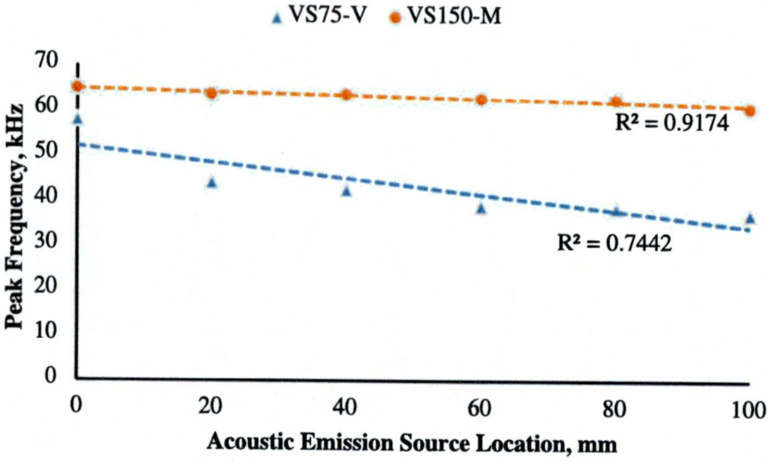

Fig. 16.12 The correlation between the peak frequency and the acoustic emission source location

was at the source location of 100 mm with the value of 4.91×10^6 eu and 1.64×10^6 eu, respectively. This phenomenon is closely related to the wave attenuation effect.

(ii) The peak frequency is also reduced as the acoustic emission location increased. The highest peak frequency in average was found at the source location of 0 mm for both sensors with the value of 57.84 kHz and 64.78 kHz for VS75–V and VS150–M, respectively.

(iii) The average acoustic emission energy for sensor VS75–V is higher than that of VS150–M. Meanwhile, it is contrast for the peak frequency where the VS75–V represents the lower value than the sensor VS150–M.

(iv) The relationship between the energy, frequency and location of pencil lead breaks indicates that the sensor VS150–M represents a good correlation compared to the VS75–V.

Acknowledgements The authors would like to express their thanks to Universiti Teknologi MARA, Cawangan Pulau Pinang and Universiti Kebangsaan Malaysia (Gran Impak Penyelidikan: DIP-2019-015) for providing financial support towards the publication of this manuscript.

References

1. ASTM E976–15 (2015) Standard guide for determining the reproducibility of acoustic emission sensor response. ASTM International, West Conshohocken, PA
2. Noorsuhada MN (2016) An overview on fatigue damage assessment of reinforced concrete structures with the aid of acoustic emission technique. Construct Build Mater

3. Md Nor N, Ibrahim A, Muhamad Bunnori N, Mohd Saman H, Mat Saliah SN, Shahidan S (2014) Diagnostic of fatigue damage severity on reinforced concrete beam using acoustic emission technique. Eng Failure Anal 41:1–9
4. Soffian Noor MS, Md Nor N, Abd Rahman N, Abdullah S, Tahir MS (2021) Evaluation of severely damaged reinforced concrete beam repaired with epoxy injection using acoustic emission technique. Theor Appl Fract Mech 112 (2021)
5. Zhang X, Li B (2020) Damage characteristics and assessment of corroded RC beam-column joint under cyclic loading based on acoustic emission monitoring. Eng Struct 205
6. Md Nor N, Ibrahim A, Muhamad Bunnori N, Mohd Saman H (2013) Acoustic emission signal for fatigue crack classification on reinforced concrete beam. Construct Build Mater 49:583–590
7. Mayr SI, Stanchits S, Langenbruch C, Dresen G, Shapiro SA (2011) Acoustic emission induced by pore-pressure changes in sandstone samples. Geophysics 76(3)
8. Wang Y, Deng H, Deng Y, Chen K, He J (2021) Study on crack dynamic evolution and damage-fracture mechanism of rock with pre-existing cracks based on acoustic emission location. J Petrol Sci Eng 201
9. Zhang Z, Liu X, Zhang Y, Qin X, Khan M (2021) Comparative study on fracture characteristics of coal and rock samples based on acoustic emission technology. Theor Appl Fract Mech 111 (2021)
10. Xing Y, Zhang G, Luo T, Jiang Y, Ning S (2019) Hydraulic fracturing in high-temperature granite characterized by acoustic emission. J Petrol Sci Eng 178:475–484
11. Winnera RA, Lub G, Prioul R, Aidagulov G, Bunger AP (2018) Acoustic emission and kinetic fracture theory for time-dependent breakage of granite. Eng Fract Mech 199:101–113
12. Ge Z, Sun Q (2018) Acoustic emission (AE) characteristics of granite after heating and cooling cycles. Eng Fract Mech 200:418–429
13. Guo TY, Wong LNY (2020) Microcraking behaviour of three granites under mode I loading: Insights from acoustic emission. Eng Geol 278
14. Kallimogiannis V, Saroglou H, Tsiambaos G (2017) Study of cracking process in granite. Proc Eng 191:1108–1116
15. Zhou XP, Zhang JZ, Qian QH, Niu Y (2019) Experimental investigation of progressive cracking processes in granite under uniaxial loading using digital imaging and AE techniques. J Struct Geol 126:129–145
16. ISRM (1981) International Society for Rock Mechanic. Pergamon
17. Md Nor N, Muhamad Bunnori N, Ibrahim A, Mohd H, Shahidan S, Mat Saliah SN (2013) An Investigation of an Acoustic Wave Velocity in a Reinforced Concrete Beam from Out-of Plane and in Plane Sources. Modeling and Measurement Methods for Acoustic Waves and for Acoustic Microdevices. Intech Open
18. ASTM E976 (2010) Standard guide for determining the reproducibility of acoustic emission sensor response. ASTM Int 5
19. Dong L, Wang J, Li X, Peng K (2018) Dynamic stability analysis of rockmass: a review. Adv Civil Eng 2018:1–22
20. He MC, Miao JL, Feng JL (2010) Rock burst process of limestone and its acoustic emission characteristics under true-triaxial unloading conditions. Int J Rock Mech Min Sci 47(2):286–298
21. Vallen Systeme (2009) Acoustic emission system AMSY-5 system description

Chapter 17
Seismic Excitations of Prefabricated Reinforced Concrete Building Performance: Case Study of Kolej Delima Universiti Teknologi MARA, Cawangan Pulau Pinang

Mohd Samsudin Abdul Hamid, Norul Mas Diyana Ahmad, Kay Dora Abd Ghani, Siti Hafizan Hassan, and Nurulzatushima Abdul Karim

Abstract Structures can deflect under large earthquake excitation. Until now there was still lack of study to model non-seismic designed reinforced concrete structure under this conditions. This study investigated the load versus displacement and determined the building performance in terms of work and energy behaviour response under different seismic excitation data for different earthquake excitations. A ten storey building were modelled under four past earthquake records: Kunak, Bukit Tinggi, El Centro North–South and Pacoima Dam. When subjected to Pacoima Dam earthquake, the building exhibits largest pushing and pulling load which are 16,610 and 12,330 kN. Under Pacoima Dam earthquake also this building shows largest displacement compare to other three cases which are 518.4 mm for pushing displacement and 624.4 mm for pulling displacement. Apart from that, energy dissipated was also higher hence reduces the strength of the structure. This situation occurs due to the larger peak ground acceleration of Pacoima Dam earthquake which is 1.19 g. In addition, under Pacoima Dam earthquake, Kolej Delima UiTMPP would undergo partial collapse hence leads to severe damage to the structure. In overall, it was found that the building was sustained under low to medium excitations; however, it will collapse if under high excitations. This study was found important to predict the maximum behaviour of non-designed earthquake reinforced concrete structure.

Keywords Excitation · Hysteresis loop · Ruaumoko 2D · Seismic · Work and energy behaviour

---

M. S. A. Hamid (✉) · K. D. A. Ghani · S. H. Hassan · N. A. Karim
School of Civil Engineering, Universiti Teknologi MARA, Cawangan Pulau Pinang, 13500 Permatang Pauh, Pulau Pinang, Malaysia
e-mail: samsudin85@uitm.edu.my

N. M. D. Ahmad
Mentrol Global Holdings Sdn Bhd, No. 4 Jalan Penggawa U1/75, Taman Perindustrian Batu Tiga, Shah Alam 40150, Selangor, Malaysia

S. Abdullah et al. (eds.), *Structural Integrity Cases in Mechanical and Civil Engineering*, Structural Integrity 23, https://doi.org/10.1007/978-3-030-85646-5_17

17.1 Introduction

Malaysia is very close to areas that have experienced strong earthquakes, including Sumatra and the Andaman Sea, while Sabah and Sarawak are located close to the earthquake zone of South Philippines and North Sulawesi. Therefore, the odds of an earthquake striking Peninsula Malaysia cannot be ruled out. Based on some of the findings, strong earthquakes occur at zones where tectonic plates collide at the Andaman Sea, the South China Sea, the Sulu Sea and the Sulawesi Sea. When a strong earthquake strives off the west coast of northern Sumatra on 2012 with a magnitude of 9.1 Richter scale, it can unleash a tsunami that can end up at the coastlines of Perlis, Kedah, Penang, Perak, Selangor, Sabah and Sarawak. The earthquake in Turkey, Taiwan, India, Iran, China, New Zealand and Japan has caused considerable loss of life and extensive damage to structures [1]. Hence, early precaution should be implementing to reduce the loss of life and extensive damage of the structure due to the impact of the earthquake in the future. Past records showed that six (6) earthquakes had been occurred in Sabah which is at Tambunan, Kota Marudu, Kudat, Beluran, Kunak and Keningau, and two (2) earthquakes happened in Belaga, Sarawak. However, an earthquake that is believed to be the strongest earthquake so far was in year 1976 in Lahad Datu, Sabah, with a magnitude of 5.8 on the Richter scale. Based on the above statement, it is possible that in the future the impact of an earthquake with larger magnitude can occur in Malaysia.

Industrialized Building Systems (IBS) is also defined as a construction technique where the components are manufactured in a controlled environment either on or off site and prefabricated reinforced concrete is also known as one of the system use in IBS structure [2]. This study was developing to examine the building using IBS under different earthquake excitation and analysed after impact of the building due to excitations in term of displacement, hysteresis loop, and mode shape.

17.2 Literature Review

The impact of an earthquake is controlled primarily by population exposed at each shaking intensity level and how vulnerable that population is to building damage at each intensity level [2]. In addition, it is also due to the distribution and severity of shaking which is caused by the degree of seismic resistance of the local building. Furthermore, it was found that based on the hypothesis of sea-floor spreading during the past few decades, plate tectonics was developed. The six main plate tectonics in the world which are Pacific Plate, American Plate, African Plate, Antarctic Plate, India Plate and Eurasian Plate. Besides that, an earthquake is a term use to describe radiated seismic energy caused by a slip or by magmatic activity or by volcanic or other sudden stress changes in the Earth. In addition, he also mentions that earthquake is a sudden slip on a fault as well as the resulting ground shaking [3].

Other than that study has stated the seismic performance parameters such as lateral strength capacity, stiffness, ductility and equivalent viscous damping were evaluated using the measured hysteresis loops from experimental work. The finding shows that in designing structure due to earthquake, it is important to pay special attention to joint detailing of precast structures components while constructing the precast buildings. The design concepts and detailing of the jointing precast structural components contribute significantly to the overall seismic performance of precast buildings under earthquakes excitations. A lot of structural damages occurred during earthquakes due to not using the current seismic code of practice and poor detailing especially at the jointing system [4].

The excitation angle or angle of incidence is the angle in which the horizontal seismic components are applied with respect to the principal structural axes during a time history analysis that reflected the performance of the structure under seismic excitations [5]. Furthermore, result shows the structure experience pounding during strong earthquakes due to insufficient or no separation distance between them from its investigation. It was found that the critical accelerations in earthquake excitations being computed by internal shear force at the base of the structure at each time step is maximized under constraints on ground motion. A numerical example presents to show the efficiency of critical excitation method in determining the maximum internal shear force and base moment under variety of constraints [6].

The time-history analysis is another option; how the seismic response can be calculated. A model for this calculation is imposed by up to three different accelerograms which describe the behaviour for each direction. Usually in this method it was used as non-linear analysis. In order to compare this method with two linear methods it is necessary to use synthetic accelerograms. These accelerograms describe the response spectrum. Eurocode 8 defines, that for the determination of the design seismic action, an average value from at least seven non-linear calculations has to be used [7]. By fever calculations, the worst result has to be chosen. Calculation of motion equations is the main principle of this method, and it is calculated by the modal analysis or by the direct integration. Modal analysis allows the calculation with only linear material behaviour. The direct integration is possible to use material non-linearities.

Study had done and point out that the fundamental requirement for seismic performance was to meet with 'no collapse' performance level as well as 'damage limitation' performance level. 'No collapse' performance level required that the structure retains its full vertical load bearing capacity after an earthquake. In order to protect life, there should also be sufficient residual lateral strength and stiffness during strong aftershocks. 'Damage limitation' performance level required that the cost of damage and other limitations should not be too high compared to the total cost of the structure [8]. Referred to Li and Kurkani [9], seismic performance of wide beam-column joint performance was dominated by the torsional behaviour of transverse beams. While Cardone et al. [10] believe that the seismic respond of earthquake does not change significantly although being governed by the effective stiffness and effective damping of the isolation system at the design displacement typically associated to shear strains of the order of 100–150%.

According to Construction Industry Development Board (CIDB), IBS are systems in which structural components are prefabricated or manufactured in a factory on or off worksite, then transported and assembled into a structure with minimal additional site works [11]. Study by Hamid et al. [12] stated that it would be an admirable move to use IBS techniques for the country as the conventional on-site construction methods have long been criticised for imposing rigorous labour, human health and safety risk as well as causing significant environmental destruction. Furthermore, IBS projects can reduce the use of concrete and reinforce for the structural components; it is more profitable compared to the non-IBS projects [13]. IBS technique and sustainability in construction are very beneficial to the people particularly and the country as a whole. In agreeing to Hamid et al. [12] has mentions that it can happen to countries like Malaysia where due to the shortage of labours, the construction industry is forced to import huge number of unskilled workers, who could unfortunately cause some problems such as delays in construction progress or completion of the projects, incompetency, wastages and social problems.

17.2.1 Hysteresis Loop and Energy Behaviour of the Reinforced Concrete Structure

Based on the research that had been done by Segupta and Li [14], they had found that the structural wall with limited transverse reinforcement possesses poor energy dissipation characteristics when subjected to repeated cyclic deformation. Hence, it will result in pinched hysteresis loop with possible sudden loss of lateral capacity as well as significant degradation. At lower displacement levels, the loops became smaller indicating lower energy dissipation while at higher displacement level, the loops became bigger indicating higher energy dissipation. It is also mention that the peak load and stiffness of the hysteresis loop is affected by analytical parameters, stiffness ratio and system natural frequency with their changing magnitudes. Hysteresis stiffness, smoothness of the hysteresis loop and the sharpness of the transition from the initial to asymptotic slope will be influenced by a few parameters. Hysteresis loop obtained from the quasi-static is analysed to determine effective horizontal stiffness and equivalent viscous damping.

Furthermore, the strain energy depends on the displacement of all degree of freedom of the structure as well as the values of all forces on the structure. They also added that near-fault ground motion has unique properties. It causes structure to dissipate earthquake input energy in a few large deformation cycles compares to far-fault ground motion. Structural damage cause by far-fault earthquake was contributed by low-cycle fatigue effects and maximum deformation. On the contrast, most of the damage caused by near-fault earthquake was the result due to few plastic cycles and immediate energy demand associated with severe pulse effects [15]. Figure 17.1 shows the hysteresis energy demand in largest yield excursion for far-fault earthquake motion and near-fault earthquake motion.

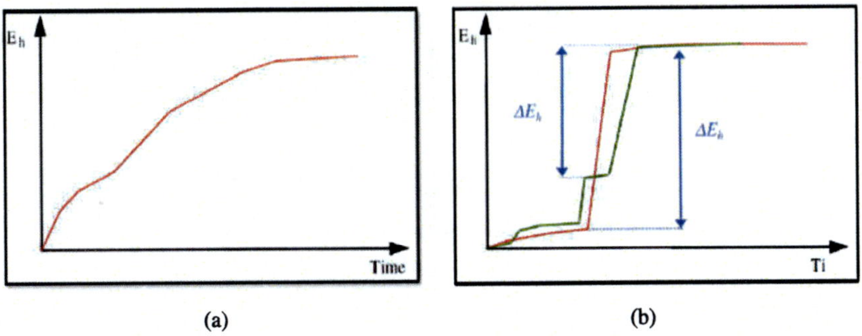

Fig. 17.1 **a** Hysteresis energy demand in largest yield excursion for far-fault motion. **b** Hysteresis energy demand in largest yield excursion for near-fault motion [15]

17.3 Methodology

Figure 17.2 shows the overall flow of analysis of Kolej Delima UiTMPP under seismic excitations by using Ruaumoko 2D. The reinforced concrete structure was designed by adopting British Standard as a code of practice because it followed the original design for Kolej Delima UiTMPP. The process of seismic excitation implementation for the building was applied after reinforced concrete design was completed because it required the complete designed for dynaplot analysis. Dynaplot analysis was phased to determine the performance of building under seismic response based on historical earthquake record. Furthermore, four different cases of previous earthquake excitations chosen in this study was based on the different region and frequency which are low, medium and strong. Two main results were the main focused in this study which are hysteresis loop relationship and work and energy behaviour respond.

The reinforced concrete structural designed has been done for Kolej Delima UiTMPP to BS8110 as code of practice for the designing process. The structural element parameters for the building was following the full scale dimension of the building that has been constructed. The designing process was took placed in this study as redesign process to reconfirm the structural performance under normal ultimate load for British Standard 8110. Based on the design, further analysis has been used to be analysed under Ruaumoko 2D software under four (4) different earthquake records with different excitations level between low to high. The past earthquake records were choosing based on the earthquake zone and level of earthquake. Therefore, four cases that has been choose for this study were Kunak Sabah, Bukit Tinggi Pahang, El-Centro California and Pacoima Dam California.

Kolej Delima UiTMPP was remodel in ESTEEM 8 into model prototype in order to produce its reinforced concrete structural system. Kolej Delima has identical floor plan from the first floor until the ten floor of the building. This building was using spun pile as foundation system with a diameter of 500 mm. As for the beams, the dimensions used for the elements in this building are as stated in Table 17.1. Figure 17.3

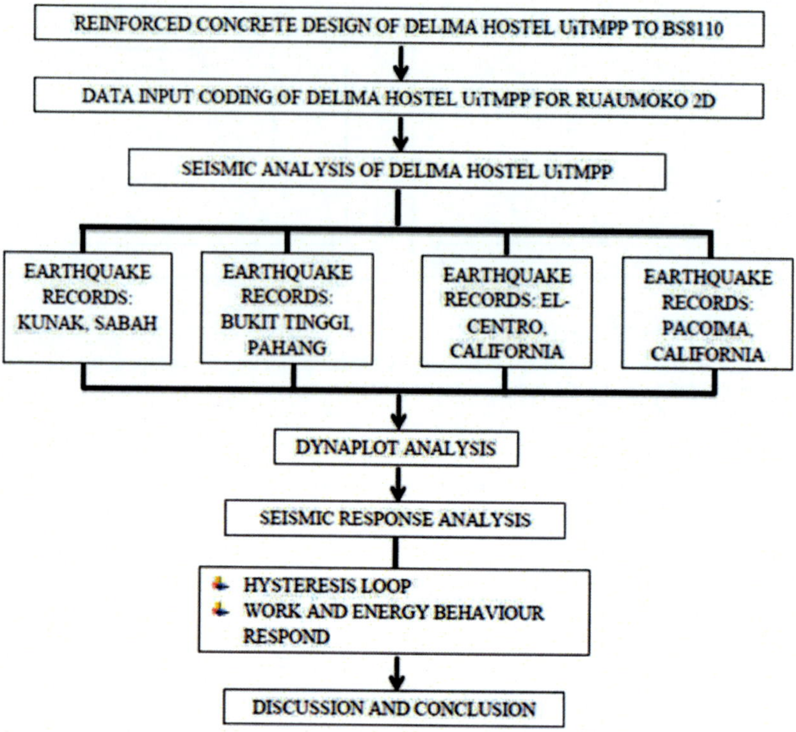

Fig. 17.2 Research methodology overall flow chart

Table 17.1 Structural elements dimensions

Elements	Size
Slab	200 mm
Beam	200 mm × 500 mm
Column	400 mm × 500 mm
	200 mm × 500 mm
	200 mm × 200 mm
Foundation	500 mm
Floor height	3000 mm

shows the picture of the existing Kolej Delima UiTMPP that has been choose as a case study for this research, and Fig. 17.4 shows the 3D Modelling by using ESTEEM 8 for the purpose of reinforced concrete structural design.

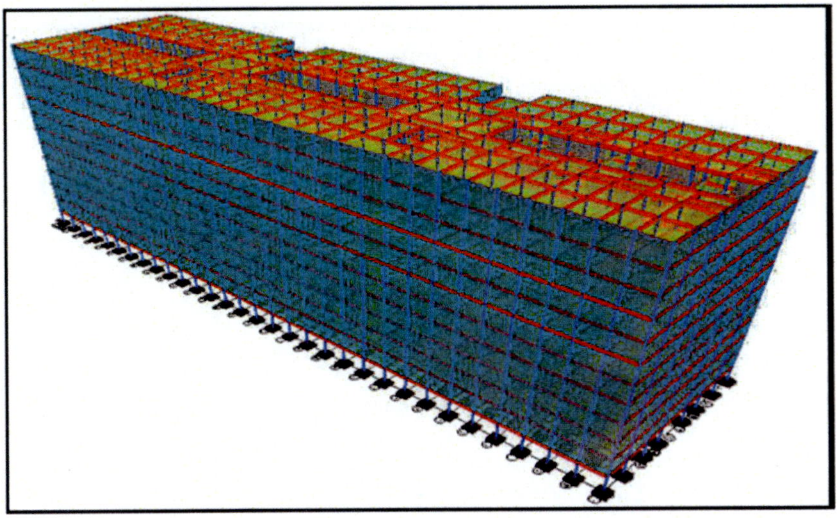

Fig. 17.3 Front view of Kolej Delima UiTMPP which constructed using IBS

Fig. 17.4 3D view model prototype of Kolej Delima UiTMPP using ESTEEM 8

17.4 Results and Discussion

The results in this study was observed through the seismic performance of the reinforced concrete building under different seismic excitation and hysteresis loop that show load over displacement performance of the building. From the Ruaumoko 2D software, a sub programme call Dynaplot will be run to determine hysteresis loop for this structure. Apart from that, other data also can be generated from this programming such as excitation graph, displacement versus time graph, work or energy versus time graph as well as mode shape of free vibration. This process was done to take out all the data from seismic analysis that had been done from analysis Ruaumoko 2D.

17.4.1 Seismic Performance of Delima Hostel UiTMPP Under Earthquake Excitations

Seismic analysis was done to observe the performance of Kolej Delima UiTMPP under different seismic respond. Past earthquake records were used for the analysis which is Kunak, Bukit Tinggi, El-Centro and Pacoima. Earthquake at Kunak, Sabah, occurred on 28 May 2012 with the PGA of 0.0065 g while earthquake at Bukit Tinggi, Pahang, with a PGA of 0.000464 g occurred on 4 December 2007. Apart from that, earthquake that happened on 18 May 1940 at El-Centro, California, has a PGA of 0.348 g and earthquake that occurred at Pacoima, California, on 13 January 2001 has a PGA of 1.19 g. The specific detail information is shown in Table 17.2. These past earthquake records were selected due to its different seismic region. Kunak and Bukit Tinggi were situated in low seismic region, while El-Centro and Pacoima were located at medium seismic region. The main reason that the past earthquake record in high seismic region was not applied in this study because of the possibility for Malaysia to encounter with such larger magnitude of the earthquake is unlikely to happen.

Table 17.3 shows a summary of the result for all the earthquake cases. Based on Table 17.2 there were no pushing load for Kunak and Bukit Tinggi cases at all nodes due to the peak ground acceleration for both instances were too low which recorded 0.0065 g and 0.000464 g respectively. Despite that, Kunak and Bukit Tinggi exhibit load at pulling direction but the value was too small. Pacoima Dam earthquake contributes the largest pushing and pulling load at all nodes compare to the other

Table 17.2 Past earthquake record adopted

Earthquake	PGA (g)	Date of event	Magnitude (Scale Richter)	Category based on frequency[a]
Kunak, Sabah	0.0065	28 May 2012	4.46	Very low
Bukit Tinggi, Pahang	0.000464	4 December 2007	3.0	Very low
El-Centro, California	0.348	18 May 1940	7.1	Major
Pacoima Dam, California	1.19	13 January 2001	6.6	Strong

[a]Categories are based on observation of the United Stated Geological Survey (USGS) [16]

Table 17.3 Summary of the result analysis

Earthquake	Node	Pushing (+ve)		Pulling (−ve)		Energy	
		Load (kN)	Displacement (mm)	Load (kN)	Displacement (mm)	Min (kJ)	Max (kJ)
Kunak (KUNAK.EQF)	353	0	0.2004	0.06979	0.4175	(1) 1.337E-07 (2) 1.427E-07 (3) 0. 2532 (4) 0.5064	(1) 1.337E-07 (2) 1.427E-07 (3) 0. 2532 (4) 0.5064
	193	0	0.2653	0.1419	0.423		
	33	0	0.08117	0.1419	0.119		
Bukit Tinggi (BUKITTINGGI.EQF)	353	0	0.2004	0.06979	0.6723	(1) 3.274E-07 (2) 4.026E-07 (3) 0.2532 (4) 0.5064	(1) 3.274E-07 (2) 4.026E-07 (3) 0.2532 (4) 0.5064
	193	0	0.2653	0.1419	0.6589		
	33	0	0.08117	0.1419	0.1832		
El-Centro (EL40NSC.EQB)	353	1872	212	2431	464.8	(1) 0.0 (2) 0.0 (3) 0.2532 (4) 0.5064	(1) 35.021 (2) 35.561 (3) 36.132 (4) 36.385
	193	3804	91.07	4941	350.7		
	33	3805	29.77	4942	109		
Pacoima Dam (PACMSW.EQB)	353	8168	518.4	6065	624.4	(1) 0.0 (2) 0.0 (3) 0.2532 (4) 0.5064	(1) 314.630 (2) 317.830 (3) 322.530 (4) 322.780
	193	16,600	464.7	12,330	476.8		
	33	16,610	269.5	12,330	244.5		

three cases. This situation occurs because Pacoima Dam earthquake has bigger peak ground acceleration which is 1.19 g compare to other earthquake cases. In addition, Pacoima Dam earthquake also has the largest pushing and pulling displacement due to its larger peak ground acceleration value.

It is important that all the structural component was securely connected together because when push and pull are happening against one another during earthquake, the connection was strong enough to transfer the earthquake force hence maintain the integrity of the structure. In the energy column, number 1, 2, 3 and 4 represent relative kinetic energy, kinetic energy plus damping work, kinetic energy plus damping work plus strain energy and applied work done on the structure respectively. As the energy increase, the strength of the structure will decrease. This happens because all the strength in the structure elements was used to dissipate the energy produce from the peak ground acceleration of an earthquake.

The category was determined based on the value of peak ground acceleration for each earthquake cases and was classified based on observation of the United Stated Geological Survey (USGS). The most critical structural element was found in column especially at the beam-column joint. That area was experienced the worst damage during the earthquake events. Based on the analysis that had been done, there would be no damage to the Kolej Delima UiTMPP building when this structure was analysing under Kunak and Bukit Tinggi earthquake due to the low peak ground acceleration values. But, for Pacoima Dam case, Kolej Delima UiTMPP building would lead to partially collapsed hence the structure would experience with severe damage. In addition, the damage causes by the earthquake was solely based on the PGA value and not the magnitude of the earthquake. This is because, PGA produce a vibration that shake the structure hence producing the damages to the structure.

Table 17.3 shows the overall summary of the pushing and pulling behaviour that reflected the values of energy dissipation of the structural elements during the simulation work based on different earthquake excitation. Hysteresis curves were produced during the analysis which were used to determine the strength and stiffness of the specimens or models and evaluated the general behaviour of the members [17]. The hysteresis loop curves were representing by the pushing and pulling load values shown in the table below with positive and negative values and reflected the displacement values of the structural systems. Higher acceleration of the past earthquake record apply to the specimen produced highest displacement of the structural system. Therefore, the results found that Pacoima Dam produced highest displacement of the structure that cause severe damage to the structural system from top floor to the lower floor of Kolej Delima UiTMPP. Therefore, it is important to strengthen the joint system of the building especial on slab to beam and beam to column connection by implementing seismic design to Eurocode 8 in order to avoid damage of the building during the vibration cause by earthquake. The location that produced higher displacement was found at the top floor due to sway behaviour which is excitation or vibration after effect from the earthquake. It shows the stiffness of the structural elements reducing during the earthquake that may lead to cause failure of other part of the buildings.

The energy dissipation results show in Table 17.3 were also reflected the pushing and pulling behaviour of the structure under seismic performance. In overall analysis of the energy dissipation, increasing of energy dissipation from the structural elements reflected the overall performance under seismic excitation where severe damage took place when the peak ground acceleration was increasing. Furthermore, based on the simulation work done the performance of the building under local history of earthquake which is Kunak and Bukit Tinngi, the building remains save with minor failure that did not cause functional problems of the main structural elements. But, under El-Centro and Pacoima Dam earthquake excitation the energy dissipated from the structural elements shows that the performance of the building decreasing where it reflected the decreasing of the structural strengths. Besides that, it was found that the critical location of the structural failure that produced red zone with highest displacement and energy dissipation were at the connection joint of the elements node 33, node 193 and node 353 where it is at the beam column connection joint.

Figure 17.5 shows the hysteresis loop of Kolej Delima UiTMPP under 4 different earthquake excitation. Under Kunak Earthquake it was found that the pulling load recorded was 0.1419 kN with the maximum pushing displacement of 0.2653 mm and pulling displacement of 0.423 mm. Under Kunak Earthquake, the building was not affected by its excitation because the maximum displacement was found at level

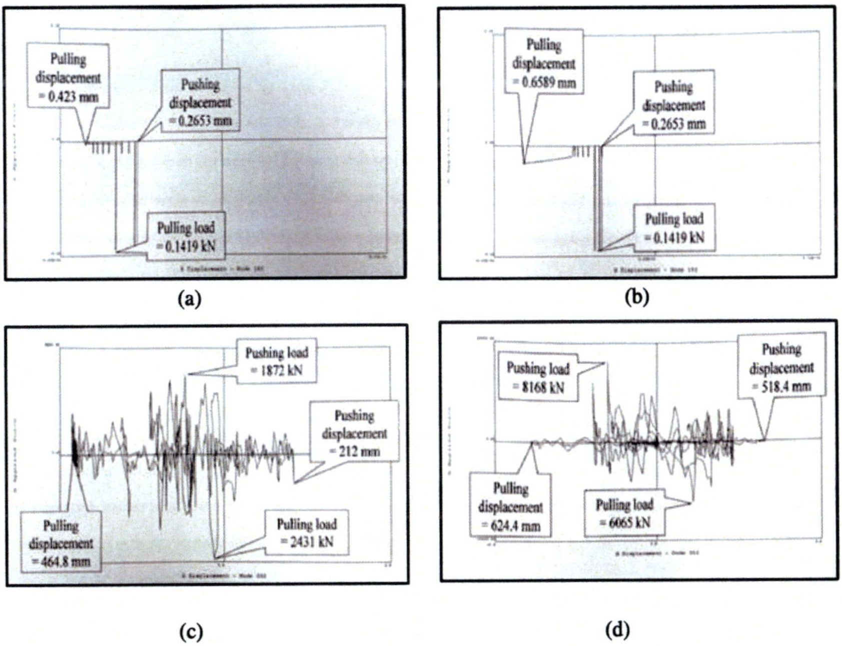

Fig. 17.5 Hysteresis behaviour of Kolej Delima UiTMPP under different seismic excitations. **a** Kunak Earthquake; **b** Bukit Tinggi Earthquake; **c** El-Centro Earthquake; **d** Pacoima Dam Earthquake

6 of the building while the other floor produced very low displacement that not cause any failure of the structural elements. Furthermore, under Bukit Tinggi earthquake excitation, the pulling load recorded was same as Kunak earthquake which is 0.1419 kN but the maximum pushing displacement was 0.2653 mm and pulling displacement was 0.6589 mm. Besides that, the performance under El-Centro and Pacoima produce large displacement under pulling and pushing load because the earthquake excitation was medium to high.

Furthermore, in overall the structure was inversely proportional to the position of the node. The higher the node, the smaller the load on the structure but for displacement it was directly proportional to the positions of the nodes. Displacement value for a node was increasing at the position of the node increased. In overall performance under seismic, it was found that the pushing and pulling load for earthquake under El-Centro and Pacoima Dam were higher compared to the Kunak and Bukit Tinggi Earthquake. This is because earthquakes from El-Centro and Pacoima Dam have larger peak ground acceleration values which is about 0.348 g to 1.19 g under medium seismic region of earthquake excitation. Based on the results, different failure behaviour were occurs under two seismic regions with low and medium seismic region. Under low seismic excitations, there were no failure occur on the structural elements but under medium seismic regions the building face minor failure and some major failure at the structural elements because the building was genuinely not design under seismic behaviour. The type of failures of the structure that was normally occurs were soft story failure, major crack, reduction of strength, bending of reinforcement, buckling of reinforcement and others. All those failures should be solving in order to avoid the lifetime of the building is below the design lifetime that was normally applied for 50 years or 100 years.

17.4.2 Seismic Performance of Delima Hostel UiTMPP Under Earthquake Excitations

Figure 17.6 shows the overall performance of Kolej Delima UiTMPP under four different earthquake excitation that has been choose for this study. It was found that small sway mode occurs at the building under Kunak and Bukit Tinggi excitation and almost cannot be seen by using naked eyes because of the Peak Ground Acceleration of both earthquakes below 0.01 g with very small excitation recorded as shown in Fig. 17.6a, b. Therefore, under both condition of earthquake Kolej Delima UiTMPP structural system was not failed and affected by this earthquake. It is considered as low region earthquake.

Under El-Centro Earthquake excitations, the building found side sway at lower story as shown in Fig. 17.6c. The behaviour of mode shape showing the structural movement and displacement under this excitation although the Peak Ground Acceleration is only 0.348 g. El-Centro Earthquake was considered as medium earthquake. Therefore, it can be concluded that under medium earthquake excitations, Kolej

Fig. 17.6 Mode shape under earthquake excitations **a** Kunak earthquake; **b** Bukit Tinggi earthquake; **c** El-Centro earthquake; **d** Pacoima Dam earthquake

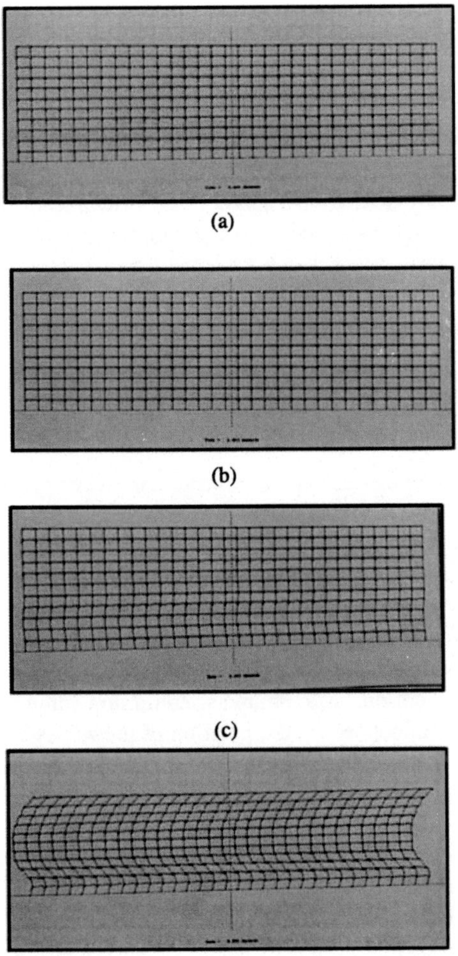

(a) (b) (c) (d)

Delima UiTMPP structure was found affected with failure of soft story especially at column-beam joint. This type of failure is one of the common type of structural failure due to an earthquake. Based on study by [18], it was stated that in some reinforced concrete buildings, especially at the ground floor, walls may not be continuous along to height of building for architectural, functional, and commercial reasons. While ground floor generally encloses with glass window instead of brick infill walls, partition walls are constructed above from this storey for separating rooms for the residential usage. This situation causes brittle failures at the end of the columns. In mid-rise reinforced concrete buildings, the most common failure mode is soft-storey mechanism, particularly at the first storey. Failure can be concentrated at any storey called as weak storey in which the lateral strength changes suddenly between adjacent stories due to lack of or removing of partition walls or decreasing of cross section of columns.

Based on above statement, it was proved that the structural performance of Kolej Delima UiTMPP under Pacoima Dam Earthquake excitations produced severe damage of the structural system as shown in Fig. 17.6d. Uneven displacement with sway effect of the building for each floor shows large displacement of the structure due to high excitations of this earthquake. Pacoima Dam earthquake produces 1.190 g Peak Ground Acceleration where it is considered as high region earthquake that is mostly more than 6.5 Richter Scale. Overall, Kolej Delima UiTMPP was not able to sustain under medium earthquake excitations that is about 3.5 Richter Scale.

17.5 Conclusion

In a nutshell, an earthquake can occur at any place in the world. The differences between all the places are that whether the magnitude and peak ground acceleration of earthquake for each case are low, medium or high. Apart from that, the magnitude and peak ground acceleration is also influenced by the location of seismic region of the earthquake for each country in the world. For the conclusion, all the objectives for this research were achieved. The hysteresis loops were generated at selected nodes to see the response of the building when subjected to different earthquake excitations. Maximum and minimum value of force and displacement for each node is influenced by the position of the nodes and recorded 624 mm displacement and 8168 kN under the largest earthquake excitation. Force of a structure is inversely proportional to the position of the node. The higher the node, the smaller the force of the structure, but for displacement, it is directly proportional to the position of the nodes. Displacement value for a node will be an increase as the position of the node increase. In overall, the larger the peak ground acceleration (PGA) of an earthquake, the larger the value of force and displacement of the structure. As for work and energy behaviour response, all four parameters that were analysed with different seismic region data of earthquake give a logic result as mention in Sect. 4. During analysis process using Ruaumoko 2D with all four cases, the structure does not experience with total collapsed but may have major damage to the elements of the structure. This is because, the higher the energy produces, the lower the strength of the structure hence leads to the damage of the structure. As the earthquake excitation is higher, the energy dissipated is also higher. Hence, it can be concluded that peak ground acceleration is directly proportional to the energy dissipated. Since the construction of the wall of Kolej Delima UiTMPP is using shear wall in the state of normal brick wall, this building can resist more energy during earthquake events. Shear wall has the ability to absorb vibration due to earthquake compared to normal brick wall. Shear wall is constructed with steel reinforcement in the wall which when connected with the beam gives a better connection between these two elements. Overall, it is crucial for a designer to add a factor of safety for earthquake during designing process to enhance the performance of building during an earthquake event.

Acknowledgements We would like to acknowledge the Universiti Teknologi MARA, Cawangan Pulau Pinang for funding this research publication.

References

1. Mortezaei A, Ronagh HR (2012) Sharif University of Technology plastic hinge length of FRP strengthened reinforced concrete columns subjected to both far-fault and near-fault ground motions. Scientia Iranica 19(6):1365–1378
2. Wald DJ, Jaiswal KS, Asce AM, Marano KD, Bausch D (2011) Earthquake impact scale. Nat Hazard Rev 12(3):125–139
3. Bangash MYH (2011) Earthquake resistant building: dynamic analyses, numerical computations, codified methods. Case studies and examples. Springer, Berlin Heidelberg, German
4. Hamid NH, Azmi IF, Awang H (2019) Seismic response of precast and IBS buildings under earthquake attacks. In: IOP conference series: earth and environmental science, volume 365, international conference on agricultural technology, engineering and environmental sciences 21–22 August 2019, Banda Aceh, Indonesia
5. Polycarpou P, Papaloizou L, Komodromos P, Charmpis DC (2015) Effect of the seismic excitation angle on the dynamic response of adjacent buildings during pounding. Earthq Struct 8(5):1127–1146
6. Kamgar R, Rahgozar R (2015) Determination of critical excitation in seismic analysis of structures. Earthq Struct 9(4)
7. EN 1998–1 (2004) Eurocode 8: design of structures for earthquake resistance: Part 1: general rules, seismic actions and rules for buildings
8. Booth E, Lubkowski Z (2009) Basic seismic design principles for buildings. Spon Press, 270 Madison Ave, New York
9. Li B, Kulkarni SA (2010) Seismic behavior of reinforced concrete exterior wide beam-column joints. J Struct Eng 136(1):26–36
10. Cardone D, Gesualdi G, Nigro D (2011) Effects of air temperature on the cyclic behavior of elastomeric seismic isolators. Bull Earthq Eng 9:1227–1255
11. Construction Industry Development Board (2005) Manual of IBS Content Scoring System (IBS Score), IBS Digest (January–March 2005). Construction Technology Unit, Technology Development Division, CIDB Malaysia
12. Hamid ZA, Kamar KAM, Zain MZM, Ghani MK, Rahim AHA (2008) Industrialised Building System (IBS) in Malaysia. Curr State R&D Initiatives Malaysian Construct Res J (MCRJ) 1(2):1–11
13. Yunus R, Yang J (2011) Sustainability criteria for industrialised building systems (IBS) in Malaysia. Proc Eng 14:1590–1598
14. Sengupta P, Li B (2014) Hysteresis behavior of reinforced concrete walls. J Struct Eng 140:1–18
15. Amin S, Beheshti-aval SB, Ala M (2014) New energy-based approach to predict seismic demands of steel moment resisting frames subjected to near-fault ground motions. Eng Struct 72:182–192
16. Gupta BL, Gupta A (2010) Principles of earthquake resistant design of structures and tsunami. Standard Publishers Distributors, New Delhi, India
17. Kucukgoncu H, Altun F (2020) The seismic behaviour of RC exterior shear walls used for strengthening of intact and damaged. Bull Earthq Eng 18:3683–3709
18. Yon B, Sayin E, Onat O (2017) Earthquakes and structural damages. In: Earthquakes—tectonics, hazard and risk mitigation. Intech

Chapter 18
The Effects of Peat Modified Asphalt Binder Concentrations on Viscoelastic Properties

Ahmad Suliman B. Ali, Mohd Idrus Mohd Masirin, Abdalrhman Milad, Allam Musbah Al Allam, and Nur Izzi Md Yusoff

Abstract In this study, the viscoelastic properties of Peat partials used to modify asphalt binders are investigated. The tests were implemented in the current study using a dynamic shear rheometer (DSR). The obtained results from the rheological tests were represented by both complex moduli (G*) and viscosity (δ) under different conditions by Isochronal Plot, Black Diagram Plot, and Surface free energy SEF. The rheological properties of modified types of binder showed better performance than unmodified binder PNpt.M.A 0 at intermediate temperatures based on the results of DSR. Different proportions of Parit Nipah Peat (PNpt) to original asphalt binder had improved. The influence of Parit Nipah Peat (PNpt) on the original binder was significantly more significant than the unmodified asphalt binder or original asphalt binder. It was found that the addition of PNpt had positive effects and the addition of modifier with PNpt.M.A 5 revealed the best results. In conclusion, the Parit Nipah Peat modified asphalt reduced temperature susceptibility and increased the hardness of asphalt. PNpt.M.A 5 showed the most substantial effect among all samples, according to the results.

Keywords Modified asphalt binder · Parit Nipah Peat (PNpt) · Rheological properties · Dynamic shear rheometer (DSR) · Surface free energy · Fail temperatures

A. S. B. Ali · M. I. M. Masirin
Faculty of Civil Engineering and Built Environment, University Tun Hussein Onn Malaysia, 86400 Parit Raja, Johor, Malaysia

A. Milad (✉)
Department of Civil and Environmental Engineering, College of Engineering and Architecture, University of Nizwa, Birkat-al- Mouz, 616, Nizwa, Sultanate of Oman
e-mail: a.milad@unizwa.edu.om

A. M. Al Allam · N. I. M. Yusoff
Department of Civil Engineering, Universiti Kebangsaan Malaysia, UKM, Bangi, Selangor 43600, Malaysia

S. Abdullah et al. (eds.), *Structural Integrity Cases in Mechanical and Civil Engineering*, Structural Integrity 23, https://doi.org/10.1007/978-3-030-85646-5_18

18.1 Introduction

Nowadays, there are multiple ways to enhance the performance of highway pavements against permanent deformations. The various compositions of original asphalt are complicated. Moreover, various types of asphalt binders have been compared as prospective cohesive material through different tests [1, 2]. Asphalt possesses a high viscosity and high elasticity properties useful in pavement and highway applications [3–5]. A small percentage of asphalt additives used in the binder, which is 5% on average based on previous research, is acceptable for pavement performance by Parit Nipah Peat (PNpt) is modified in asphalt pavement which is enhancing rutting resistance on soft, yielding ground, advantage and application of product (basis of new alternative) in civil engineering [6]. Several types of asphalt modifiers can be used to improve the performance of asphalts certainly. On the same side, the rheology of viscoelastic and physical properties responsible for its mixture's behavior consists of the mineral additives and materials, shape, nature, structure, proportional ratio, and modified composition of asphalt binder and material. As demonstrated, mineral fillers in asphalt binders are one of the main body parts of the asphaltic mixture and are considered essential as they play a vital role in stiffening rigidity and toughening asphalt binder during its service life [7]. Another essential point given is that the filler acts as an active adhesion of asphalt binder, aids in the dispersion and homogenous of asphalt binder's contain in the asphaltic mixture [8], increases the elasticity of asphaltic mixture, and at the same time leads decreases permanent deformations such as rutting, fatigue and thermal cracking [9, 10].

However, the lower amount than the recommended percentages has a different amount based on the filler properties. Based on that, filler percentage content from asphalt binder's weight and high asphalt binder content can increase influence mixture sensitivity to rutting, and on the other hand, high filler percentage content from asphalt binder's weight can increase asphalt binder stiffness and improve its workability or permanent deformations in binders during initial tests [11]. Similarly, as Mohammadiroudbari et al. explained, asphalt binder should resist rutting to achieve its function, which is one reason to modify the asphalt binder before using it as a final product. In the same way [12], supported by, additives can enhance the viscoelastic properties of asphalt binder [4, 13–15].

However, natural materials of organic origin are used as modifiers to enhance some or all the properties in asphalt binder against a few types of deformation. The non-metallic and organic material should be suitable to be used in asphalt during the blending process. They should also be able to dissolve or interact with asphalt during the conventional mixing process. Organic materials used as asphalt modifiers by many researchers have been found to retain the main properties of modified asphalt binder during its service life [16, 17]. A significant number of additives are added to the mixtures to improve mixture properties or reduce cost. Asphalt modification offers one solution to overcome the deficiencies of asphalt, thereby improving asphalt mixtures [18].

The key aspect discussed for achieving excellent performance in various aspects is improved durability against rutting and resistance to permanent deformations [19, 20]. Another essential point is that modified asphalt at high temperatures was reduced by adding rice husk ash, wood sawdust ash, and Batu Pahat Soft Clay [16, 21]. In contrast, the use of excessive modifiers or inappropriate materials can increase the probability of rutting in asphalt mixtures [22]. According to Moayedi et al., peat can be derived from vegetation changed chemically and fossilized under acidic conditions [23]. It is a problematic soil due to severe problems in long-term consolidation settlements. It is also not suitable for construction, as its natural state cannot withstand any form of loading [24]. Several critical studies have been highlighted by Elghatas et al. that the chemical compositions, structural and molecular distribution of peat provide beneficial diatomite characteristics, and it has unique characteristics such as low shear strength and high compressibility [25]. The composition and structure of this type of soil are also very different from inorganic soils such as clay, sand, and gravel. Peat fibers have been applying in asphalt as a stabilizing additive and modifier for based asphalt and asphalt mixes, providing high performance to the road surface at a low cost [26].

Peat is used in asphalt to prevent drainage of bitumen; this becomes more significant for gap classified mixes like stone mastic asphalt, which have a high binder content. The addition of peat provides advantages such as increased binder content, increased aggregate binder film thickness, increased mix stability, and reduced binder drainage during transportation and laying [27, 28].

Based on the findings, it can be argued that direct mixing of peat fibers with hot aggregates and filler before adding the bitumen does not guarantee an even distribution of the fibers in the asphalt mix when mixing peat fibers with asphalt [29]. In contrast to that, as confirmed by Masirin to reach a homogeneous distribution of peat fibers in the asphalt mix, mixing should be performed only by the wet method, i.e., by blending the peat fibers in hot bitumen before adding them to the hot aggregates in the mixer [30].

This study investigates the effects of Parit Nipah Peat (PNpt) modified asphalt on binder rheological properties. Binders' viscoelastic properties were evaluated in the DSR using an oscillatory test followed by a strategic highway research program test to evaluate the modified asphalt binder properties. Similarity, assess the rutting resistance of asphalt by investigating the complex components of asphalt binder and the effectiveness of Parit Nipah Peat (PNpt) when used as a modifier in asphalt binder with different concentrations. Furthermore, the surface free energy and morphology properties of modified asphalt binder will also be evaluated.

Table 18.1 The properties of the original asphalt binder

Properties	Specification	Value
Penetration @ 25 °C	ASTM D5	96.2(dmm)
Softening point	ASTM D36	46.1(°C)
Ductility @ 25 °C	ASTM D113	> 100(cm)
Flashpoint	ASTM D-92	275(°C)
Fire point	ASTM D-92	301(°C)

18.2 Materials and Methods

18.2.1 Properties of the Original Asphalt

The asphalt binder used in this study is a soft binder with a penetration grade of 80/100. This asphalt binder was tested in the lab, and the characteristics of the original asphalt binder are listed in Table 18.1.

18.2.2 Parit Nipah Peat (PNpt)

The unique characteristics that stimulate peat in the asphalt industry are compositions, structure, and molecular distribution provide beneficial diatomite characteristics such as aloft imbibition, constancy, ability to spread, and low density [23]. In contrast, the physical properties of peat are listed in Table 18.2.

In this study, the modifier that has been used, Parit Nipah Peat (PNpt), was used as a modifier after it has passed the sieve size of 0.075 mm, which was prepared as a pan size in the blending phase [31]. PNpt was mixed in the UTHM lab using a mechanical blender with nearly 300 g of heated original asphalt binder in different proportions such as 0% or PNpt.M.A 0, 3% PNpt. M.A 3 and so on until 7% PNpt.M.A 7 and more than that negatively affect the properties, added with the mass of original asphalt binder. It was mixing performed for 60 min at a speed of 3500 rpm until a homogeneity point was reached with a temperature of 120 °C. Additives were added slowly, and mixing in the container was terminated when the binder became homogenous.

Table 18.2 Physical properties of peat

Properties	Value
Optimum moisture content, MOC (%)	132.47
Organic content (%)	78.77
Fiber content (%)	40.97
Ash content (%)	3
Liquid Limit (%)	162.11

18.2.3 Rheological Isochronal Plot

The most direct benefit of isochronal plots is comparing the complex modulus or phase angle at different temperatures collected by the DSR test. Therefore, viscoelastic data can be presented over a range of temperatures at a given frequency using an isochronal plot. The plot is defined as a simple viscoelastic function versus temperature. For the Strategic Highway Research Program SHRP testing, complex modulus and phase angle were used at a frequency of 10 rad/s 1.6 Hz to predict rutting specifications [32, 33]. This frequency was used because it corresponds to the loading time of 0.1 s with every high traffic loading cycle, which is a standard time for laboratory testing and is meant to simulate traffic loading. For rutting resistance, a high G" value and low δ are desirable. The selected viscoelastic function was plotted against temperature in which the complex moduli for the binders are plotted versus temperature.

18.2.4 Black Diagram Plot

Another fundamental curve obtainable from the frequency sweep data is known as a black diagram. It is a plot of curves from a complex modulus versus phase angle, obtained from a dynamic test, G* versus δ obtained from the DSR test. It can be defined as a curve on a graph representing the system's behavior at a constant frequency (time of loading). In this section, the physical laboratory tests were done on the original asphalt. The frequency of temperature is eliminated from plot, which allows all the dynamic data to be presented in one plot without the need to perform the shifting of plots for raw data [34]. However, Black diagram plots are used to identify and quantify inconsistencies in rheological data caused by aging, content, and non-linear effects.

18.2.5 Surface Free Energy

The surface free energy SFE test can be used to determine the moisture property of asphalt binders based on the Wilhelmy Plate method and the (SFE) values obtained from the software. Binder samples are mixed by adding peat to the asphalt and mixed by a blinder. SFE test is done with a loading rate of 0.5 ml at 25 °C of distilled water to measure the contact angle.

18.3 Results and Discussion

18.3.1 Isochronal Plots

Moreover, several properties, such as the temperature susceptibility of asphalt binder, can be interpreted or presented from this type of plot. Meanwhile, the isochronal plots of the complex modulus G* (MPa) versus temperature (°C) at the frequencies start of 1 and 10 rad are shown in (Figs. 18.1 and 18.2), respectively [4, 20]. In comparison, increased complex modulus (G*) demonstrated the ability for all Parit Nipah Peat modified asphalts PNpt.M.As and the 5% PNpt or PNpt. M.A 5 sample shows the

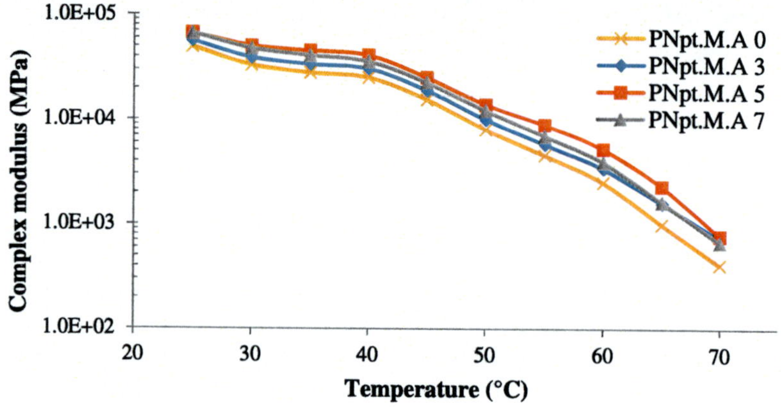

Fig. 18.1 Isochronal plots of the complex modulus G* at 1 rad

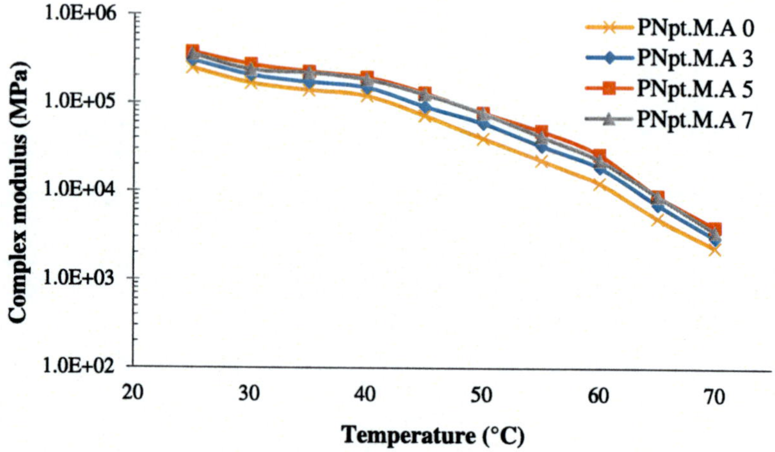

Fig. 18.2 Isochronal plots of the complex modulus G* at 10 rad

highest value in the increase in complex modulus (G*). After that, PNpt.M.A 7 decreased slightly in G*, but its value is still higher than the rest of the samples, including original asphalt PNpt.M.A 0.

Furthermore, the best-improved values compared to all samples were tested for unmodified and modified asphalt binder samples PNpt.M.As. The increase in viscosity (δ) was similar to the increase in the complex modulus (G*) for peat to control asphalt binder. This is probably due to cation exchanges where the ions and layers of PNpt's components are exfoliated and intercalated in the asphalt binder. Ammonium NH4 + and carbonic anhydrase Ca2 + in asphalt binder are the major exchangeable cations in the reaction process based on the literature review [35].

Here the laboratory investigations were conducted to evaluate both complex modulus (G*) and viscosity (δ) at the same time to compare them at different temperatures. The viscoelastic function to improve pavement modification uses different modifiers in the asphalt binder under different conditions [25, 30]. However, at high frequencies and temperatures, a high increase in complex modulus G* was noted, indicating improvement under actual conditions. The modification indices of G* at 0.1 and 10 Hz for the binders presented slightly similar trends in terms of the modification indices of viscosity with PNpt 5% showing a peak modification of 70 °C.

The modification indices at the two frequencies allow the influence of loading time or frequency and temperature to be evaluated compared to the only temperature for the viscosity results. However, the modification indices at 10 Hz are considerably lower than those at 0.1 Hz. This means that the relative performance of PNpt.M. changes at higher frequencies.

18.3.2 Black Diagram

Another fundamental curve is obtainable from the viscoelastic and frequency sweep data. Curves from complex modulus versus phase angle are analyzed and plotted in a vast extension of methods [3, 36]. Hence, the values were measured and plotted in the current form. Figure 18.3 shows that PNpt.M.A 5 has a higher complex modulus and elastic recovery performance at a low phase angle than the different percentages prepared in the lab under the same conditions. In short, it has better rutting resistance and performance compared to other modified blenders.

After the reactions, denser solid materials are melted, and a stable bonding framework is formed. Therefore, asphalt binder microstructure improves the complex shear modulus, thereby enabling PNpt modified asphalt binder to have the potential resistance toward rutting. Furthermore, the work dissipated per load cycle is calculated for rutting influence.

As explained in the results shown by [18, 30, 35, 37], the modifier was added to the original asphalt binder at different concentrations. The complex shear modulus G* of modified asphalt increased compared to the control asphalt. This means that

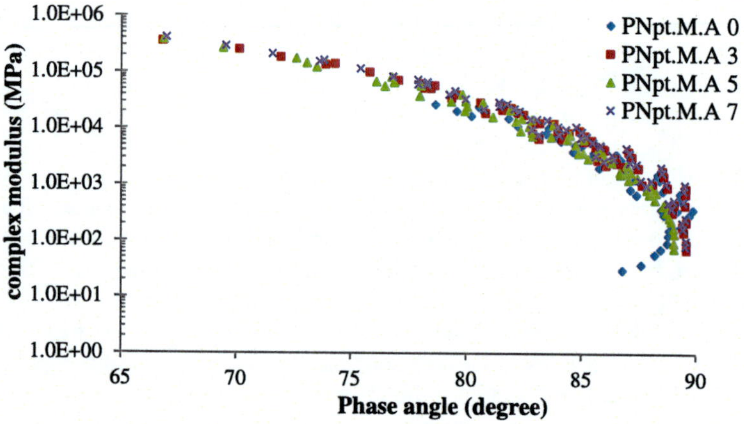

Fig. 18.3 Black Diagrams for original and PNpt modified asphalt binder

the presence of the modifier in asphalt improves the resistance toward rutting and produces the best performance.

18.3.3 Surface Free Energy

It is apparent from the graph that the values of the modified asphalt binders indicate a higher ability to minimizing the contact angel between water surface and asphalt surface, when compared with unmodified binders. The change in asphalt values depends on the types of modifiers used and their percentages. The number of modifiers in PNpr.M.0 did affect the increase in the percentages of modifiers. PNpt.M.A 7 showed the lowest value of SFE, while the unmodified asphalt binder recorded the highest SFE, the high value of SFE the big contact angel is as shown in Fig. 18.4.

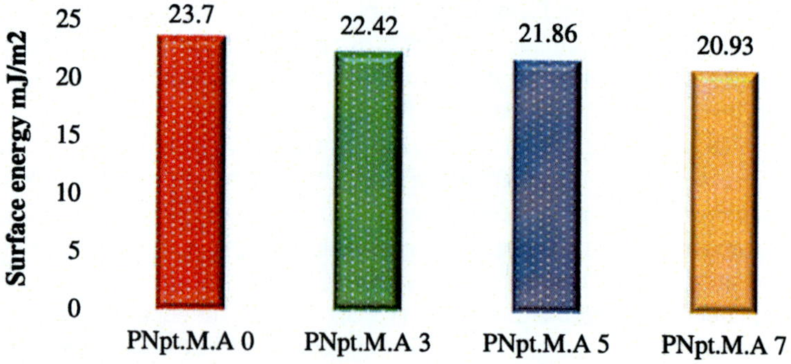

Fig. 18.4 Surface energy of asphalt binder

Additionally, the primary chemical of the modifier promote better adhesion between asphalt components. Literature indicates that amines consist of a long hydrocarbon chain and an amine group. The amine group reacts with the aggregate surface while the hydrocarbon portion (hydrophobic) is directed into the asphalt binder [31, 38, 39].

As a result, a strong bond between the modifier and asphalt binder would form due to the long hydrocarbon chain, which acted as a link between the modifier and asphalt binder surfaces and the characteristics of an organic modifier. Therefore, these factors could increase the adhesion between the asphalt binder and modifier. Thus, in most cases, modified asphalt binders will increase resistance to damage by increasing its adhesive. Similarly, the addition of a modifier to the original asphalt could also decrease the damage of the asphalt binder in most cases. A comparison of modified and unmodified asphalt to the original asphalt showed high susceptibility to moisture damage than the modified asphalt. Overall, the results from the SFE method proposed that the modified binders were significantly improved compared with that of original asphalt.

18.3.4 Electron Images of Modified Asphalt Binder

The addition of PNpt changes the internal microstructure of modified blended asphalt and that can be seen obviously and disperse without clustering for the modifier inside the asphalt and that were of some changes and effects due to the modifier to the original asphalt binder [30]. It is realized that the PNpt might have physical dispersion and chemical reactions with the asphalt. Based on the DSR outcomes, the addition of all modifiers can increase the complex shear modulus compared to the original asphalt binder as mentioned in Fig. 18.5 and also improves the resistance to rutting damages.

The morphology shows that the crystalline structure (a) appears to be straightforward and homogeneous (b). The addition of PNpt changes the internal structure of the blended asphalt [30]. This is observed from the SEM analysis. The PNpt phase

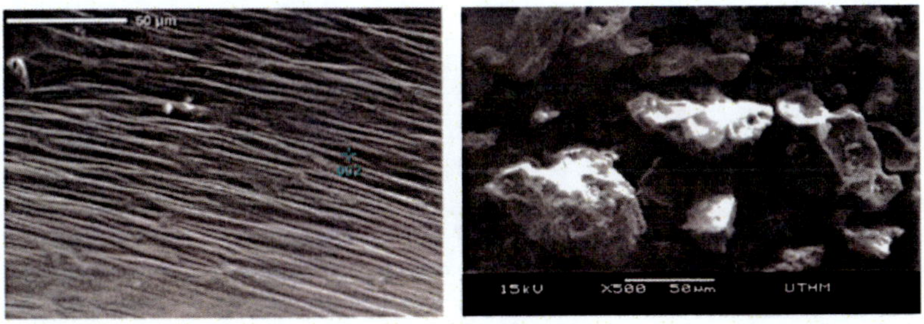

Fig. 18.5 Electron images of PNpt modified asphalt binder

appears light and rigid, while the asphalt phase appears dark. Also, significant differences in morphology can be seen. The modified asphalts with PNpt do not have phase separation, and the addition of PNpt has a significant effect on the compatibility of PNpt modified asphalts.

In the same way, the refined brown PNpt particles are dispersed in the asphalt binder. This means that the blends with PNpt can be stored at a high temperature. The SEM images of PNpt used in this research are both irregulars in shapes with fractured and harsh, sharp angles surfaces. Besides, the particles were pieced together; nevertheless, different surface [40, 41].

18.3.5 Failure Temperature

The failure temperatures for the case of unaged asphalt binder are G*/sin δ values of less than 1.0 kPa, considering all asphalt proportions [42]. In general, the purpose of computing the failure temperature in all the samples is to determine the grade of the asphalt binder [6]. Moreover, it is indicated that the original asphalt binder PNpt.M.A 0 and PNpt modified asphalt binder have failure temperatures at 64 °C or more than 85 °C, similar to the Superpave specifications.

The original asphalt binder PNpt.M.A.0 has the weakest resistance and the lowest failure temperature among all the samples, around 64 °C. Meanwhile, PNpt.M.A 3, PNpt.M.A 5, and PNpt.M.A 7 recorded failure temperatures around 66, 71, and 68 °C respectively as shown in Fig. 18.6.

Previous investigations on asphalt binder modified by different organic materials showed similar behavior where the modified asphalt binder has a higher failure temperature than original asphalt or unmodified asphalt. The specific amount of modifiers used also depends on the nature of the asphalt and its source [28].

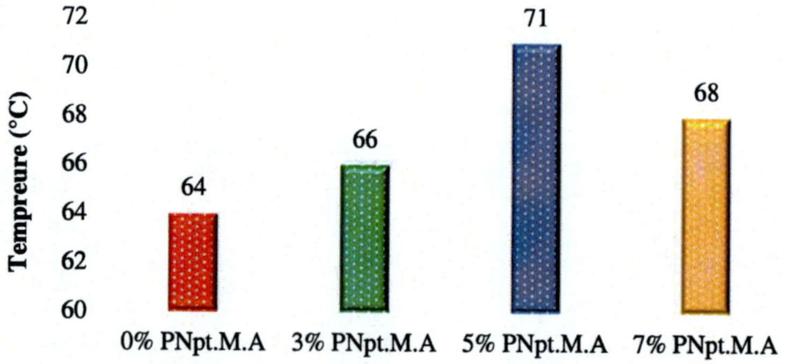

Fig. 18.6 Failure temperatures for the original and PNpt modified asphalt binder

18.4 Conclusion

In summary, asphalt compatibility allows a continuous to be established. Modification is provided by a highly elastic network which increases the viscosity, stiffness, and elastic response. However, the level of enhancement increased for the binders containing a greater quantity of PNpt. Meanwhile, the modification of asphalt as shown by the dynamic shear rheometer (DSR) results revealed that the addition of PNpt resulted in additional interactions which changed the structure of unmodified asphalt or the original asphalt binder. Therefore, the modifiers managed to improve the stiffness of the asphalt binder, which in turn improves its rutting resistance. The results of rheological testing were presented using the Isochronal Plot, Black Diagram Plot, and Surface free energy SEF. The results indicated that the addition of PNpt increased the rutting resistance by increasing the stiffness of the modified binders as a result of the black diagram plot.

Similarly, the modified asphalt binder recorded the highest waterproof surface to get a better raincoat. The modifier appears to be a compatibly homogeneous crystalline structure and by high dispersion with the asphalt. Finally, the addition of PNpt also improved the failure temperatures of the modified asphalt binder.

Acknowledgements The authors would like to express their gratitude to Universiti Kebangsaan Malaysia for the financial support for this work through Grant No. DIP-2020-003.

References

1. Milad AA, Ali ASB, Yusoff NIM (2020) A review of the utilisation of recycled waste material as an alternative modifier in asphalt mixtures. Civil Eng J 6:42–60
2. Cong P, Liu N, Tian Y, Zhang Y (2016) Effects of long-term aging on the properties of asphalt binder containing diatoms. Const Build Mater 123:534–540
3. Usman N, Masirin MIM, Ahmad KA, Ali ASB (2019) Application of recycled polyethylene terephthalate fiber in asphaltic mix for fatigue life improvement. Lect Notes Civil Eng 9:1401–1413
4. Milad A, Taib AM, Ahmeda AG, Solla M, Yusoff NIM (2020) A review of the use of reclaimed asphalt pavement for road paving applications. Jurnal Teknologi 3(82):35–44
5. Milad A, Ali ASB, Babalghaith AM, Memon ZA, Mashaan NS, Arafa S (2021) Utilisation of waste-based geopolymer in asphalt pavement modification and construction a review. Sustainability 13(6):3330
6. Muniandy R, Yunus R, Hasham S, Aburkaba E (2013) Influence of clay nanoparticles on the physical and rheological characteristics of short term aged asphalt binder. 7(11):488–498
7. Abreu LPF, Oliveira JRM, Silva HMRD, Palha D, Fonseca PV (2017) Suitability of different foamed bitumens for warm mix asphalts with increasing recycling rates. Constr Build Mater 142:342–353
8. Hu X, Wang N, Pan P, Bai T (2017) Performance evaluation of asphalt mixture using brake pad waste as mineral filler. Constr Build Mater 138:410–417
9. Miró R, Martínez AH, Pérez-Jiménez FE, Botella R, Álvarez Á (2017) Effect of filler nature and content on the bituminous mastic behaviour under cyclic loads. Const Build Mater 132:33–42

10. Pérez I, Gómez-Meijide B, Pasandín AR, García A, Airey G (2021) Enhancement of curing properties of cold in-place recycling asphalt mixtures by induction heating. Int J Pavement Eng 22(3):355 368
11. Dulaimi A, Al Nageim H, Ruddock F, Seton L (2017) High performance cold asphalt concrete mixture for binder course using alkali-activated binary blended cementitious filler. Const Build Mater 141:160–170
12. Mohammadiroudbari M, Tavakoli A, Razavi Aghjeh MK, Rahi M (2016) Effect of nanoclay on the morphology of polyethylene modified bitumen. Const Build Mater 116:245–251
13. Golestani B, Hyun B, Moghadas F, Fallah S (2015) Nanoclay application to asphalt concrete: characterization of polymer and linear nanocomposite-modified asphalt binder and mixture. Const Build Mater 91:32–38
14. Yu X, Dong F, Ding G, Liu S, Shen S (2016) Rheological and microstructural properties of foamed epoxy asphalt. Const Build Mater 114:215–222
15. Masoudi S, Abtahi SM, Goli A (2017) Evaluation of electric arc furnace steel slag coarse aggregate in warm mix asphalt subjected to long-term aging. Const Build Mater 135:260–266
16. Al Allam AM, Idrus M, Masirin M, Abdullah, ME, Kamaruddin M (2016) NH Influence of using Batu Pahat soft clay on the mechanical properties of hot mix asphalt mixture. ARPN J Eng Appl Sci 11(4):2380–2386
17. Al-Mansob RA, Ismail A, Yusoff NIM, Albrka SI, Azhari CH, Karim MR (2016) Rheological characteristics of unaged and aged epoxidised natural rubber modified asphalt. Const Build Mater (102):190–199
18. Milad A, Ahmeda AG, Taib AM, Rahmad S, Solla M, Yusoff NIM (2020) A review of the feasibility of using crumb rubber derived from end-of-life tire as asphalt binder modifier. J Rubber Res 23(3):1–14
19. Al Allam AM, bin Masirin MI, Ali ASB (2017) Influence of aging on the physical properties and chemical compositions of asphalt binder with soft clay particles. Adv Eng For 24:48–54
20. Xue Y, Wu S, Cai J, Zhou M, Zha J (2014) Effects of two biomass ashes on asphalt binder: dynamic shear rheological characteristic analysis. Const Build Mater 56:7–15
21. Jia X, Huang B, Bowers BF, Zhao S (2014) Infrared spectra and rheological properties of asphalt cement containing waste engine oil residues. Const Build Mater 50:683–691
22. Moayedi H, Nazir R, Kazemian S, Huat BK (2014) Microstructure analysis of electrokinetically stabilized peat. Measurement 48:187–194
23. Idrus M et al (Jan. 2020) Analysis of physical and microstructural properties on Parit Nipah peat particles as sustainable asphalt modifier. Mater Sci Forum 975:197–202. https://doi.org/10.4028/www.scientific.net/msf.975.197
24. Qiu J, Liu H, Lai J, Lai H, Chen J, Wang K (2018) Investigating the long-term settlement of a tunnel built over improved loessial foundation soil using jet grouting technique. J Perform Constr Facil 32(5)
25. Elghatas HMA, Bin Aman MY, Ali ASB (2019) An assessment on the physical and rheological properties of asphalt binder modified with micro bauxite powder (MBP). Lect Notes Civil Eng 19:30–37
26. Yu J-Y, Feng P-C, Zhang H-L, Wu S-P (2009) Effect of organo-montmorillonite on aging properties of asphalt. Constr Build Mater 23(7):2636–2640. https://doi.org/10.1016/j.conbuildmat.2009.01.007
27. Paul DR, Robeson LM (2008) Polymer nanotechnology: nanocomposites. Polymer 49(15):3187–3204
28. Rodríguez-Alloza AM, Gallego J, Giuliani F (2017) Complex shear modulus and phase angle of crumb rubber modified binders containing organic warm mix asphalt additives. Mater Struct 50(1):1–9
29. Kádár R, Abbasi M, Figuli R, Rigdahl M, Wilhelm M (2017) Linear and nonlinear rheology combined with dielectric spectroscopy of hybrid polymer nanocomposites for semiconductive applications. Nanomaterials 7:23
30. Masirin MIM, Al Allam AM, Ali ASB (2017) Effect of Batu Pahat Soft Clay (BPSC) concentrations on the physical and rheological properties of asphalt binder. Pertanika J Sci Technol (25):101–108

31. Yusoff NIM, Shaw MT, Airey GD (2011) Modelling the linear viscoelastic rheological properties of bituminous binders. Const Build Mater 25:2171–2189
32. Chailleux E, Ramond G, Such C, de La Roche C (2006) A mathematical-based master-curve construction method applied to complex modulus of bituminous materials. Road Mater Pavement Design 7(1):75–92
33. Cui S, Blackman BRK, Kinloch AJ, Taylor C (2014) Durability of asphalt mixtures: effect of aggregate type and adhesion promoters. Int J Adhes Adhes 54:100–111
34. Lu H, Zhang X, Knauss WG (1997) Uniaxial, shear, and Poisson relaxation and their conversion to bulk relaxation: studies on poly (methyl methacrylate). Polymer Eng Sci 18(2):211–222
35. Wetton RE, Marsh RDL, Van-de-Velde JG (1991) Theory and application of dynamic mechanical thermal analysis. Thermochim Acta 175(1):1–11
36. Zhang J, Walubita LF, Faruk ANM, Karki P, Simate GS (2015) Use of the MSCR test to characterize the asphalt binder properties relative to HMA rutting performance—a laboratory study. Construct Build Mater 94:218–227
37. Lesueur D, Gerard J-F, Claudy P, Letoffe J-M, Planche J-P, Martin D (1996) A structure-related model to describe asphalt linear viscoelasticity. J Rheol 40(5):813–836
38. Celauro C, Fecarotti C, Pirrotta A (2017) An extension of the fractional model for construction of asphalt binder master curve. Eur J Environ Civil Eng 21(1):78–93
39. Zhu J, Birgisson B, Kringos N (2014) Polymer modification of bitumen: advances and challenges. Eur Polymer J 54:18–38
40. Yao H, You Z, Li L, Shi X, Goh SW, Mills-Beale J, Wingard D (2012) Performance of asphalt binder blended with non-modified and polymer-modified nanoclay. Constr Build Mater 35:159–170
41. Shafabakhsh GH, Ani OJ (2015) Experimental investigation of effect of Nano TiO2/SiO2 modified bitumen on the rutting and fatigue performance of asphalt mixtures containing steel slag aggregates. Const Build Mater 98:692–702
42. El-Shafie M, Ibrahim IM, Rahman AMM (2012) The addition effects of macro and nano clay on the performance of asphalt binder. Egypt J Petrol 21(2):149–154

Chapter 19
Experimental Assessment of Steel Fibre Reinforced Concrete Beam Strengthened with Carbon Fibre Reinforced Polymer

Noorsuhada Md Nor, Abdul Hakeem Zulkifli, Soffian Noor Mat Saliah, Noor Syafeekha Mohamad Sakdun, Nor Z. Amin, Nor N. A. Anisah, and Azmi Ibrahim

Abstract Due to an increase of load-carrying capacity, rigidity and stability of building structure become limited to accommodate the space. Thus, most of the structural elements experienced excessive cracking and this can endanger the structure. To overcome this problem, this paper presents the experimental assessment of steel fibre reinforced concrete (SFRC) beams strengthened with carbon fibre reinforced polymer (CFRP) using acoustic emission (AE) technique. An experimental investigation was conducted to obtain the crack propagation of the reinforced concrete (RC) beam under monotonic load. Three RC beams sized 150 × 200 × 1500 mm were tested. The first beam was designed as a control beam with no steel fibre added (B0SF0C), the second beam was added with 1% steel fibre based on concrete weight (B1SF0C), and the third beam was added with 1% steel fibre and strengthened with CFRP on the soffit (B1SF1C). The relationship between AE signal strength, load and time for all beams were analysed and observed the crack modes of the beams. It was found that the B1SF1C produced the highest signal strength when reached to the localization of shear crack due to the presence of CFRP that provides higher strength and stiffness. The experimental results indicate that the SFRC beam strengthened with CFRP was the highest in strength among other beams. Furthermore, the relationship between signal strength and crack modes of the beams is good. Hence, this study is significantly prime to enhance the performance and integrity of the structure.

Keywords Steel fibre reinforced concrete · Carbon fibre reinforced polymer · Acoustic emission signal strength

N. M. Nor (✉) · S. N. M. Saliah · N. S. M. Sakdun · N. Z. Amin · N. N. A. Anisah
School of Civil Engineering, College of Engineering, Universiti Teknologi MARA, Cawangan Pulau Pinang, Pulau Pinang, Malaysia
e-mail: ida_nsn@uitm.edu.my

A. H. Zulkifli
Doka Formwork Sdn. Bhd, Kota Damansara, Petaling Jaya, Selangor, Malaysia

A. Ibrahim
Faculty of Civil Engineering, Universiti Teknologi MARA, Shah Alam, Selangor, Malaysia

S. Abdullah et al. (eds.), *Structural Integrity Cases in Mechanical and Civil Engineering*, Structural Integrity 23, https://doi.org/10.1007/978-3-030-85646-5_19

19.1 Introduction

The service life of reinforced concrete (RC) structures is limited because of the deterioration, aging, increase in load demands and changes in use of the structures [1]. The use of steel fibre in the concrete mix for construction industry has advantages in order to increase the service life of the RC structures. Steel fibre reinforced concrete (SFRC) is a combination of steel fibre and normal reinforced concrete blended into one composite. There has been an explosive growth of interest in civil engineering application. It has been used extensively for maintenance purpose. It is due to the fact that the steel fibre in the RC can delay the crack development [2, 3]. The steel fibre also can maintain the bond between concrete even though the steel bar corroded. The mechanical properties of the concrete can be improved [4, 5]. It is because the inclusion of steel fibre in the concrete mix at a certain dosage level can increase the characteristic strength and mechanical properties of the RC structures [6]. Altun et al. [7] found that steel fibre added into RC beam with dosage 30 kg/m^3 give higher toughness compare to RC beam having the same conventional reinforcement without steel fibre dosage. The optimum dosage should be within the range of 1.0–2.5% by absolute volume. However, a dosage smaller than 1.0 and beyond 2.5% become ineffective due to the physical difficulties in providing a homogeneous distribution of the steel fibres within the concrete causing an appreciable drop in the compressive strength as compared to the plain concrete. It indicates that the inclusion of steel fibre in RC mix is limited to a certain range. Hence, it needs for the use of other fibres to strengthen the SFRC performance such as carbon fibre reinforced polymer (CFRP).

The use of CFRP is now widely used for repairing and retrofitting to prolong service life of the concrete structure [1, 8]. The CFRP offers many advantages when compared to conventional steel as reinforcement such as high specific strength and stiffness [9]. According to Naaman et al. [10], the RC beam strengthened with CFRP has good rigidity after the initial crack with the adequate bonding. The use of CFRP for strengthening of RC beam has also been investigated by Yun et al. [1] and Lee and Lee [9]. The CFRP has also been applied on the masonry surface by Ghiasi et al. [11]. They found that the CFRP is low weight to strength ratio and flexibility in application have made the CFRP an efficient solution for strengthening purposes. From the review, it is found that the application of CFRP to strengthen the SFRC beam surface has not been widely explored and reported. It indicates that the generic study of the hybrid SFRC strengthened with CFRP is needed.

When the hybrid SFRC strengthened with CFRP subjected to loading, the acoustic emission (AE) wave would be developed. The study of AE mechanisms is vital in order to investigate the characteristic and behaviour of the hybrid SFRC beam strengthened with CFRP. Moreover, its damage can be evaluated. The progression of damage can be evaluated by analyzing the AE parameters such as signal strength. The signal strength is "the measured area of the rectified AE signals with units proportional to volt-seconds" [12], which normally include the absolute area of both the positive and negative envelopes [13]. It relates to the relative energy, which relates to the amount energy released by the specimen [14]. It is also a function of the amplitude and duration of the signal [14]. The AE signal strength has been used extensively for assessment of crack [15–20]. The signal strength has been used for assessment of fatigue crack of RC beam [15, 16, 18]. Md Nor et al. [18] stated that the signal strength increases as the load increases. De Rosa [20] affirmed that the AE activity increases as the stress which induced the occurrence of crack in the material increases. Hence, this paper presents the damage evaluation of hybrid SFRC beam strengthened with CFRP on the soffit using AE signal strength.

19.2 Methodology

For the determination of the acoustic emission trend, the process of the flow is presented in Fig. 19.1. A total of three specimens were prepared. In this study, the three-point flexural tests were performed to investigate the strength of beam specimens in conjunction with AE monitoring. To investigate the behaviour of the beam specimen, the development cracks were observed. Next, the data collected from the AE sensor to record the crack activities was then analysed. Hence, the determination of the AE trend and the crack pattern of the specimen was carried out.

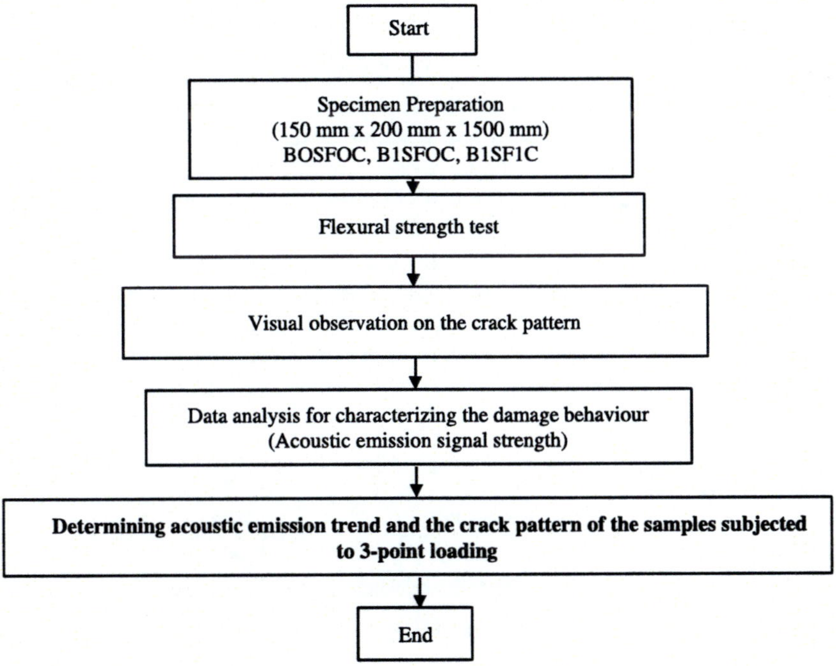

Fig. 19.1 Process flow for determining the acoustic emission trend of the samples' behaviour

19.2.1 Preparation of Materials

Three RC beam specimens were prepared. The first specimen was a control RC beam which no steel fibre added in the RC beam and labelled as B0SF0C. The second specimen was added with 1% steel fibre from the total concrete weight and labelled as B1SF0C. The third specimen was added with 1% steel fibre in the RC beam strengthened with one layer of CFRP on the soffit and labelled as B1SF1C. The size of the RC beam specimens was 1500 mm length by 150 mm width by 200 mm depth, generally used in flexural tests. The concrete mix was designed with grade C40. For the reinforcement bar, the size used was 4T12 for tension bars and 2H8 for hanger bars. The size of the stirrup was H8 with the spacing of 150 mm centre to centre. The concrete mix was designed by weight of cement: water: fine aggregate: coarse aggregate of 1.00: 0.45: 2.07: 2.74, respectively [21]. To improve the workability of the concrete, 1.0% of superplasticiser was added to the concrete mix. The steel fibre used in this study was hooked-end of 50 mm length and 0.75 mm diameter. The tensile strength of the steel fibre was 1250 N/mm2 which has bright and clean wire appearance.

19.2.2 Installation of the CFRP on the SFRC Beams

The CFRP used in this study was supplied by SikaWrap. It was a unidirectional woven carbon fibre fabric that was used for dry application. Table 19.1 shows the physical properties of the CFRP. The CFRP was installed on the soffit of the beam specimen using an adhesive. The epoxy with the tensile strength of 30 N/mm2, elongation at break of 0.9% and tensile E-modulus of 4500 N/mm2 was used throughout the study.

Before setting the CFRP, the beam surface was cleaned using sandpaper to remove impurities. The epoxy was prepared where the epoxy part A and part B were mixed and stirred with the ratio of 4:1 by weight. The epoxy was then applied onto the beam surface using the clean brush, and the CFRP was laid on it. The CFRP was rolled using plastic impregnation roller parallel to the fibre direction until the epoxy squeezed out and the epoxy distributed evenly over the fabric surface. To investigate

Table 19.1 The physical properties of the CFRP

Properties	Value
Area weight	230 g/m^2 $\pm$ 12 g/m^2
Fibre density	1.80 g/cm^3
Fabric design thickness	0.127 mm
Tensile strength	4900 N/mm^2
Tensile E-modulus	230,000 N/mm^2
Elongation at break	2.1%
Laminate thickness	1 mm per layer

the material properties, all the beam specimens were air-cured for seven days prior to test at room temperature.

19.2.3 Test Setup and Acoustic Emission Monitoring

AE technique was used to evaluate the crack propagation of the RC beam specimens when subjected to monotonic load to failure. In the AE hardware, the threshold level of 40 dB and the AE wave velocity of 4000 m/s were set [21]. Four sensors VS75–V with the frequency range of 20–100 kHz were fixed on the RC beam specimens using high vacuum grease [22]. According to Yepez Roca [23], the high vacuum grease able to act in good contact with the specimen surface and the sensor. Then, the sensors were fixed to the surface of the RC beam specimen using the magnetic clamps that hold on the 2.2 mm steel plates that glued on the specimen. Two sensors were fixed on the top, and the rest were fixed on one side of the beam specimen as shown in Fig. 19.2.

The two sensors on the top were located at 500 mm from the edge of the beam specimen and named as CH6 and CH7. Meanwhile, the two sensors on the side were located at 300 mm from the edge of the beam and named as CH5 and CH8.

All the sensors were then connected to the pre-amplifier with the gain of 34 dB using cable. Then, the cable was connected to the ASIP-2 board. From the ASIP-2 board, it was connected to a computer for AE data acquisition. In this study, only AE signal strength collected from all channels CH5, CH6, CH7 and CH8 was analysed to evaluate the crack performance that occurred on the RC beam specimen. The AE signal strength was investigated to assess the behaviour of the crack when the

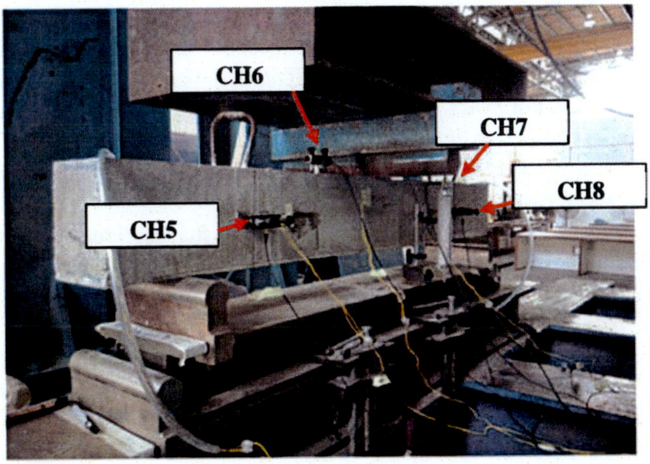

Fig. 19.2 A setup of the instruments on the beam specimen

beam subjected to monotonic load to failure. A comparison between all channels was performed.

The monotonic load was performed using a servo-hydraulic testing machine on RC beam in the force-control mode associated with AE monitoring as depicted in Fig. 19.2. The constant rate of 0.02 mm/s was applied. Prior to the test, the beam specimen was supported with the effective length of 1200 mm.

19.2.4 Observation of the Crack Propagation

The grid lines were drawn on one surface of the beam specimen. The crack propagation that appeared on the beam surface was mapped on graph paper which has the same grid lines as on the beam surface. Since both sides of the beam have almost the same crack patterns, cracks only on one side are presented here [21]. Another side surface was occupied due to the instalment of AE sensors, strain gauges and transducers. However, all data collected from the strain gauges and transducers are not presented in this paper.

In order to have a precise crack observation, a torchlight was used. During mapping crack on the graph paper, the time was recorded, and the load applied was noted. The crack was numbered as 1 onwards.

19.3 Results and Discussion

19.3.1 Observation of the Crack Pattern at Ultimate Loads

For crack observation, torchlight was used to observe the occurrence of hairline cracks on the beam surface. The crack appeared on the beam surface which mapped on the graph paper. For beam specimen B0SF0C, more wide cracks were generated when the beam subjected to the maximum load of 107.8 kN as presented in Fig. 19.3a. From the crack pattern, it is clearly seen that serious damage has occurred to the beam on the flexural and shear crack regions.

However, for beam specimen B1SF0C, it seems that the crack pattern demonstrated a phenomenon the main reason including steel fibre of 1% in the concrete. According to Aggelis et al. [24], the addition of steel fibre in the concrete leads to higher absorbance of fracture energy. Figure 19.3b shows the crack pattern when the load reached the ultimate load. At the mid-span of the beam found that the wide crack developed from the bottom. However, at the shear region, small cracks were developed and propagated. At the same time, the inclusion of steel fibre in the beam specimen is found bridging the cracks and restrains them. Calogero et al. [25] confirmed that the inclusion of steel fibre in the concrete delayed the formation and

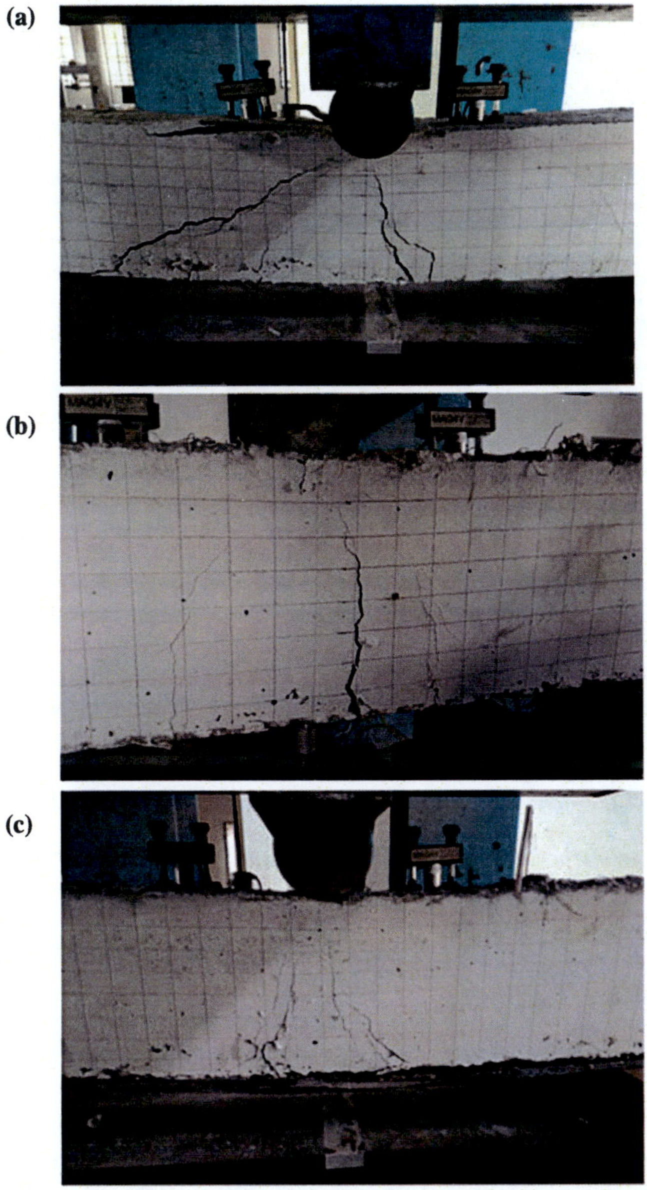

Fig. 19.3 Crack pattern on the beam specimen (a) B0SF0C, (b) B1SF0C, and (c) B1SF1C

growth of cracks. From the monotonic test to failure, it was found that the maximum load of this beam was 144.9 kN.

The addition of 1% steel fibre in the concrete mix and strengthened with the CFRP on the soffit of the beam reduced the formation of crack as well as the flexural and shear regions as shown in Fig. 19.3c. This beam is designated as B1SF1C, and the ultimate load of this beam is 153.1 kN. It is found that there is good agreement between the beam types and the crack pattern at the ultimate load. The ultimate load of the beam increases as the strengthening material provided to stiffen the beam. Moreover, the addition of 1% steel fibre improved the performance of the beam specimen compared to the control beam specimen. However, when the beam with the addition of 1% steel fibre strengthened with CFRP in the transverse direction, extra stiffness is provided in the beam compared to other beams.

19.3.2 Signal Strength of the Control Beam (B0SF0C)

Figure 19.4 shows the relationship between signal strength, load and time for beam specimen B0SF0C. The beam was designed with no steel fibre in the concrete mix and no CFRP attached on the soffit of the beam. Four channels of CH5, CH6, CH7, and CH8 were analysed. From this analysis, it is found that the CH7 collected the highest AE data of 100,697 numbers with low signal strength. This is because the CH7 is positioned on the top of the beam. CH6 collected the second highest AE data of 76,587 numbers with high signal strength. The CH6 was also positioned on the top of the beam specimen. CH 5 collected the third highest AE data of 68,248 numbers with low AE signal strength. Meanwhile, CH8 collected the lowest AE data

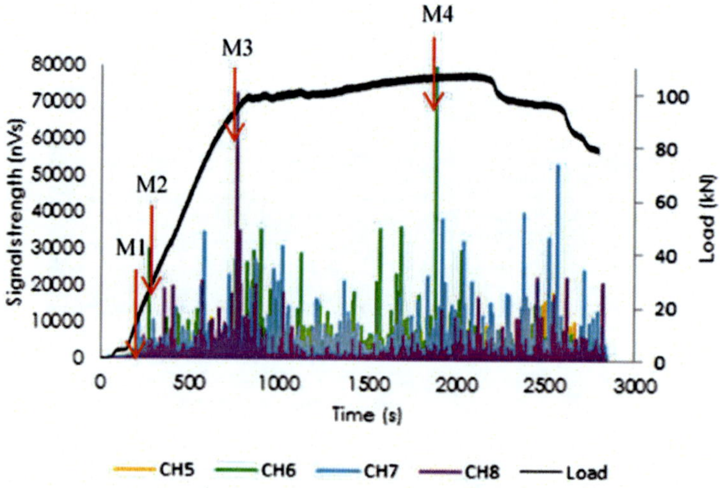

Fig. 19.4 Signal strength, load with respect to time for beam specimen B0SF0C

of 40,687 numbers with high signal strength. The CH8 was positioned on the side of the beam surface 300 mm from the edge. From these analyses, it is confirmed that the position of the sensor would affect the collection of the AE data but not necessarily capture the highest signal strength.

Two points of highest signal strength were noticed which obtained from CH8 and CH6. Those points represented the behaviour of the crack developed in the beam specimen B0SF0C. Hence, four crack modes were identified namely mode 1 (M1), mode 2 (M2), mode 3 (M3) and mode 4 (M4). M1 indicates the nucleation of crack which is unable to be seen through naked eyes. At time 90 s, the highest signal strength was noticed when the beam subjected to a load of 12 kN. The signal strength is 2040 nVs captured by CH7. No crack has appeared on the beam surface. As the load increased to 24.7 kN, the first crack was observed on the beam surface at time 272 s which induced a signal strength of 29,900 nVs. The first crack was noted as crack mode 2, M2 and the signal was collected from CH6. The flexural cracks were localized as the first crack progressively propagated beyond the mid-depth of the beam and named as M3. This crack generates high signal strength of 72,600 nVs from CH8 at time 760 s. Just after the flexural crack localized, the shear crack propagated diagonally from a new crack on the beam specimen and the tip of the flexural cracks. The shear cracks were then more localized when reached 106 kN at the time 1900s. This crack was designated as M4, which was generated 89,800 nVs from CH6.

It can be inferred that when the beam is elastic, a low signal strength produced. However, when reached the plastic limit, as the concrete unable to resist the load, it produced high signal strength. Hence, the signal strength can be used to identify the condition of the beam specimen.

19.3.3 Signal Strength of the 1% SFRC Beam (B1SF0C)

Figure 19.5 shows the relationship between signal strength, load and time for beam specimen B1SF0C. The beam was designed with the inclusion of 1% steel fibre in the concrete mix and no CFRP attached on the soffit of the beam. Four channels of CH5, CH6, CH7, and CH8 were analysed. From Fig. 19.5, it is found that the CH7 collected the highest AE data of 265,191 numbers with low signal strength. This is because the CH7 is positioned on the top of the beam. CH6 collected the second highest AE data of 159,672 numbers with low signal strength. The CH6 was also positioned on the top of the beam specimen. CH 8 collected the third highest AE data of 59,707 numbers with the highest AE signal strength. Meanwhile, CH5 collected the lowest AE data of 30,512 numbers with low signal strength. The CH8 was positioned on the side of the beam surface 300 mm from the edge. From these analyses, it is confirmed that the position of the sensor would affect the collection of the AE data but not necessarily capture the highest signal strength.

In Fig. 19.5, two points of highest signal strength were noticed which obtained from CH8. Those points represented the behaviour of the crack developed in the

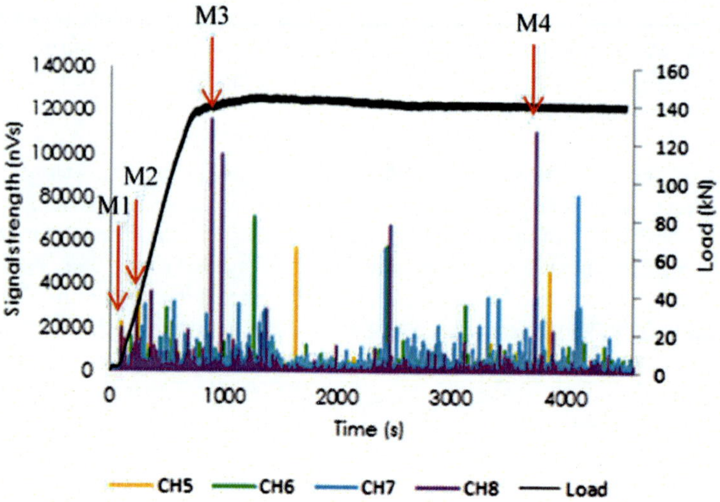

Fig. 19.5 Signal strength, load with respect to time for beam specimen B0SF1C

beam specimen B1SF0C. Hence, four crack modes were identified namely mode 1 (M1), mode 2 (M2), mode 3 (M3) and mode 4 (M4). M1 indicates the nucleation of crack which is unable to be seen through naked eyes. At time 100 s, the high signal strength was noticed when the beam subjected to a load of 14 kN. The signal strength is 19900 nVs from CH8. At this crack mode, no crack has appeared on the beam surface. As the load increased to 34 kN, the first crack, noted as crack mode M2, was observed on the beam surface at time 312 s which induced a signal strength of 30,600 nVs. The signal strength for M2 was collected by CH7. The flexural cracks were localized as the first crack progressively propagated beyond the mid-depth of the beam and noted as crack mode M3. It generates high signal strength of 116,000 nVs from CH8 at time 883 s. Just after the flexural crack localized, the shear crack propagated diagonally from a new crack on the beam specimen and from the tip of the flexural cracks. The shear cracks were then more localized when reached 141 kN at time 3745 s. This crack was designated as M4. The signal strength is 111000 nVs which was collected from CH8.

A similar phenomenon was found for beam B0SF1C and B0SF0C where the beam is elastic, a low signal strength produced. However, when reached the plastic limit, as the concrete unable to resist the load, it produced high signal strength. Hence, the signal strength can be used to identify the condition of the beam specimen.

19.3.4 Signal Strength of the 1% SFRC Beam Wrapped with CFRP (B1SF1C)

Figure 19.6 shows the relationship between signal strength, load and time for beam specimen B1SF1C. The beam was designed with the inclusion of 1% steel fibre in the concrete mix and one layer of CFRP attached to the soffit of the beam. It was found that the CH7 collected the highest AE data of 302,568 numbers with low signal strength. Meanwhile, CH6 collected the second highest AE data of 283,050 numbers with high signal strength. The CH6 was also positioned on the top of the beam specimen. CH8 collected the third highest AE data of 65,446 numbers with the highest AE signal strength. Meanwhile, CH5 collected the lowest AE data of 58,640 numbers with the highest signal strength. From these analyses, it is confirmed that the position of the sensor would affect the collection of the AE data but not necessarily capture the highest signal strength.

When load subjected to the beam of 36.9 kN, the nucleation of crack, M1 was developed with a high signal strength of 3580 nVs at time 321 s. The AE signal was captured by CH8. The first crack, M2 has appeared on the beam surface when a load of 60.4 kN subjected to the beam at 385 s. The signal strength was captured by CH6 with the value of 12,500 nVs. The flexural cracks, M3 were generated and localized when CH6 captured high signal strength of 70,200 nVs at time 910 s. The load was 142.9 kN.

The formation of shear crack, M4 was generated when the load reached 147.4 kN produced a signal strength of 140,000 nVs at time 1123 s. Then, at time 1801s, the CFRP was delaminated and generated a high signal strength of 234,000 nVs. The

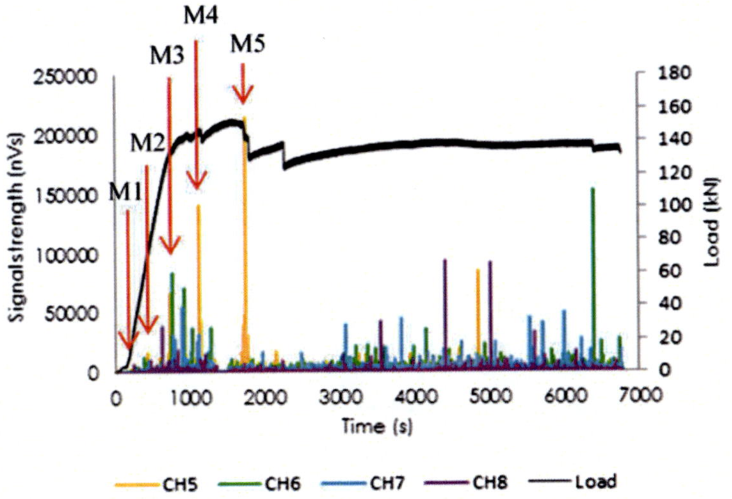

Fig. 19.6 Signal strength, load with respect to time for beam specimen B1SF1C

high signal strength was collected from the CH6. However, when the delamination occurred, the load was reduced to 143.9 kN.

19.3.5 Signal Strength with Respect to Damage Modes for All Beam Specimens

From the graph of signal strength with respect to damage modes as presented in Fig. 19.7, it is clearly seen that the signal strength increases as the beam specimens approaching failure. Muralidhara et al. [26] stated that the AE energy increased up to 80% when reached to failure stage. Based on the AE signal strength, the integrity of the structure can be sufficiently quantified [27, 28]. The beam B0SF0C was represented the lowest signal strength compared to other beam specimens. The lowest signal strength was noticed when M1 to M3 was developed in the B1SF1C. However, it produced high signal strength when reached to the localization of shear crack. This is because the high stiffness in the transverse direction owing to the CFRP and shear of the isotropic CFRP provides strengthening confinement.

There is a good agreement between the AE signal strength and the damage modes, to the extent that the constitutive signal strength captured for the control beam, 1% SFRC beam and 1% SFRC beam amplified with CFRP can predict the damage that occurs to the beam.

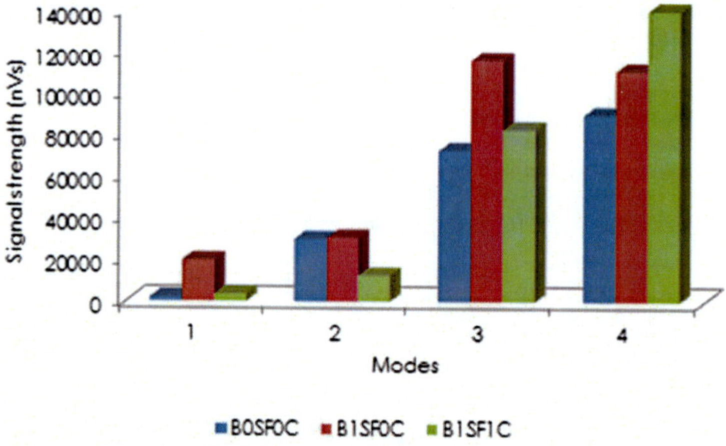

Fig. 19.7 Signal strength with respect to damage modes for all beam specimens

19.4 Conclusion

Based on the observation of the crack propagation for all beam specimens and based on the observation of the crack propagation for all beam specimens and the relationship between AE signal strength and damage modes, it is found that the beam B0SF0C produced a wider crack at the ultimate load with the lowest signal strength compared to other beam specimens. For beam specimen B1SF0C, the formation of crack pattern seems reduced but small cracks were developed and propagated at the shear region. This occurs when the inclusion of 1% steel fibre in the concrete delayed the formation of cracks. The signal strength produced higher compared to beam specimen B0SF0C.

The results obtained from beam specimen B1SF1C which was designed with the inclusion 1% of steel fibre in the concrete mix and layered with CFRC on the soffit demonstrated that the crack propagation is reduced more as well as the formation of cracks in the flexural and shear region. The relationship between AE signal strength and damage mode was noticed to have the highest signal strength because of the high stiffness on transverse direction provided by CFRP layer. From this study, it proved that the application of CFRP layer to strengthen SRFC beam surface can help increase the strength characteristic of the beam and very good for rehabilitation purposes.

Acknowledgements The financial support of this conference's fee by the grant RAGS/1/2014/TK02/UITM//2 with the reference no. of 600-RMI/RAGS 5/3 (227/2014) from Ministry of Education (MOE) is gratefully acknowledged. Special thanks to Universiti Teknologi MARA (UiTM) Pulau Pinang. The authors also thank to all assistant engineers in heavy structural laboratory and concrete, Faculty of Civil Engineering, Universiti Teknologi MARA Pulau Pinang.

References

1. Yun HD, Choi WC, Seo SY (2010) Acoustic emission activities and damage evaluation of reinforced concrete beam strengthened with CFRP sheets. NDT Int 43:615–628
2. Gholamhoseini A, Khanlou A, MacRae G, Scott A, Hicks S, Leon R (2016) An experimental study on strength and serviceability of reinforced and steel fibre reinforced concrete (SFRC) continuous composite slabs. Eng Struct 114:171–180
3. Chang D, Chai WK (1995) Flexural fracture and fatigue behavior of steel-fiber-reinforced concrete structures. Nucl Eng Des 156:201–207
4. Patil SP, Sangle KK (2016) Tests of steel fibre reinforced concrete beams under predominant torsion. J Build Eng 6:157–162
5. Mat Saliah SN, Megat Johari, MA, Md Nor N (2008) Effects of steel fiber on properties of concrete. In: Proceedings of the international conference on science & technology: applications in industry & education. 869–875
6. Ruano G, Isla F, Pedraza RI, Sfer D, Luccioni B (2014) Shear retrofitting of reinforced concrete beams with steel fibre reinforced concrete. Constr Build Mater 54:646–658
7. Altun F, Haktanir T, Ari K (2017) Effects of steel fiber addition on mechanical properties of concrete and RC beams. Constr Build Mater 21(3):654–661

8. Lee YO, Yun HD, Seo S (2011) Acoustic emission monitoring and fracture process of reinforced concrete beams strengthened in flexure with CFRP. Adv Mater Res 163–167:2581–2584
9. Lee JK, Lee JH (2002) Nondestructive evaluation on damage of carbon fibre sheet reinforced concrete. Compos Struct 58:139–147
10. Naaman A, Park SY, Lopez MDM (1999) Repair and strengthening of reinforced concrete beams using CFRP laminates. The University of Michingan, Department of Civil and Environmental Engineering
11. Ghiassi B, Verstrynge E, Lourenco PB, Oliveira DV (2014) Characterization of debonding in FRP-strengthened masonry using the acoustic emission technique. Eng Struct 66:24–34
12. ASTM E 1316—06 (2006) Standard terminology for non-destructive examinations. ASTM Int 5
13. Ziehl PH (2000) Development of a damage based design criterion for fiber reinforced vessels. University of Texas, Ph.D
14. Xu J (2008) Nondestructive evaluation of prestressed concrete structures by means of acoustic emissions monitoring. PhD thesis, Auburn Alabama
15. Md Nor N, Ibrahim A, Muhamad Bunnori N, Mohd Saman H, Mat Saliah SN, Shahidan S (2015) Fatigue crack inspection and acoustic emission characteristics of precast RC beam under repetition loading. Appl Mech Mater 773–774:1022–1026
16. Md Nor N, Mat Saliah SN (2016) Acoustic emission signal strength for damage evaluation of reinforced concrete beam. ARPN J Eng Appl Sci 11(4):2622–2626
17. Wan Ahmad WNA, Md Nor N (2016) A study on damage assessment of RC beam wrapped with carbon fibre sheets using parameters of acoustic emission signal. ARPN J Eng Appl Sci 11(4):5391–5396
18. Md Nor N, Ibrahim A, Muhamad Bunnori N, Mohd Saman H, Mat Saliah SN, Shahidan S (2014) Diagnostic of fatigue damage severity on reinforced concrete beam using acoustic emission technique. Eng Fail Anal 41:1–9
19. Degala S, Rizzo P, Ramanathan K, Harries KA (2009) Acoustic emission monitoring of CFRP reinforced concrete slabs. Constr Build Mater 23:2016–2026
20. Gostautas RS, Ramirez G, Peterman RJ, Meggers D (2005) Acoustic emission monitoring and analysis of glass fibre-reinforced composites bridge decks. J Bridg Eng 10(6):713–721
21. Md Nor N, Ibrahim A, Muhamad Bunnori N, Mohd Saman H (2013) Acoustic emission signal for fatigue crack classification on reinforced concrete beam. Constr Build Mater 49:583–590
22. Noorsuhada MN (2016) An overview on fatigue damage assessment of reinforced concrete structures with the aid of acoustic emission technique. Constr Build Mater 112:424–439
23. Yepez Roca LO (1997) Acoustic emission examination of high strength prestressed concrete girders. Master of Science in Engineering Thesis, The University of Texas
24. Aggelis DG, Soulioti D, Barkoula NM, Paipetis AS, Matikas TE, Shiotani T (2010) Acoustic emission monitoring of steel-fiber reinforced concrete beams under bending. J Acoust Emiss 28:23–40
25. Calogero C, Lidia M, Papia M (2003) Effectiveness of stirrups and steel fibres as shear reinforcement. Cem Concr Compos 777–786
26. Muralidhara S, Prasad BKR, Eskandari H, Karihaloo BL (2010) Fracture process zone size and true fracture energy of concrete using acoustic emission. Construct Build Mater 479–486
27. Shahidan S, Muhamad Bunnori N, Md Nor N, Basri SR (2012) Health index evaluation on acoustic emission signal for concrete structure by intensity analysis method. Adv Mater Res 403–408:3729–3733
28. Mat Saliah SN, Md Nor N, Abd Rahman N, Abdullah S, Tahir MS (2021) Evaluation of severely damaged reinforced concrete beam repaired with epoxy injection using acoustic emission technique. Theor Appl Mech 112

Chapter 20
Simulation of Structural Analysis on Fatigue Crack (Isotropic) in FRP Laminate

Roslin Ramli, Mohd Hisbany Mohd Hashim, Suhailah Mohamed Noor, Anizahyati Alisibramulisi, and Muhammad Azrie Husainy Mohd Jasri

Abstract Fiber Reinforced Polymer (FRP) is a composite material that consists of a polymer matrix embedded with high strength fibers, such as glass, aramid, and carbon. FRP laminate to be improved on the matrix cracking which the damage between fiber in the layer parallel. However, the studies on this fatigue crack initiation, including on matrix cracking under fatigue loading. This paper aims to emphasize a study on the effect of fiber angle orientation on fatigue strength of FRP, which are [+45°] and [−45°]. The method of laminate is isotropic with the same material properties in all directions, and normal loads create only normal strains. The model of isotropic laminate simulation that will be used is [+45°/−45°] as the angle orientation. The model size is 250 × 25 × 8 mm in rectangular bar according to the ASTM D 3309 and ABS rules for thickness 8 mm part of the hull. The simulation of the fatigue test by using the Ansys software. The results were presented in the form of simulation results, showing that [+45°] orientation has a higher fatigue strength as a function of angle orientation. The benefit of result on isotropic in FRP laminate for the naval architecture and can be a guideline for the boat building industry.

Keywords Fatigue crack initiation · Isotropic · Fiber reinforced polymer

R. Ramli (✉) · M. A. H. M. Jasri
UniKL Malaysian Institute of Marine Engineering Technology (UniKL-MIMET), Jalan Pantai Remis, 32200 Lumut, Perak, Malaysia
e-mail: roslin@unikl.edu.my

R. Ramli · M. H. M. Hashim · A. Alisibramulisi
Faculty of Civil Engineering, Universiti Teknologi MARA (UiTM), 40450 Shah Alam, Selangor, Malaysia

S. M. Noor
Faculty of Civil Engineering, Universiti Teknologi MARA (UiTM), 13500 Bukit Mertajam, Pulau Pinang, Malaysia

S. Abdullah et al. (eds.), *Structural Integrity Cases in Mechanical and Civil Engineering*, Structural Integrity 23, https://doi.org/10.1007/978-3-030-85646-5_20

20.1 Introduction

The Fiber Reinforced Polymer (FRP) is a composite material that consists of a polymer matrix embedded with high strength fibers, such as glass, aramid, and carbon. The main benefits of composite materials are high strength and stiffness, with low density [1–4]. In previous years, some of the researches have focused on composite materials particularly glass fiber to identify the fatigue strength and failure mechanisms of the material.

Several factors affect the behavior of the material, mostly due to material design and environmental effects. From the perspective of environmental impact on the strength of the material, humidity levels and temperature influence the adhesive strength of the fiber-matrix interface and the strength of a polymer matrix [5]. The material properties and the behavior under fatigue loading are not just affected by the material design, for example, the delamination, matrix failure, and fiber fracture, while other factors can also give significant effects [6–8]. The most important factors can be loading rate, load frequency, thickness, fiber stack sequence, and fiber angle orientation.

Fatigue is the degradation of material properties that will happen when the material is testing under cyclic loading. This will be causing damages and cracks in the material. Fatigue occurs when the material is subject to a variable level of stress over a period of time.

However, the studies on this fatigue crack initiation, including on matrix cracking under fatigue loading. FRP laminate to be improved on the matrix cracking which the damage between fiber in the layer parallel [9]. This paper aims to emphasize a study on the effect of fiber is angle orientation on fatigue strength of FRP, which are [+45°] and [−45°].

20.2 Material

The laminate of the fiber will use the isotropic method. The isotropic is the same material properties in all directions and normal load. The sequence of laminate is nine layers which are tissue mat, chopped strand mat fiberglass, and woven roving. The resin is unsaturated polyester, and the hardener is catalyst Methyl Ethyl Ketone Peroxide. The example of angle orientation and stack sequence an as shown in Figs. 20.1 and 20.2.

20.3 Methodology

The method has few steps to complete the simulation. The first step is the parameters of the specimen with the model of isotropic laminate simulation that will be used is [+45°/−45°] as the angle orientation as shown in Table 20.1. The model size is 250

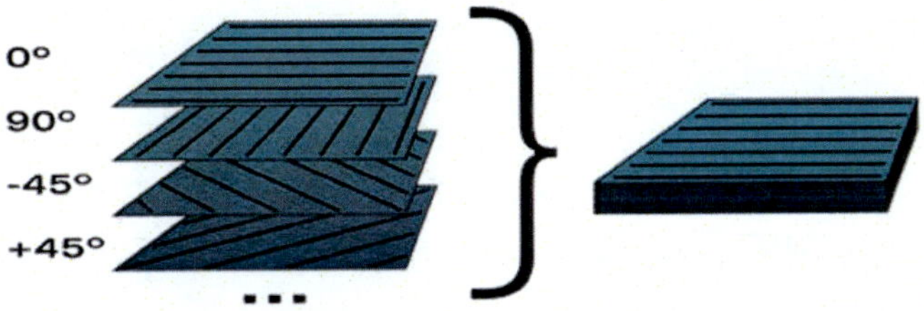

Fig. 20.1 Example of isotropic laminate

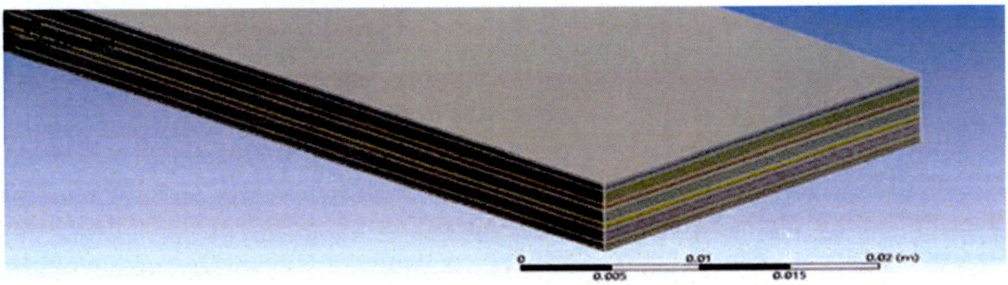

Fig. 20.2 Nine layers of fiberglass laminate

Table 20.1 The angle of orientation and lamination sequences

Specimen	Angle orientation	Lamination sequence
+45°	[0/0/ + 45/0/ + 45/0/ + 45/0/0]	T/CSM/WR/CSM/WR/CSM/ WR/ CSM/T
−45°	[0/0/−45/0/−45/0/−45/0/0]	T/CSM/ WR/CSM/WR/CSM/WR/CSM/T

× 25 × 8 mm in rectangular bar according to the ASTM D 3309 and ABS rules for thickness 8 mm part of the hull.

The simulation of the fatigue test was studied by using the Ansys software. Start with the workbench to create the specimen layer by layer according to the types of fiberglass and size. Once complete, assemble according to the angle orientation and stack sequences. Add the material data of the specimen +45° and −45° as shown in Table 20.2.

Table 20.2 Material data for specimen +45° and -45°

Material data	Specimen + 45	Specimen −45
Load (kN)	30	30
Tensile strength (Pa)	1.3536×10^8	9.4306×10^7
Young modulus	2.82×10^{10}	3.6272×10^{11}
Possion's ratio	0.23	0.23
Bulk modulus (Pa)	1.747×10^{10}	2.239×10^{11}
Shear modulus (Pa)	1.146×10^{10}	1.475×10^{11}

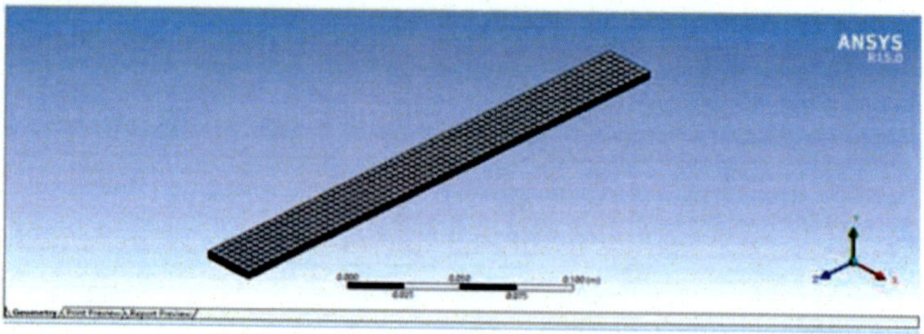

Fig. 20.3 Meshing

The static structural has been chosen to do this simulation. Setup the meshing, fixed support, force, and solution. The meshing nodes are 33,876 with 4347 elements as shown in Fig. 20.3. The force is the same in every layer which is 30 k Newton. The solution results are directional deformation, equivalent elastic strain, maximum principal elastic strain, equivalent stress, and maximum principal stress. Nevertheless, the results on the nine layers of fiberglass laminate also will be considered for this study.

20.4 Results and Discussions

The specimens stacking sequence and orientation use either tissue, chopped strain mat or woven roving fiberglass with [+45°/−45°] angle orientation laminate together by using ANSYS laminate glass characteristic material data. The simulation results in the aftereffect of mechanical properties of the laminate fiber after been loads 30 k Newton. The detailed results as shown in Tables 20.3 and 20.4, respectively, are of +45° and −45°.

The directional deformation results for +45° and −45° show that minimum and maximum occurred on the CSM 450. The result for +45° shows the minimum deformation as 2.8904×10^{-4} m and the maximum deformation as 2.8904×10^{-4} m.

Table 20.3 Result of the + 45°

	Minimum	Maximum
Directional deformation (m)	$-2.8904X10^{-4}$	$2.8904X10^{-4}$
Equivalent elastic strain (m/m	$5.6283X10^{-2}$	0.1012
Maximum principal elastic strain (m/m)	$6.5919X10^{-2}$	0.10124
Equivalent stress (Pa)	$1.5863X10^{9}$	$2.8538X10^{9}$
Maximum principal stress (Pa)	$1.9494\ X10^{9}$	$2.8756\ X10^{9}$

Table 20.4 Result of the –45°

	Minimum	Maximum
Directional deformation (m)	-2.2472×10^{-5}	2.2472×10^{-5}
Equivalent elastic strain (m/m	4.3758×10^{-3}	7.8708×10^{-3}
Maximum principal elastic strain (m/m)	5.125×10^{-3}	7.8708×10^{-3}
Equivalent stress (Pa)	1.5863×10^{9}	2.8538×10^{9}
Maximum principal stress (Pa)	1.9494×10^{9}	2.8756×10^{9}

However, the result for −45° shows the decrease to the minimum deformation as −2.2472 × 10^{-5} m and the maximum deformation as 2.8904 × 10−4 m.

The results of equivalent elastic strain and equivalent stress for +45° and −45° has occurred minimum on CSM 450 and maximum occurred on the tissue mat. The result of +45° minimum equivalent elastic strain is 5.6283 × 10^{-2} m/m and the maximum is 0.1012 m/m with the minimum equivalent stress same results for +45° and −45° 1.5863 × 10^{9} Pa and maximum 2.8538 × 109Pa. However, the results for equivalent elastic strain −45 decrease with the minimum 4.3758 × 10^{-3} m/m and the maximum 7.8678 × 10^{-3} m/m.

Figure 20.4 shows the result for +45° and −45° equivalent stress. From the minimum to maximum, it shows that the fatigue crack tends to propagate initially along the specimen surface.

The minimum and maximum occurred on the tissue mat for both angle orientation +45° and −45°. The result of maximum principal elastic strain for +45° with minimum 6.5919 × 10^{-2} m/m and a maximum of 0.10124 m/m, with the −45° decrease of minimum result is 5.125 × 10^{-3} and the maximum result is 7.8708 × 10^{-3} m/m. Whereas, the maximum principal stress for +45° and −45° is the same, which is the minimum 1.9494 × 10^{9} Pa and the maximum 2.8756 × 10^{9} Pa.

Figures 20.5 and 20.6 shows the result for +45° and −45° maximum principal elastic strain. The minimum to maximum shows the strain on the specimen surface.

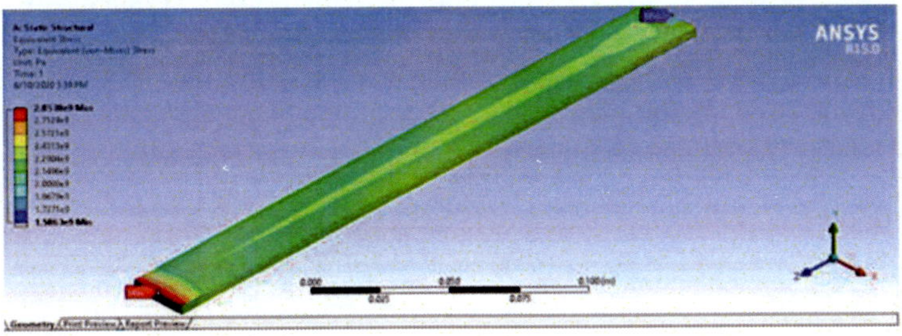

Fig. 20.4 +45 and −45 equivalent stress

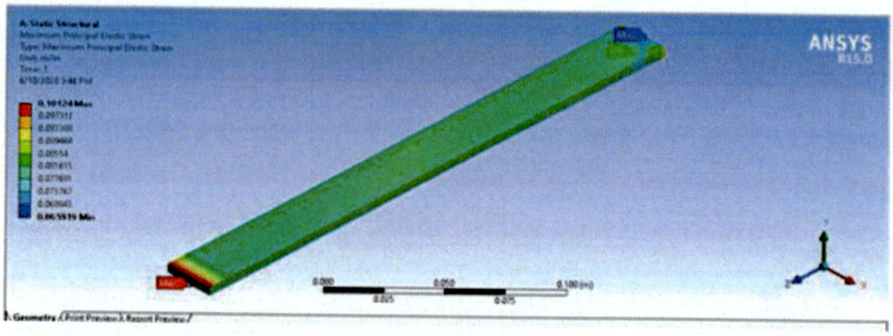

Fig. 20.5 +45 maximum principal elastic strain

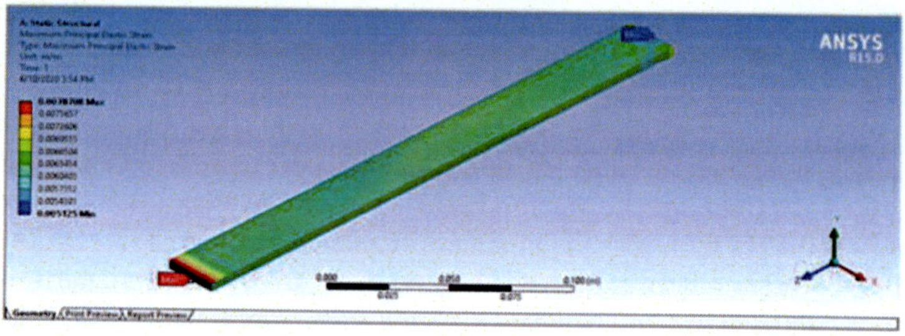

Fig. 20.6 −45 maximum principal elastic strain

The +45° has a higher maximum principal elastic strain which is 0.10124 m/m. It occurred on the tissue mat which is the first layer of the specimen. The −45° has the lowest maximum principal elastic strain which is 7.8708×10^{-3} m/m. It also occurred on the tissue mat which is the first layer of the specimen.

The overall results for are shows that initiation cracks will have on the first and second layer of laminate. The first layer is a tissue mat, and the second layer is a chopped strand mat (CSM 450) for −45° angle orientation and stack sequence. The difference is minimum and maximum value overtime on the maximum elastic strain.

The minimum value over the time results for +45° is 6.5919×10^{-2} m/m and the maximum value over the time results is 0.10124 m/m. The minimum value over the time results for −45° is 5.125×10^{-3} m/m and the maximum value over the time results is 7.8708×10^{-3} m/m. Both results, show the initiation cracks occur on the surface of the specimen −45° in a short time than the +45°.

The result of +45° and −45° can verify with the crack initiation of the experimental data. The fatigue test was performed using the Shimadzu Machine. The experimental results show that the specimen +45° has the highest cycle count, which is 225584. The −45°has the lowest cycle count, which is 109012. The difference between the two specimens is 116572.

The length of the crack initiation for the angle orientation on +45° is 1.8825 mm and the angle orientation for −45° is 4.7813 mm. The results of the crack initiation

Table 20.5 Fatigue experimental results

	Specimen +45°	Specimen −45°
Cycle count	225,584	109,012
Crack initiation–length (mm)	1.8825	4.7813
Crack initiation–width (mm)	0.0458	0.0677
Crack initiation–depth (mm)	0.005	0.12
Elapsed	6 h 33 min	3 h 01 min
Types of cracking	Matrix	Matrix

are shows that the +45° is the higher strength compared to the −45° lower strength with the difference of the length is 2.8358 mm.

The width of crack initiation for the angle orientation +45° is 0.0458 mm and the angle orientation on −45° is 0.0677 mm. The results of the crack initiation are shows that the +45° is the higher strength compared to the −45° lower strength with the difference of the width is 0.0219 mm.

The depth of the angle orientation on +45° is 0.005 mm and the depth of the angle orientation on −45° is 0.12 mm. This shows angle orientation +45° initiation cracks will have on the first layer of laminate with tissue mat. The angle orientation of −45° that has occurred on the first layer is tissue mat and the second layer is chopped strand mat (CSM 450) with the thickness 0.1 mm (Table 20.5).

Figures 20.7 and 20.8 shows the crack initiation length of the angle orientation + 45° and −45° crack initiation on the surface of the laminate by using Dino-Lite. The results of the fatigue test are the same as the Ansys simulation, which is the cracking initiation that occurred on the surface of the specimen.

The +45° crack initiation length is 1.8825 mm, which occurred on the surface of the specimen. Thus, the −45° crack initiation length is 4.7811 mm, which also occurred on the surface of the specimen. The −45° is the longest crack on the surface

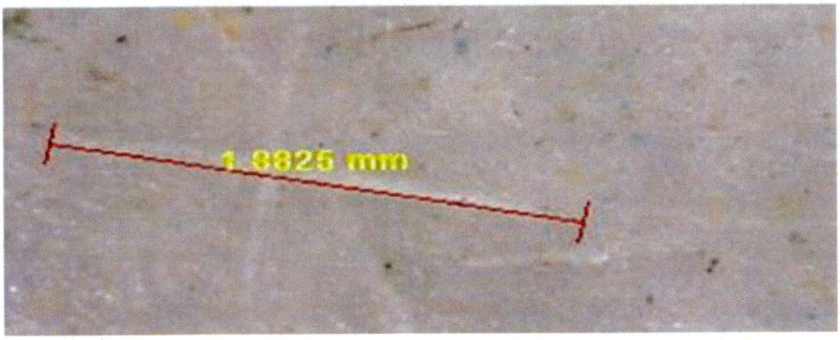

Fig. 20.7 +45° Crack initiation length

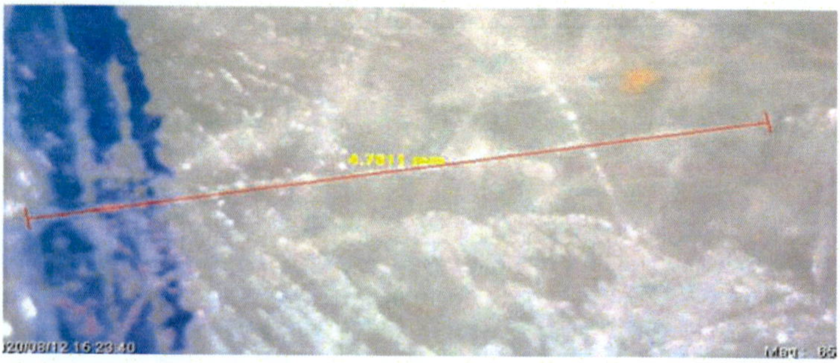

Fig. 20.8 -45° Crack initiation length

of the specimen, compared to +45° which is the shortest cracking that occurs on the specimen.

Figures 20.9 and 20.10 shows the crack initiation width of the angle orientation + 45° and −45° crack initiation on the surface of the laminate by using Dino-Lite. The results of the fatigue test are the same as the Ansys simulation, which is the cracking initiation that occurred on the surface of the specimen.

The +45° crack initiation width is 0.0458 mm, which occurred on the surface of the specimen. Thus, the −45° crack initiation width is 0.0677 mm, which also

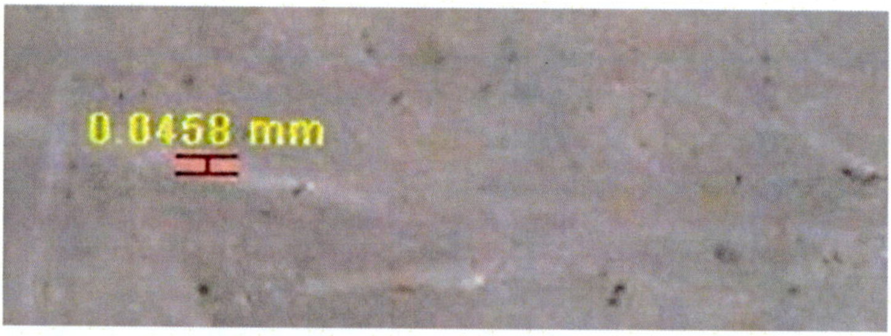

Fig. 20.9 −45° crack initiation width

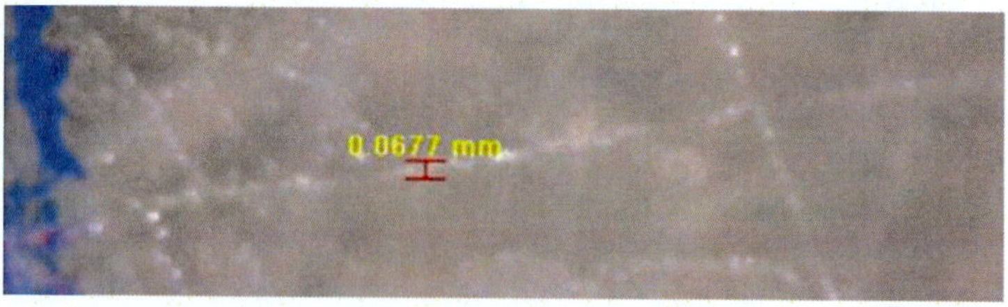

Fig. 20.10 −45° crack initiation width

occurred on the surface of the specimen. The −45° is the biggest crack on the surface of the specimen compared to +45°, which is the smallest cracking that occurs on the specimen.

20.5 Conclusion

In this simulation, it can be concluded that the +45° and −45° have different strengths, the strain, and the time of initiation cracking. The +45° orientation has a higher fatigue strength as a function of angle orientation. The result of +45° for the maximum principal elastic strain is 0.10124 m/m and the maximum principal stress is 2.8756 GPa.

The simulation results are proved by the experiment with the cracking initiation on the nine layers of fiberglass laminate. The cracking initiation has occurred on the first layer and the second layer of the laminate. The minimum occurs on the tissue mat and the maximum occurs on the chopped strand mat (CSM 450). Other than that, the result of the cycle counts for the angle orientation +45° more than the angle orientation −45°.

The outcome of this study shows that the angle orientation +45° method is the best specimen. This method is used for improving the strength of the fiberglass boat. The benefit of result on isotropic in FRP laminate is for the naval architecture and can be a guideline for the boat building industry.

Acknowledgements The authors would like to express their gratitude and thanks to UniKL-Malaysian Institute of Marine Engineering Technology (UniKL-MIMET) for funding this research paper under the Research and Innovation section (R&I) and to her main supervisor and co-supervisor at UiTM. Lastly, regards and blessings are offered to all those who supported the completion of the paper in any respect.

References

1. Iqbal AA, Arai Y (2015) Study on low-cycle fatigue behavior of cast hybrid metal matrix composites. Int J Automot Mech Eng
2. Ismail AE, Aziz CA, Aswadi M (2015) Tensile strength of woven yarn kenaf fiber reinforced polyester composites. J Mech Eng Sci
3. Roslan SAH, Hassan MZ, Rasid ZA, Zaki SA, Daud D, Aziz S, Sarip S, Ismail Z (2015) Mechanical properties of bamboo reinforced epoxy sandwich structure composites. Int J Automot Mech Eng
4. Mazumdar SK (2001) Composites manufacturing: materials, product, and process engineering.
5. Zakaria KA, Jimit RH, Ramli SNR, Aziz AA, Bapokutty O, Ali MB (2016) Study on fatigue life and fracture behaviour of fibreglass reinforced composites. J Mech Eng Sci
6. Hosoi A, Kawada H (2018) Fatigue life prediction for transverse crack initiation of CFRP cross-ply and quasi-isotropic laminates. Materials 11:7
7. De Carvalho NV, Chen BY, Pinho ST, Ratcliffe JG, Baiz PM, Tay TE (2015) Modeling delamination migration in cross-ply tape laminates. Compos Part A Appl Sci Manuf 71:192–203

8. Naik NK, Meduri S, Chandrasekher Y (2001) Polymer matrix woven fabric composites subjected to low-velocity impact: part III–Effect of incident impact velocity and impactor mass. J Reinf Plast Compos
9. Masuelli MA (2013) Introduction of fibre-reinforced polymers—polymers and composites: concepts, properties and processes, in fiber reinforced polymers. Technol Appl Concr Repair

Chapter 21
Managing Damages Recovery: Adopting Green Road Operation and Maintenance Index Criteria

Jeffryl Azniel Adzar, Rozana Zakaria, Eeydzah Aminudin, Dayalan Rainoo Raj, Rozelawati Ishak, Mohamad Faizal Sahadan, Ridzuan Mohd Radzi, Mohamad Hafizudin Syafiq Abd Rashid, Vikneswaran Munikan, Siti Mazzuana Shamsudin, and Sherliza Zaini Sooria

Abstract Transport infrastructure such as road networks was imprinted as a significant contributor to the sustainability Triple Bottom Line. Road networks cover millions of kilometers globally and are identified as the most common traveling options. However, due to its sensitivity to climatic conditions, it has progressively contributed to road asset deterioration. Therefore, road Operation and Maintenance (O&M) is performed to ensure satisfactory road functions and perform its intended design. From the O&M perspective, balancing of road damages from climate change and the rehabilitation work's impact can be accomplished by sustainability assessment in a road rating system. Presently, the O&M phase is refuted and not featured by numerous international green rating systems. The study aims to supplement the pHJKR (Road) rating system with the O&M index indicator, aiming to limit road damages. This paper highlighted research methods, including pilot study, comparative review, focus group discussion, carbon emission survey, and expert validation. Results from the study identified an improvement is required for pHJKR (Road) due

J. A. Adzar · R. Zakaria (✉) · E. Aminudin · D. R. Raj · R. Ishak · M. F. Sahadan · R. M. Radzi · M. H. S. A. Rashid
School of Civil Engineering, Universiti Teknologi Malaysia, Johor Bahru, Malaysia
e-mail: rozana@utm.my

J. A. Adzar
e-mail: jeffryl@jkr.gov.my

V. Munikan
Civil Engineering Department, Universiti Pertahanan Nasional Malaysia, Kuala Lumpur, Malaysia

S. M. Shamsudin
Faculty of Architecture Planning and Surveying, Universiti Teknologi MARA, Shah Alam, Malaysia

S. Z. Sooria
Public Work Department, Kuala Lumpur, Malaysia

S. Abdullah et al. (eds.), *Structural Integrity Cases in Mechanical and Civil Engineering*, Structural Integrity 23, https://doi.org/10.1007/978-3-030-85646-5_21

to the inadequacy of O&M criteria and carbon criteria. As a result, a new green road rating system for the O&M phase is proposed to allow the criteria responsive to road damages reduction; thus, the risks can be predicted by the green assessment beforehand.

Keywords Sustainable · Green road · Operation and maintenance · Damage · Recovery

21.1 Introduction

Infrastructure facilities' contribution to the national economy is highlighted as a part of the essential component in the development framework. However, due to repercussions on a country's surrounding environment and socio-economic, substantial measures need to be comprehended to delicately curb the development's impact. The transportation sector was imprinted as a significant contributor and inflicted risks to the sustainability Triple Bottom Line. Apart from the impact of the road construction, the roads Operation and Maintenance (O&M) environmental effectis identified as one of the liabilities contributing to road damages. The life cycle of a road can elongate to a staggering of 40 years before the refurbishment phase. The outstretched phase of O&M in the road life cycle is foresighted as a dominant contributor to road sustainability.

The assessment to scale-yard road's sustainability by employing the road rating system reduces the need for road repair by diminishing reconstruction and rehabilitation works. In Malaysia, the inadequacy of the green road rating system is obvious in contrast to the vertical or building rating system. The *Penarafan Hijau JKR* (pHJKR) developed by PWD Malaysia is an assessment tool for non-tolled road projects, covering the planning, design, and construction phase, excluding the O&M phase. Hence, the purpose of this study is to supplement the existing pHJKR by proposing green criteria for a non-tolled road in the O&M phase as an index indicator, aiming to limit the effects of road damages originating from construction and rehabilitation activities.

21.1.1 Infrastructure

The demands for public services to complement the population's basic living necessities are supplemented by global infrastructure development. Power, telecommunication, water supply, sanitation and sewerage, solid waste collection and disposal, and piped gas are essential services insisted by the communities. Pertinent to that, public works infrastructure, such as roads, dam and canal works, railways, urban transport, ports and waterways, and airports [1], are persistently developed globally. Road networks cover millions of kilometers globally and play a vital role in connecting all the utilities with its elaborate yet specific network throughout a country. The importance of roads is beyond doubt in ensuring physical infrastructures operate and perform their primary function without disruption. Therefore, it is noteworthy to preserve road's design objectives by executing necessary actions to increase its service cycle [2].

However, with the increasing dependency on the transportation system, it has inflicted risks to the sustainability Triple Bottom Line, namely, the economy, social, and environment in a variety of consequences including congestion, accidents, fuel costs, excessive energy use, and pollution emission [3]. It is noted not just for the development but also the operation of the roads itself is learned to influence the three-basis element of sustainability [4]. Admittedly, according to ASSHTO [5], in promoting transportation sustainability objectives, Triple Bottom Line needs to be equally considered as it is crucial to the efficiency and effectiveness of a transportation system.

The generic development phases of roads consist of planning, design, construction, and O&M, whereby it requires the engagement of various stakeholders and technical expertise from owners, architects, engineers, planners, contractors to public users. The road planning, design, and construction phase is the most focused phase in the road life cycle [6], while the operational phase that includes road maintenance is regularly side-lined. Although planning, design, and construction represent the outcome of the road physically, the maintenance treatment application during the O&M phase contributes to the improvement in safety and ride comfort for the road users [7].

21.1.2 Operations and Maintenance (O&M)

Sullivan et al. [8] defines O&M as the decisions and actions in property and equipment control and upkeep. From the road network's perspective, managing deteriorating road facilities minimized the life cycle cost while maintaining existing road assets in acceptable condition [9]. Generally, O&M has different terms and definitions based on its functions. Operation is essential in ensuring roads function satisfyingly for the road users' needs, while maintenance is essential in maintaining the road performance with its previous intended design [10]. Varying between 30 and 40 years, O&M is remarked as the most extended period of a road life cycle [10] and requires maintenance to provide the best performance of road safety and practical facilities to accommodate the community's daily activities. Maintenance activities are required to ensure infrastructure remains in serviceable condition to end-user and are categorized into three groups, which cover preventive, corrective, and rehabilitation [11, 12]. The preventive maintenance comprises of scheduled and rigorous activities [8], including vegetation control, sign replacement, culvert and drain clearing, resealing, resurfacing, regrading (gravel roads), overlay, and reconstruction [13].

The performance of roads relies upon the demands and the type of vehicles utilizing the facilities. As Sánchez-Silva et al. [14] mentioned, over time, the degradation of a particular structure is unavoidable due to the variations of demands. The alteration to local socio-economic implicates new demands, and often, it was not anticipated during the planning stage of road development. These have influenced the vehicles' increment on the road resulting in acceleration of road deteriorations. These effects induce various types of socio-economic damages, such as traffic congestion, accidents, and time delays [15].

Therefore, maintenance management for roads in fair condition compared to rehabilitation works when in poor condition is vital to ensure road facilities consistently in better condition [16]. Preservation works need to be planned and performed precisely to sustain or restore roads' overall condition [17]. An insufficient maintenance program will not only affect the road operation but consequently induce various social problems [18], where in most cases, the maintenance backlog in the road sector contributes to cost implications due to vehicle operation, and maintenance resulted from travel time due to the increased road roughness [13]. For instance, in Australia, road accidents' increment contributed to an estimated cost of $27 billion annually [19].

21.1.3 Maintenance Impact on Climate Change

Previous study reveals that the maintenance works impacted the environment, hence enlisted as one of the activities affecting global climate change. According to UNFCC, in developed countries, measures in dealing with climate change is targeted in the energy sector and followed by the transportation sector [20]. The effects of maintenance, including greenhouse gas or GHG generation, eventually contribute to the world's rising climate change issues. Espinoza et al. [21] conquered the statement by stating that the impact of road construction and rehabilitation is very much associated with GHG emission increment in the atmosphere. Furthermore, the maintenance and rehabilitation frequency in preserving road usability is identified as carbon dioxide emissions contributors [15].

There are two types of energy consumed in the maintenance operation which are direct energy from on-site construction or maintenance works, and indirect energy from construction products and services [22]. The excessive reliance on materials, such as earth, stone, lime, cement, and asphalt, has emerged in road construction as the leading area for energy consumption and carbon emissions [23]. As part of road infrastructure assets, pavement preservation and rehabilitation are recognized as significant GHG emission contributors; therefore, it is crucial to consider environmental friendly pavement management [16]. Hence, predictive maintenance needs to be executed before poor road conditions to prevent the road from deteriorating at a rapid pace [24]. Besides physical damages to the road infrastructure, the maintenance works indirectly induce community safety risks and can temporarily stonewall economic activities.

21.1.4 Climate Change's Impact

Road rehabilitation's effects on climate change spheres back to road damages due to the alteration of climate change impact. Climate changes pose severe implications to public road infrastructure's operational and structural performance, resulting in a significant reduction in pavement service life [25–28]. The damages trigger cost increment due to intensified maintenance work and demand for climate resilience actions in not only road design but also in management [25, 29].

Climate change's impacts on the transportation system signify by the extreme weather effects such as changes in temperature, precipitation, sea-level rise, and hurricanes, whereby besides temperature influences on the pavement, 75% of road damages caused by flooding is due to increased precipitation [25, 30, 31]. The effects instigated by extreme weather, including damaged road lighting and traffic lights, water and sewerage structure disruptions, displaced maintenance hole/inspection covers, collapsed lighting, utility poles, traffic lights and road signs, displaced guardrails, flooded and damages of a sidewalk or walkway structure and foundation, urgently demand corrective maintenance [32]. These impacts can be minimized by several measures, including a better understanding of climate change effects and long-term maintenance program; however, due to climate changes vulnerability and cost increment, these efforts hinge on funding availability [33].

21.1.5 Sustainable Maintenance

To neutralize the sphere of effects from climate change and maintenance works, sustainable maintenance is requisite, where according to Bengtsson and Lundström [34], effective maintenance focuses on safety, environmental care, quality, and reliability/availability where needed. While Franciosi et al. [35] suggested that maintenance-related impacts on sustainability should be considered in maintenance management. Corresponding to Zhang and Mohsen [36] study, sustainable pavement maintenance is defined as an activity that reasonably uses labor, money, and natural resources; hence, reducing negative impacts on the surrounding environment, pavement users, workers, and neighborhood people while restoring pavement performance which eventually extends its service life. For sustainable maintenance, performance indicators are employed to measure and monitor the actual and desired maintenance's sustainable performance [37].

21.1.6 Sustainability Rating System

From the O&M perspective, balancing road damages from climate change and the rehabilitation work's impact on climate change can be accomplished by sustainability assessment in a green road rating system. The government, especially in developing countries, aims to administer the severe environmental impact on road's infrastructure by diminishing road repair's needs in construction and rehabilitation by utilizing the assessment scheme [15, 24].

Sustainability performance measurement or performance indicators is an option and it is implied for environmental, economic, and social performance monitoring and works as performance communicators to stakeholders [38]. The objective of having a rating system as a sustainable performance indicator is to allow designers and contractors to raise sustainability as a priority in their planning and execution of a project [39]. Considering the economic, social, and environmental impacts, a set of indicators is necessary when planning a comprehensive and sustainable transportation development [40].

The sustainable performance measurement for infrastructure is essential and should be considered in the design, construction, and O&M phases [41]. Implementing and incorporating sustainable measures in the maintenance and rehabilitation phase contributed to significant environmental and economic benefits [16], apart from triggering impacts in different parts of the pavement cycle [2]. For these reasons, recognizing and quantifying these impacts can be accomplished by implementing a rating system [42].

21.1.7 Malaysia's Green Rating System

In Malaysia, the Committee of Green Road and Committee of Green Rating, both associated with road sustainability assessment [43], have established the Public Works Department of Malaysia Green Rating Scheme (*Penarafan Hijau JKR*—pHJKR) to assess the application of green value in road projects [44]. This tool is currently applied to a new project for the Federal road valued more than RM50 million, and the one-off certification awards road planning, design, and construction phases with an exemption for the O&M [44]. Table 21.1 indicates the available rating system utilized by PWD Malaysia for roads, buildings, and infrastructure project certifications.

Referring to the PWD Director-General Instruction No.2/2020, the implementation phase of green certification for PWD projects is according to Table 21.2. In this matter, pHJKR is applied for road projects in the design and construction phase only.

Table 21.1 Green certification by project type and cost

Project type	Project cost	Type of certification
Buildings	>20 Mil	PWD standard/MyCREST
Roads	>50 Mil	pHJKR road sector
Infrastructure	>100 Mil	Sustainable INFRASTAR

Table 21.2 Certification phase by green rating tools

Certification/phase	Buildings		Roads/infrastructure		
	PWD standard	MyCREST	GPSS	pHJKR	Infra STAR
Design	√	√	√	√	√
Construction	√	√	√	√	√
O&M	√	√	X	X	X

21.1.8 Road Development in Malaysia

Road constructions undeniably one of the dynamic progress sectors catering to the towering development's demands, and the soaring needs for global transportation infrastructure by 2030 are projected to be almost $90 trillion [45]. Simultaneously, the estimation for Asia's transportation infrastructure investment suggests a staggering $8.4 trillion by 2030 [46]. In Malaysia, as of January 2020, 362 ongoing road projects (excluding highways) are under implementation by PWD Malaysia either in the planning, design, construction, or O&M phase with an overall cost of more than RM28.6 billion [47]. These projects consist of 177 projects in the O&M phase, 93 projects in construction, and 92 projects in the planning and design phase [47]. From the breakdown of projects by states, areas, project costs, and implementation stages, projects valued <RM 50 Mil seem to be dominating the list while roads in urban areas are modestly higher than rural areas.

Figure 21.1 indicates the overall statistics for road networks in Malaysia, while as reflected in Fig. 21.2, the Federal roads portioned to 7% of the overall road network in Peninsular Malaysia. In terms of paved road length in Peninsular Malaysia, Federal roads constituted 17,852.79 km; State roads consist of 227,646.71 km while Highways with 1,956.75 km [48]. The statistic between Federal and State roads (non-tolled road) with the Highway (tolled road) revealed the unproportioned road network

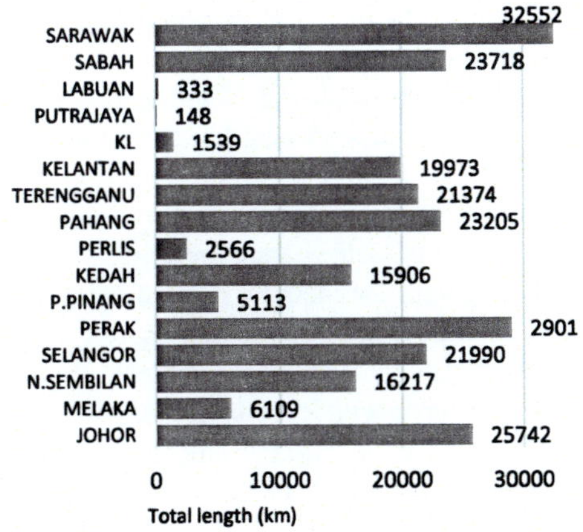

Fig. 21.1 Overall road network in Peninsular Malaysia

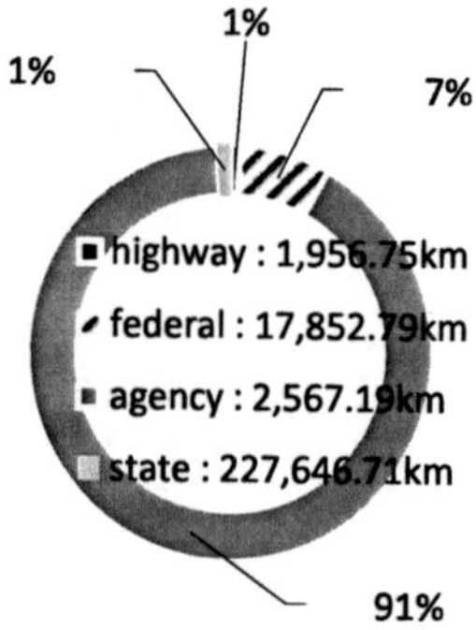

Fig. 21.2 Comparison between road network length in Malaysia with a clear indication of State road as the 'primary' road in terms of length.

21.2 Problem Statement

Presently, there are numerous international and regional green rating systems developed to evaluate sustainability. However, most of these rating systems focus more on buildings' assessment, contributing to limited roadwork assessment numbers. Among the established rating systems, only a handful assessed green roads while others concentrate mainly on highways and general civil infrastructure. Apart from the road-centric rating system's unavailability, the O&M phases are refuted and not featured by many road rating systems. However, the study showed the O&M consequent to a more significant impact on communities; nevertheless, the planning, design, and construction phase is undeniably spotlighted by many rating systems for road development sustainability performance. Admittedly, the carbon emission's impact on global climate change is vital and requires the government's policies and guidelines in undertaking climate change issues. Maintenance activities are acknowledged as contributing sources of carbon emissions in transportation infrastructure development and should be incorporated as green road assessment indicators. Nonetheless, carbon assessment criteria were only included in several rating systems as a baseline for future carbon reduction strategies. The scarcities of carbon criteria have hampered the aspiration for a comprehensive green road rating system that responds to climate change initiatives.

Along with the incorporation of O&M and carbon assessment criteria, the green road rating system's indicators should correspond to regional geographic climate, population, local resources, and local community's needs. Undoubtedly, many developing countries with less comprehensive rating system adopted an international rating system for their local assessments; however, creating a specific rating system that complements the local environment and geographic conditions will ensure a holistic assessment based on the country's geopolitics perspective. Malaysia's local green road rating system pHJKR (Road) applied to new federal road projects valued more than RM50 million in planning, design, and construction. These project selection criteria impeded the objective of assessing roads holistically, whereby most projects were discounted from the pHJKR (Road) assessment. For this reason, a comprehensive local green road rating system which constituted O&M and carbon assessment criteria was a requisite for complete road life cycle inclusive of all relevant criteria and credit.

21.3 Purpose of the Study

This study aims to establish pHJKR green road rating criteria in O&M phase for all hierarchies of non-tolled roads in Malaysia. The scorecards will also include carbon footprint measurements from routine maintenance activities. The following are the objectives carried out to achieve the goal of this research:

1. To investigate the applicability of Malaysia's green road rating system pHJKR.
2. To determine relevant criteria for road O&M assessment.
3. To identify the source and measurement of carbon emission in road maintenance activities.
4. To develop a sustainability index for pHJKR O&M's scorecard.

21.4 Research Scope

This study focuses on operational non-tolled Federal roads connecting states and districts and State roads maintained by the Public Works Department. The data collection is limited to Federal and State road maintenance under concessions companies in Peninsular Malaysia. This research also analyzes the carbon source and emission measurement in routine maintenance activities where the maintenance works are scheduled with definite work cycles. Malaysia's local green road rating system, pHJKR (Road), was employed as a benchmark and the focus for this study to determine the road's O&M green assessment criteria.

21.5 Methodology

This research adopted a non-experimental descriptive research design applying quantitative data collected from the questionnaire survey and data audit survey forms. The multiple analysis performed in this research design enabled the researcher to evaluate the data gathered, leading to the development of O&M criteria in the sustainability assessment scorecard. The research subject consists of professional practitioners from the Public Work Department (PWD), road maintenance concession companies, roadwork contractors, consultants, and other experts from various Departments and Ministries which direct and indirectly involve in road development in Malaysia.

The planned objectives were accomplished by several research methods, including Literature Review, Pilot Study, Comparative Review, Focus Group Discussion (FGD) Workshop, Carbon Data Survey, Carbon Emission Audit Online Survey, and Expert Validation Survey. The collected data from all the research procedures were analyzed using the IBM SPSS software data management tool to conduct frequency analysis and factor analysis.

21.6 Results and Discussion

21.6.1 Pilot Study

The pilot study's objective is to investigate the applicability of Malaysia's green road rating system—pHJKR (Road) by employing pHJKR (Roads) Ver. 2.0 scorecard to actual projects in the operation and maintenance phase. In this procedure, pilot investigation was conducted in an assessment session with the representatives of 12 PWD districts from 9 states in Peninsular Malaysia to gather information on the application, acceptance, and issuance of the pHJKR (Road) rating system.

Only 19% or 4 out of 21 projects achieved a minimum rating as Potential Certification from the assessment result, while 81% or 17 projects achieved below the lowest pHJKR (Road) rating. The unavailability of planning and design information during the project's operational stage has resulted in lower assessment scores. In contrast, the criteria in the construction stage collected the majority of points

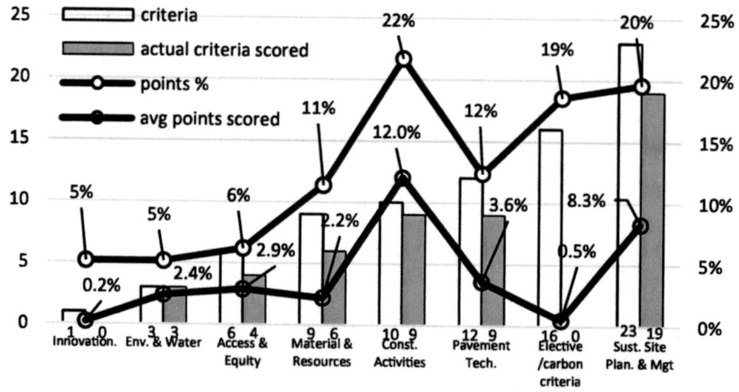

Fig. 21.3 pHJKR scoring by total criteria and points

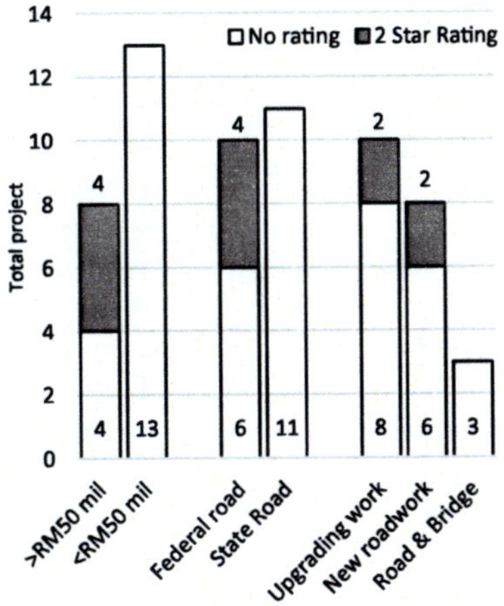

Fig. 21.4 pHJKR project rating

(Fig. 21.3), which relates to the same respondent who supervised and managed the project during the construction and operational stage.

All four projects (Fig. 21.4) with the minimum rating are in the category of above RM50 mil and Federal road. The pilot investigation results confirmed that pHJKR (Road) criteria were designed to evaluate projects in the specified categories, which are above RM50 mil, Federal road category, in planning, design, and construction stages but not tailored for the O&M phase. Conclusively, since the assessment was executed for operational road projects, the insufficient O&M criteria were clearly reflected in the assessment result.

21.6.2 Comparative Review

Comprehensive content analysis was performed by compiling established international and regional transportation rating system for review. The review was intended to understand the transportation sustainability concept and parameters, current best practice, assessment method, and appropriate green road green assessment criteria. The analysis was exploited as a guide and theory to create relevant criteria and credits for green O&M assessment of roads in Malaysia.

From the review of the established green road rating systems (*with O&M assessment criteria*), ten generic categories were identified, and the average percentage from total criteria related to the categories were ranked. The categories are Management, Environment/ecology, Social and Safety, Material, Energy, Construction activities, Water and Waste, Technology/Innovation, Economy, and others.

Table 21.3, pHJKR (Road) admittedly requires improvement, especially on the assessment phase where evaluation is conducted only for planning, design, and construction. The need for the O&M phase to be included in the assessment is

Table 21.3 pHJKR (Road) comparison to generic categories average %

Category	Main criteria	pHJKR (%)	Avg. (%)	Phase	Carbon criteria
Management	Sustainable site planning and management	17.39	18.61	P, D, C	No
Environment/ecology	Environment and water	4.35	15.10		
Social and safety	Access and equity	8.70	13.69		
Material	Material and resources	13.04	9.83		
Energy	Innovation	4.35	9.79		
Construction activities	Construction activities	30.43	9.78		
Water and waste	Environment and water	4.35	8.79		
Other	0	0.00	6.46		
Technology/innovation	Pavement technologies	17.39	4.98		
Economy	0	0.00	2.99		

vital to ensure sustainability is highlighted in every road development phase. The lack of carbon assessment criteria reflects the limitation of pHJKR (Road) and the criteria unavailability aligned with the inadequacy of carbon criteria in other rating systems studied. Although there are imperfections in pHJKR (Road), based on the common categories and average weightage distinguished from comprehensive content analysis, pHJKR (Road) criteria and its weightage appear to be relevant and averagely correlated with other rating systems. Table 21.4 reveals the concurrence of pHJKR (Road) with the basic conditions and theories in developing a comprehensive transportation rating system summarized from the literature study.

21.6.3 Focus Group Discussion—FGD

The process was initiated by inviting 50 participants from various organizations. The focus group discussion was organized in a workshop format and participated by 50 road experts from contrasting engineering backgrounds (Table 21.5) that contributes to a more reliable discussion. The experts were split into five groups, and questionnaires were answered by the experts according to their shared experiences in road development phases.

Table 21.4 Concurrence of pHJKR with the basic conditions and theories of a comprehensive transportation rating system development

No.	Theory	pHJKR (road)	Remarks
1	Planning and management	Yes	Included in main criteria (*Sustainable planning and management*)
2	Project lifecycle	No	O&M excluded
3	Maintenance work	No	Maintenance criteria not included
4	Economic elements	No	No direct economic criteria
5	Environment/ecology	Yes	Included in main criteria (*Environment and water*)
6	Localized	Yes	Developed by Malaysia PWD
7	Access and equity	Yes	Included in main criteria (*Access and equity*)
8	Green construction	Yes	Included in main criteria (*Construction activities*)
9	System and policy	No	Policy limited to certain project selection criteria
10	Energy efficiency	No	Energy not measured
11	Material and resource	Yes	Included in main criteria (*Material and resource*)
12	Standard and technology	Yes	Included in main criteria (*Pavement technologies and innovation*)
13	Indicators developm	Yes	Developed in 2012 (*currently at Revision no.3*)
14	Acceptance principles	Yes	Revision/improvement required

Table 21.5 Experts attended to FGD workshop

Organization	Nos.	Background	Nos.	Experience	Nos.
JKR road department	15	Highway engineering	30	<5 years	14
Government agency	10	Environmental engineering	6	6–10 years	16
JKR road project team	10	Structural/civil engineering	10	11–15 years	5
Road maintenance contractor	10	Business/infrastructure services	3	16–20 years	2
JKR environmental branch	2	Construction management	1	>20 years	13

Table 21.6 indicates the 7 Main criteria, 26 Sub-criteria, and 169 criteria from various sources, collected and listed in questionnaire survey forms for expert discussion in the workshop. While Fig. 21.5 reflects the factor analysis outcome conducted

Table 21.6 Criteria for expert validation in FGD

Main criteria	∑ criteria	∑ sub-criteria
1. Sustainable planning/management	4	22
2. Pavement technologies	1	10
3. Environment and water	6	34
4. Access and equity	4	32
5. Construction activities	3	27
6. Material and resources	6	37
7. Innovation	2	7
Total	26	169

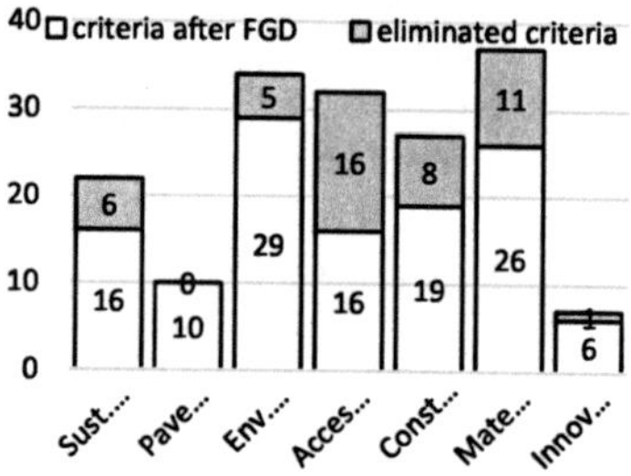

Fig. 21.5 Σ criteria post-FGD analysis

from the FGD workshop results, where 122 out of 169 criteria enlisted in the discussion were verified as relevant for O&M assessment. In comparison, 47 criteria were eliminated due to either identified as irrelevant by the experts or analysis results showing disputing opinion between experts.

21.6.4 Carbon Data and Measurement

A carbon data availability survey and carbon emission audit were conducted to gather as much information on carbon emission sources and measurement ability for road maintenance work. Eleven road maintenance companies from 11 states involving 74 districts all over Malaysia were approached for the data gathering. The companies represent the four regional areas assigned for Federal and State road maintenance, including the Northern region, Central region, Southern region, and Eastern regions.

Table 21.7 indicates the carbon source identified from the survey, which includes embodied carbon from material consumed, fuel consumption from material and waste transportation, fuel usage from machinery employed in maintenance works, vehicle fuel usage from routine inspection and staff commuting, energy usage from electricity, water, and paper usage in the office.

Table 21.7 Emission source from routine maintenance discovered from data availability survey

R01-Pothole patching	Material—pavement production
	Fuel usage—material delivery, types of machinery and equipment for patching works and waste disposal's transport
R02-Maintenance of road shoulder	Material—production
	Fuel usage—types of machinery and equipment for works
R03-Grass cutting	Fuel usage—machinery for grass cutting work
R04-Maintenance of road furniture	Material—cleaning agent and water usage
	Fuel usage—machinery and equipment for cleaning/repair works
R05-Maintenance of bridges and culverts	Fuel usage—machinery and equipment for cleaning/repair works
R07-Drainage	Fuel usage—types of machinery and equipment for cleaning/repair works
RI-Routine inspection for roads	Fuel usage—the vehicle for inspections
Office-Staff	Electrical usage—kWh, Water—M^3, or gal—Paper/-sheets or rim, fuel usage—vehicle for staff commuting
Overall	Carbon emission calculation and reporting

Figure 21.6, the total carbon emission was derived from routine maintenance activities by road maintenance companies in different districts. Results implicate that grass cutting activities stand out with a total annual emission of 133.987 tCO_2/km for Federal road maintenance, while drainage cleaning works contribute a total annual emission of 61.056 tCO_2/km for State road maintenance. The emission data concurred with the previous study, which stated direct energy from on-site construction or maintenance and materials usage is recognized as the source of energy consumption and carbon emissions. The carbon emission result also reflects the maintenance and rehabilitation works' impact on the environment as previously

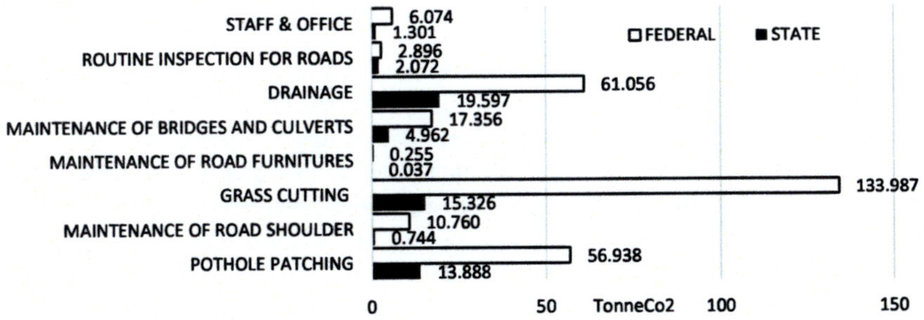

Fig. 21.6 Total carbon emission per year per kilometer for Federal and State road

discussed. The data from the study were later linked and integrated with O&M criteria as carbon assessment criteria in the assessment scorecard. Simultaneously, the average total emission is benchmarked as the baseline carbon emission for the activities prior to a further detailed study on actual emission from the maintenance works.

21.6.5 Expert Validation

The validation process was conducted using questionnaire survey through online survey and expert validation survey distributed by emails and other internet platforms. The instrument used to validate criteria scoring and weightage in the assessment scorecard was developed from the previous focus group discussion workshop and carbon data survey. The procedures commenced with the development of a questionnaire survey form encompassing 122 criteria and scores, developed from the analysis output of Focus Group Discussion. 239 respondents from various engineering backgrounds were involved in the validation survey.

Contrasting with the scoring and weightage validation survey, the subsequent validation is more meticulous and entails an expert outlook to examine the significance of the 47 eliminated criteria and 13 carbon emission criteria. Forty respondents from various organizations were invited to provide feedback in the survey as expert panels. From the analysis, all the respondents agreed with all the criteria scoring and weightage and concurred with the eradication of the eliminated criteria. The survey analysis results are the scorecard for the O&M assessment, complete with score points and carbon assessment criteria, as indicated in Table 21.8.

21.7 Conclusion

Operation, and maintenance of road induced similar impact as other construction works on the social, economic, and natural environment. Even though implementing maintenance activities should steer clear road damages, the maintenance works are partly responsible for the environmental effects. Therefore, sustainable practice in maintenance works is opted for deterring such impact while assuring road is operated in favor of the surrounding communities. To blend sustainability with road O&M, the development of a rating system is imminent. This study established an assessment index as a sustainable indicator for the non-tolled road where it caters for much-needed O&M criteria. The inadequacy of formal sustainability elements in road O&M induces impacts on the environment, leading to carbon emission concerns. Undoubtedly, the carbon assessment criteria incorporated in the assessment index precede the responsiveness of the rating system to sustainability. The pHJKR (Road) selection criterion is limited to new projects for Federal roads, valued more than

Table 21.8 Operation and maintenance scorecard derived from expert validations

Criteria	Code	Total points
Sustainable Planning and Management		
Maintenance works	SMMW	46
Noise mitigation plan	SMNM	
Staff and workers	SMWO	
Maintenance funding	SMFN	
Construction Activities		
Air quality	CAAQ	55
Waste management	CAWM	
System and policy *Carbon criteria + baseline*	CASP	
Environment and Water		
Wildlife protection	EWWP	73
Natural environment *Carbon criteria + baseline*	EWNE	
Trees and plants preservation	EWTP	
Storm water *Carbon criteria + baseline*	EWSW	
Water conservation	EWWC	
GHG emissions *Carbon criteria + baseline*	EWGE	
Pavement Technology		
Pavement *Carbon criteria + baseline*	PTPV	32
Material and Resources		
Fuel consumption	MRFC	72
Energy efficiency *Carbon criteria + baseline*	MREE	
Reuse and recycle	MRRR	
Local and alternative material	MRLA	
Environmentally friendly products and materials	MREP	
Access and Equity		
Public and community	AEPC	39
Pedestrian comfort *Carbon criteria Carbon criteria + baseline*	AEPD	
Safety and health	AESH	
Innovation		
Standards and technology	INST	17
Gr. Cleaning *Carbon criteria + baseline*	INGC	
$\sum$ main criteria: 7	$\sum$ sub-criteria	$\sum$ points
$\sum$ criteria: 26	122	334

RM50 mil, and in the planning, design, and construction phase, hence disappoints the aims to expand numbers of certified green roads in Malaysia.

Results from the pilot study on existing pHJKR (Road) assessment confirmed the ability of the scorecard to assess Federal road projects valued more than RM50 mil. However, only 19% of the projects scored the minimum 2-Star rating due to the insufficient O&M criteria. From the pHJKR (Road) comparative study and rating system review, generic O&M categories, including basic conditions and theories of a comprehensive transportation rating system, were identified, and checked for concurrence. Although with a few imperfections, pHJKR (Road) was found to be aligned with other established rating systems internationally and regionally. The questionnaire survey consists of 7 Main criteria, 26 Criteria, and 169 Sub-criteria validated by experts in the FGD workshop, and from the result analysis, 122 Sub-criteria verified as relevant for O&M assessment while 47 Sub-criteria was eliminated due to disputing views during the workshop. Subsequently, the carbon emission data determined from routine maintenance revealed that the largest emission is from grass cutting works, which emit 133.987 tCO_2/km annually for Federal road maintenance, while drainage cleaning works contribute 61.056 tCO_2/km annually for State road maintenance. Eventually, all the criteria, scoring, and weightage were validated in an expert validation survey, which resulted in the development of an O&M assessment scorecard to complement the existing pHJKR (Road). The enhancement of pHJKR (Road) with proposed O&M criteria allows for a more responsive road damage management and, by comprehensively utilizing the assessment, reduces risks of road deteriorations by climate change.

Acknowledgements This work was financially supported by Collaborative Research Grant PY/2018/07G93, Research Management Center, and research team members Green PROPMT, CREATE JKR, UiTM, and UPNM. Thus, directly it is supported by Ministry of Education, Malaysia.

References

1. Word Bank (1994) World Development Report 1994: Infrastructure for development: executive summary. World Bank, Washington, DC
2. Montoya-Alcaraz M, Mungaray-Moctezuma A, García L (2019) Sustainable road maintenance planning in developing countries based on pavement management systems: case study in Baja California México. Sustainability. 12(1):1–4
3. Clevenger CM, Ozbek ME, Simpson SP, Atadero R (2016) Challenges in developing a transportation sustainability rating system that meets the preferences of a department of transportation. J Transp Eng 142(4):04016005
4. Griffith A, Bhutto K (2008) Improving environmental performance through Integrated Management Systems (IMS) in the UK. Manag Environ Qual Int J 19(5):565–578
5. ASSHTO (2011) AASHTO—a policy on geometric design of highways and streets. American Association of State Highway and Transportation Officials, Washington, DC
6. Oswald Beiler M, Waksmunski E (2015) Measuring the sustainability of shared-use paths: development of the green paths rating system. J Transp Eng 141(11):04015026
7. Ćirilović Stanković J, Mladenović G, Queiroz C (2019) Impact of CO_2 emissions on low volume road maintenance policy: case study of Serbia. Transp Res Board 2673(12):747–755

8. Sullivan GP, Pugh R, Melendez AP, Hunt WD (2010) Operations & maintenance best practices: a guide to achieving operational efficiency. United States Department of Energy, Oak Ridge, Tennesy
9. Sarkar AK (2011) Development of a sustainable rural roads maintenance system in india: key issues. Transp Commun Bull Asia Pacif (81):36–51
10. Akbiyikli R, Eaton D (2006) Operation and maintenance (O&M) management in PFI road projects in the UK. In: Boyd D (ed) Proceedings of 22nd annual arcom conference 2006. Association of Researchers in Construction Management, Birmingham, UK, pp 393–402
11. Davis J, Brikke F (1995) Making your water supply work. IRC International Water and Sanitation Centre, p 29
12. Torres-Machi C, Pellicer E, Yepes V, Chamorro A (2017) Towards a sustainable optimization of pavement maintenance programs under budgetary restrictions. J Clean Prod 148:90–102
13. Harvey MO (2012) Optimising road maintenance. In: International transport forum discussion paper, No. 2012–12, Organisation for Economic Co-operation and Development (OECD), International Transport Forum, Paris
14. Sánchez-Silva M, Frangopol DM, Padgett J, Soliman M (2016) Maintenance and operation of infrastructure systems: review. J Struct Eng 142(9):F4016004
15. Ho Choi J (2019) Strategy for reducing carbon dioxide emissions from maintenance and rehabilitation of highway pavement. J Clean Prod 209:88–100
16. Ruiz A, Guevara J (2020) Environmental and economic impacts of road infrastructure development: dynamic considerations and policies. J Manag Eng 36(3):1–16
17. FHWA, Memo: guidance on highway preservation and maintenance (2016). https://www.fhwa.dot.gov/preservation/memos/160225.pdf. Accessed 1 Apr 2020
18. Shen L, Wu Y, Zhang X (2011) Key assessment indicators for the sustainability of infrastructure projects. J Constr Eng Manag 137(6):441–451
19. BITRE, Who's using the roads: variations in usage by drivers, BITRE (2014). https://www.bitre.gov.au/publications/2014/files/is_053.pdf. Accessed 1 Apr 2020
20. United Nations Climate Change Secretariat: Yearbook of Global Climate Action 2019. UNFCCC, Bonn, Germany (2019)
21. Espinoza M et al (2019) Carbon footprint estimation in road construction: La Abundancia-Florencia case study. Sustainability 11(8):1–13
22. Huang L, Krigsvoll G, Johansen F, Liu Y, Zhang X (2018) Carbon emission of global construction sector. Renew Sustain Energy 81(June 2016):1906–1916
23. Zhang Z, Gao X, Wang J, Ji X (2019) Prediction model for energy consumption and carbon emission of asphalt surface construction. In: IOP conference series earth environment 2019. Science 330:022052
24. Ruiz A, Guevara J (2020) Sustainable decision-making in road development: analysis of road preservation policies. Sustainability 12(3)
25. Melvin AM et al (2016) Climate change damages to Alaska public infrastructure and the economics of proactive adaptation. In: Proceeding of national academy of science, USA, vol 114, no 2. PNAS, Tempe, AZ, E122–E131
26. Abdul Rahman H (2018) Climate change scenarios in malaysia: engaging the public. Int J Malay-Nusantara Stud 1(2):55–77
27. Guest G, Zhang J, Maadani O, Shirkhani H (2020) Incorporating the impacts of climate change into infrastructure life cycle assessments: a case study of pavement service life performance. J Indus Ecol 24(2):356–368
28. Qiao Y et al (2020) Assessing impacts of climate change on flexible pavement service life based on falling weight deflectometer measurements. Phys Chem Earth (July):102908
29. Qiao Y, Santos J, Stoner AMK, Flinstch G (2020) Climate change impacts on asphalt road pavement construction and maintenance: an economic life cycle assessment of adaptation measures in the State of Virginia, United States. J Indus Ecol 24(2): 342–355
30. Oforah A, Emenike G, Ogbuji C (2020) Impact of climate change on transport infrastructure in Port Harcourt City, River State, Nigeria. Int J Innov Res Adv Stud 7(3):11–15

31. Hemed A, Ouadif L, Bahi L, Lahmili A (2020) Impact of climate change on pavements. In: E3S Web Conference 2020, vol 150, p 01008
32. Diakakis M, Boufidis N, Salanova Grau JM, Andreadakis E, Stamos I (2020) A systematic assessment of the effects of extreme flash floods on transportation infrastructure and circulation: the example of the 2017 Mandra flood. Int J Disaster Risk Reduct 47(September):101542
33. Lawrence J, Blackett P, Cradock-Henry NA (2020) Cascading climate change impacts and implications. Climate Risk Manag 29(August 2019):100234
34. Bengtsson M, Lundström G (2018) On the importance of combining 'the new' with 'the old'—one important prerequisite for maintenance in Industry 4.0. In: Procedia manufacturing, vol 25. Elsevier, Stockholm, Sweden, 118–125
35. Franciosi C, Lambiase A, Miranda S (2017) sustainable maintenance: a periodic preventive maintenance model with sustainable spare parts management. IFAC-Pap Line 50(1):13692–13697
36. Zhang Y, Mohsen JP (2018) A project-based sustainability rating tool for pavement maintenance. Engineering 4(2):200–208
37. Franciosi C, Voisin A, Miranda S, Riemma S, Iung B (2020) Measuring maintenance impacts on sustainability of manufacturing industries: from a systematic literature review to a framework proposal. J Clean Prod 260:121065
38. CH2M HILL and Good Company (2009) Transportation and sustainability best practices background. In: AASHTO transportation and sustainability peer exchange, pp 1–32
39. Reeder L (2011) Guide to green building rating systems. Wiley, New Jersey
40. Litman T (2019) Well measured—developing indicators for sustainable and livable transport planning. Victoria Transport Policy Institute, Canada
41. Lounis Z, McAllister TP (2016) Risk-based decision making for sustainable and resilient infrastructure systems. J Struct Eng 142(9):F4016005
42. Hubbard SML, Hubbard B (2019) A review of sustainability metrics for the construction and operation of airport and roadway infrastructure. Front Eng Manag 6(3):433–452
43. Jabatan Kerja Raya [JKR] (2016) Polisi Pembangunan Lestari JKR 2016–2020. Cawangan Alam Sekitar & Kecekapan Tenaga, K.Lumpur
44. JKR (2018) Manual of Penarafan Hijau JKR—pHJKR—Road Sector Version 3.0. Public Works Department Malaysia, Kuala Lumpur
45. World Resource Institute (2016) The sustainable infrastructure imperative. The New Climate Economy, Washington, DC
46. Bank AD (2017) Meeting Asia's infrastructure needs. Asia Development Bank, Manila, Philippines
47. PWD: SKALA 2020. https://www.jkr.gov.my. Accessed 1 Apr 2020
48. PWD Malaysia (2018) Road Statistics 2018 Edition, 2018th ed. Road Facilities Maintainance Branch, PWD Headquarters, Kuala Lumpur

Chapter 22
Building Defects Assessment Framework for Malaysian Construction Projects

Siti Hafizan Hassan, Mohd Samsudin Abdul Hamid, Mohd Faisal Zaini, Syahrul Fithry Senin, Zulfairul Zakariah, and Mohamad Zain Hashim

Abstract The building industries of Malaysia have grown steadily, supported by ongoing infrastructure projects. However, some of the construction projects were poorly built in terms of quality. Building defects are not new in the Malaysian construction industry, and as a developed country that grows rapidly, therefore, this problem cannot be ignored, because it will generate a negative image toward the future construction industry. This study aimed to determine the most common types of defects that occurred in buildings, and the contributing factors to building defects in the construction project are also determined. In addition, the effects and solutions of building defects in the construction industry are examined. By identifying the objectives, the building defects framework can be implemented and practiced more effectively. The data were analyzed using SPSS version 23 and the relative importance index. The result shows that cracks were produced. The highest-ranked type of building defects occurred with the relative importance index value of 77.3%, and poor workmanship is the most common factor that contributed to building defects in the construction industry with the relative importance index value of 81.3%. These defects also affect the appearance and safety of the building, which brings discomfort and dissatisfaction to the occupants living in the building. Therefore, the solutions are to overcome the defects with regular monitoring and supervision of workmanship, provide the workers with proper training, and ensure the usage of materials, especially the quality.

Keywords Construction defects · Assessment framework · Construction projects

22.1 Introduction

A building is a structured design constructed by engineers for many purposes. The main functions of a building are a place for people to live, work, do activities, store things, and others. As stated in [1], a building is a place to accommodate and work

S. H. Hassan (✉) · M. S. A. Hamid · M. F. Zaini · S. F. Senin · Z. Zakariah · M. Z. Hashim
School of Civil Engineering, Universiti Teknologi MARA Cawangan Pulau Pinang, Permatang Pauh, 13500 Pulau Pinang, Malaysia

S. Abdullah et al. (eds.), *Structural Integrity Cases in Mechanical and Civil Engineering*, Structural Integrity 23, https://doi.org/10.1007/978-3-030-85646-5_22

together, or for an organization to carry out its work. Moreover, the building also provides protection to the occupants from weathering and the conditions of the temperature. Also, various kinds of buildings such as commercial buildings, residential buildings, industrial buildings, educational buildings, government buildings, and so on can be distinguished by their function, several floors, and purposes [1].

Unfortunately, nearly every building has flaws and defects; whether it is a newly developed building or the old one. Several building defects usually occur in parts such as building roofs, walls, floors, toilets, windows, and doors [2]. These defects also give an uncomfortable view of sight to the occupants. Building defects range from minor cosmetic issues such as erosion of mortar joint, peeling paint, to the growth of mold due to exposure to hot and wet conditions. These issues also own minor defects where it does not harm the safety of the occupant in the building but cause harm to the property owner in the form of cost of repair and extra expenses. Meanwhile, major defects such as cracks on the beam, foundation failure, etc. due to overload cause more damage to the structure of the building, which could harm the occupant. Depending on the functions, system types, and materials used, various constructed facilities generate different defect types and require different levels and quality types [3]. Therefore, one of the objectives of this research is to identify the most common types of defects that occurred in the building. These building defects occurred in construction projects due to various contributing factors such as poor materials quality, poor workmanship, error design, corruption, water intrusion, lack of maintenance, external environmental condition, limited time, and cost. This defect problem occurs when the work that does not comply with the contract's express descriptions or requirements. An expert, such as an engineer or an architect, is the one responsible for the supervision, improper design, monitoring materials, and workmanship. Maintenance is therefore needed to prevent these types of defects. In Malaysia, building condition survey is underrated and disregarded by owners, managers, and expertise [4]. Furthermore, if these problems continue, not only the parties involved in the construction project will generate a negative image toward the future construction industry but also will be facing numerous procedures and regulations before being awarded a construction project. These problems cannot be ignored. Therefore, one of the objectives of this study is to determine the consequences and possible solutions to building defects.

22.2 Methodology

Based on this study, one of the methods is a questionnaire. Other than that, to verify the result of the questionnaire, interviewing with an expert was one of the methods to strengthen the result data as well as obtain the information related to the study based on their experience and opinions. The selection of the methods needs to make sure that all the objectives can be achieved. Figure 22.1 shows the flowchart of this study.

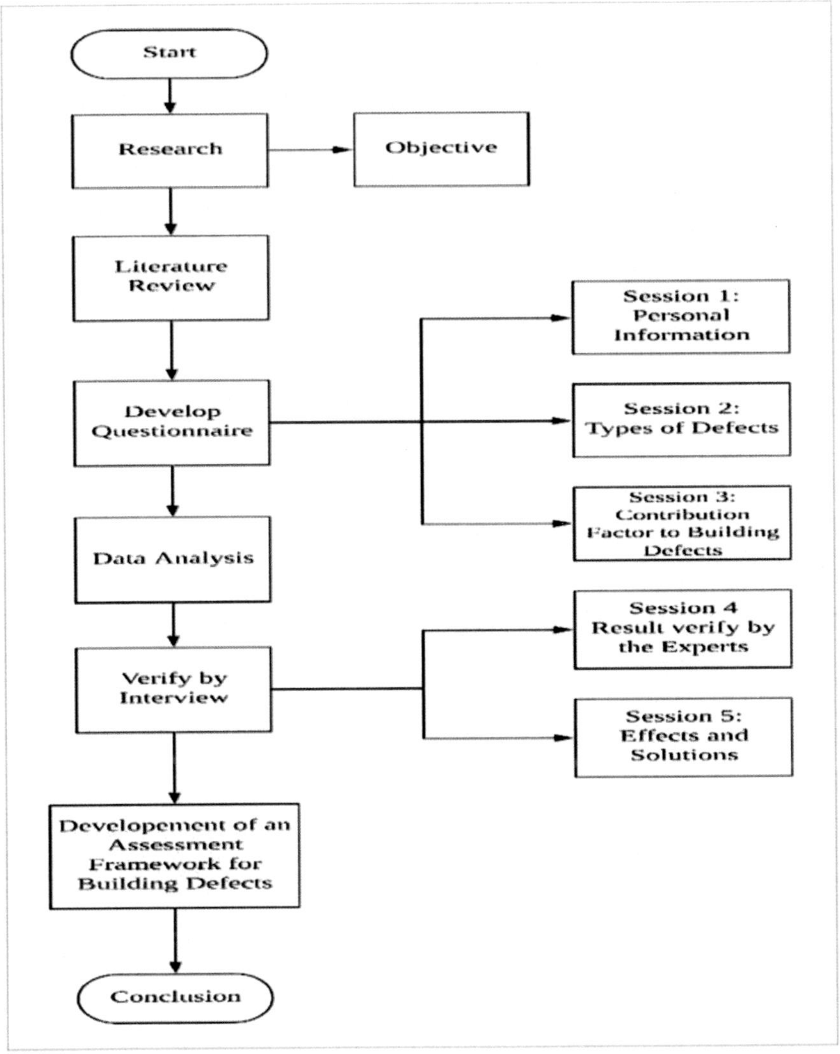

Fig. 22.1 Flowchart methodology

22.2.1 A Subsection Sample

The questionnaire forms were developed using the Likert scale often used interchangeably, with a rating scale from one to five (Table 22.1).

The questionnaire consists of three sessions: personal information, common types of defects, and common contributing factors to building defects. Personal information is used to obtain the respondents' profiles such as age, gender, working experience,

Table 22.1 Likert scale [5]

Scale	Description
1	Less occurred/less agree
2	Slightly occurred/slightly agree
3	Occurred/important/agree
4	Considerably occurred/considerably agree
5	Mostly occurred/very agree

position in the company, etc. The second session is for the respondents to score from one to five for the most common types of defects. The third session is for the respondents to score from one to five for the most contributing factors to building defects. By using this questionnaire, the objective is possibly achieved.

22.2.2 Data Analysis

IBM SPSS Version 23 Software. In this research study, after the results of the questionnaires were collected, descriptive statistics is used to determine the frequency, mean, etc. for every variable in the questionnaire by using International Business Machines (IBM) Statistical for Social Science (SPSS) statistic software version 23 platform that offers advanced statistical analysis, text analysis, etc. The data will be represented in the form of a table and bar chart.

Relative Importance Index (RII). The questionnaire data were analyzed using the relative importance index (RII) method to determine the relative importance among the types of defects that occurred and the contributing factors to building defects. The highest value of RII indicated the most common defect that occurred or agreed by the respondents. Equation (22.1) describes the relative importance index.

$$RII = \frac{\sum M}{ÆN} = \frac{5N_1 + 4N_2 + 3N_3 + 2N_2 + 1N_1}{5N} \tag{22.1}$$

where W = is the weights given to each factor by the respondents (1–5); N_1 = less occurred/agree; N_5 = mostly occurred/agree; A = is the highest weight (i.e., 5 in this case), N = is the total number of respondents.

Verify by Interview. This interview session involved the respondents who are willing to help verify the data analysis as well as examine the effects of building defects and solutions to building defects in the construction industry. Based on the analyzed data, the result will be verified by the expert. By interviewing with the expert we would be able to verify the information and facts based on their knowledge related to the study field. In addition, the experts were asked several questions regarding the effects of the building defects and the possible solution to defects in the construction industry.

The answer was based on their viewpoint, opinion, and experience related to building defects in the construction industry. The interview session involved all the parties responsible, such as the project engineer, senior engineer, consultant, contractor, and site supervisor. By doing so, the objectives of this study can be achieved.

Develop the Defect Assessment Framework. An assessment framework for building defects in Malaysian construction industry is developed. A lucid chart is used to develop the framework. The lucid chart is a web-based proprietary platform used to allow users located in multiple locations to collaborate with their colleagues in drawing, revising, and sharing charts and diagrams. The assessment framework will visualize the results, explore connections between contribution factors to building defects and the types of defects, effects, and solutions of building defects in the construction industry.

22.3 Results and Discussion

The data were collected at Sarawak. The targeted respondents are engineers, site engineers, project engineers, site supervisors, etc. A total of 30 usable questionnaires were collected and five respondents were interviewed.

22.3.1 Respondents Profile

Position of the Respondents. The respondents are project managers, senior engineers, resident engineers, site engineers, etc. Figure 22.2 states the total number of respondent's positions in their company. The highest number is the senior engineer which is 8, followed by the site engineer with a total of 6, which is the second-highest total out of 30 respondents. The lowest total is 1 for the site supervisor and officer supervisor.

Working Experience of the Respondents. Table 22.2 shows the total of respondent's working experience such as <5 years, between 5 and 10 years, and >10 years. The highest total of respondents with <5 years of working experience is 15, although a total of five respondents has working experience of >10 years.

Interviewee's Profile. These sessions involved one project manager, 2 site engineers, and one resident engineer. The list of expert's details is given in Table 22.3.

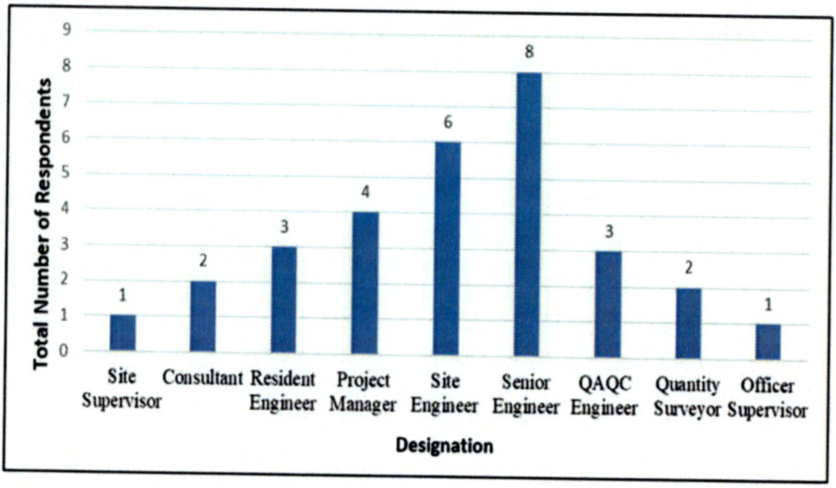

Fig. 22.2 Total number of respondents and their position

Table 22.2 Years of working experience of the respondents

Years of working experience	Total
<5	15
Between 5 and 10	10
>10	5

Table 22.3 List of interviewee's details

Position	Company	Experience (years)
Site supervisor	Junda Realty Sdn. Bhd	5
Resident engineer	JKR Miri Sarawak	4
Site engineer	Kayahan Construction Sdn. Bhd	6
Project manager	JKR Miri Sarawak	16
Site engineer	Borneo Bintang Sdn. Bhd	5

22.3.2 Common Types of Defects in Construction

This subsection is to find out in more detail regarding the types of defects in a construction project, especially in Sarawak. Based on the data from the questionnaire, the data had been interpreted by using SPSS version 23 software to identify the statistical data such as frequency, mean, total sum, etc. Besides, the relative importance index (RII) of types of defects was also calculated to determine the ranking of the most common types of defects.

Descriptive Statistics of Types of Defects. The statistics in Table 22.4 shows that cracking had the highest total number of scores by the respondents with a total of 116 out of 150. Cracking had a mean of 3.8667, which is also the highest mean among the rest of the defects. The lowest score by the respondents is foundation failure with a total of 65 out of 150. Foundation failure also had the lowest mean, which is 2.1667.

Ranking for the Most Common Types of Defects. A further and detailed way to determine the most common types of defect is by using the relative importance index (RII), which shows how important the variables are to the respondents. From the result in Table 22.5, the highest rank of common types of building defect that occurred is cracking with the RII value of 0.773. Cracks can be classified as non-structural and structural. The RII value for most types of defects is more than 0.4, which means all these types of defects are most likely to occur. The higher RII value shows a bad quality of production and should be controlled and monitored during and after construction. Further precautions are being suggested by an expert in Table 22.12.

Structural cracks can be found on the wall, columns, or beam caused by overloading. One of the factors is the limited cost that leads to cut down of the cost in material and labor. Lower cost on the material will lead to poor quality of materials, and thus affect the strength of the concrete to withstand the load. Therefore, the structure will crack easily. Also, cut down on the labor cost may cause fewer workers and inexperienced workers at the construction site. Owing to lack of knowledge and insufficient workers, the workers simply handle the concrete works and formworks which will affect the concrete strength in structure.

Table 22.4 Descriptive statistics for types of defects

	N	Descriptive statistics			
		Minimum	Maximum	Sum	Mean
Cracking	30	2.00	5.00	116.00	3.8667
Dampness	30	2.00	5.00	107.00	3.5667
Peeling paint	30	1.00	5.00	104.00	3.4667
Defective plaster	30	2.00	5.00	103.00	3.4333
Rendering erosion mortar	30	2.00	5.00	102.00	3.4000
Blemishes	30	1.00	5.00	101.00	3.3667
Timber decay	30	1.00	4.00	87.00	2.9000
Roof defect	30	1.00	5.00	72.00	2.4000
Corrosion	30	1.00	5.00	72.00	2.4000
Reinforcement steel	30	1.00	4.00	65.00	2.1667
Foundation failure	30	2.00	5.00	116.00	3.8667
N	30				

Table 22.5 The rank of common types of building defects based on the relative importance index

No	Types of defect	Relative importance index (RII)
1	Cracking	0.773
2	Dampness	0.713
3	Peeling paint	0.693
4	Defective plaster rendering	0.687
5	Erosion of mortar joint	0.680
6	Blemishes	0.673
7	Timber decay	0.580
8	Roof defects	0.480
9	Corrosion of reinforced steel	0.480
10	Foundation failure	0.433

22.3.3 Contribution Factors to Building Defects in Construction

Descriptive Statistics of Contribution Factors to Building Defects. The statistics in Table 22.6 show poor workmanship had the highest total number of scores by the respondents with a total of 122 out of 150. Furthermore, poor workmanship also had the highest mean of 4.0667 compared to the rest of the contributing factors to defects. The second-highest number of sums is limited cost with a total of 117 out of 150 with a mean of 3.900. Lastly, the lowest score by the respondents is corruption

Table 22.6 Descriptive statistics of the contributing factors to building defects

	Descriptive statistics				
	N	Minimum	Maximum	Sum	Mean
Poor workmanship	30	2.00	5.00	122.00	4.0667
Limited cost	30	2.00	5.00	117.00	3.9000
Construction materials	30	1.00	5.00	114.00	3.8000
Limited time	30	2.00	5.00	114.00	3.8000
Climatic condition	30	2.00	5.00	109.00	3.6333
Design deficiencies	30	2.00	5.00	107.00	3.5667
Water intrusion	30	1.00	5.00	105.00	3.5000
Maintenance deficiencies	30	1.00	5.00	104.00	3.4667
Corruption	30	1.00	5.00	92.00	3.0667
Valid N (listwise)	**30**				

Table 22.7 The rank of common contributing factors to building defects based on relative importance index

No	Contribution factors to defects	Relative importance index (RII)
1	Poor workmanship	0.813
2	Limited cost	0.780
3	Construction materials	0.760
4	Limited time	0.760
5	Climatic condition	0.727
6	Design deficiencies	0.713
7	Water intrusion	0.700
8	Maintenance deficiencies	0.693
9	Corruption	0.613

with a total number of sums of 92 out of 150. Corruption also has the lowest mean, which is 3.0667.

Ranking for Contribution Factors to Building Defects. Based on the result from Table 22.7, the highest rank is poor workmanship with the RII value of 0.813. According to [6], the residential buildings that are newly constructed experienced defects due to the reason of poor workmanship by the contractor or developer. The problem involves lack of supervision, lack of experience, and lack of communication at the construction site. The labor workers with a lack of knowledge may simply construct the construction work without the supervise of the contractor. Therefore, the role of the contractor to supervise the activities such as concrete work and install formwork on construction site are very important. As shown below, the second-highest rank is limited cost followed by construction materials. The quality of material used in construction may drop as the cost is limited which can be influenced by the type of material chosen and the market price. Thus, the strength and durability of the structure also decrease. Climatic condition also may affect the duration of completion which led to limited time, thus is a uncontrollable constraint.

22.3.4 Expert Verification

To get more confirmation and verification regarding data analysis, the results from Tables 22.5 and 22.7 have been shown to the expert. These are some comments from their experience and point of view in Table 22.8. The result shows that most of the interviewees agreed that the workmanship is the main cause that frequently happens and is the cause of the defects in the building. As mentioned by [6], defects that occur in the new buildings were due to poor workmanship of the labor workers, lack of supervision, and usage of the materials. Also, these factors are affected by cost. As agreed by the experts, limited cost becomes one of the major causes followed by poor workmanship.

Table 22.8 Highlighted comments from interviewees on the result from Tables 22.5 and 22.7

Factors	Highlighted comments from experts
Poor workmanship	• Unskilled workers and fewer workers on site • Lack of supervision, work executed is not in a proper way, and not following the method given by the consultant • Example works such as plastering, concreting, and dismantle formwork • Cracking resulted from poor workmanship if there is no checking on the cement quality, rebar, etc. and the structure surely will crack • Improper installation of pipe which causes the leakage of water
Construction materials	• Materials not according to the specification • Poor quality of materials • The material is also affected by climatic condition
Limited cost	• Affecting poor workmanship and construction materials
Climatic condition	• Hot conditions affecting the work performance of the workers • Affect the quality of the paint

22.3.5 *Expert Verification*

The following show types of questions that were asked during the interview session with the experts.

Interview Questions

1. What are the consequences of the building defects to the parties involved?
2. What are the effects of building defects on the building?
3. What happens if no action is taken against defects?
4. What are the precautions to reduce the defects?

Expert's Answers. Based on the result from Table 22.9, as most of the experts agreed that the cost of maintenance will bring an effect on the contractor. As stated by [7], the social and economic impact of building defects may include the depreciation of assets and increased maintenance cost. Other than that, defects also affect the image and reputation of the company. The future client may not trust the contractor for tendering. This view has been supported by [8], and loyal customers are important for the reputation of companies and their overall success.

From Table 22.10, the major effect of building defects on the building is the quality of the building. When the quality is low, the aesthetic value decreases and the company may lose profit. The occupant may feel unsatisfied and uncomfortable living in the building. As stated by [9], customers are not satisfied with the building and cannot fully utilize it as the defects are on the building.

Based on Table 22.11, most experts agreed that the building will collapse if no action was taken against the defect. The collapse will happen if major defects on the beam or columns are left untreated.

According to [10], collapse is the ultimate and most serious result of structure failures, but overstressing is also evident at earlier stages through the development of

Table 22.9 Answers for consequences of the building defects to the parties involved

Interviewees	Answers
Interviewee 1	It depends on how serious the defects. If the defects can cause the collapse of the building, then the responsible parties can be sued and enter court. Therefore, it will affect the reputation of the company
Interviewee 2	The contractor will face the consequences because of ruining the image of their company. The company can be labeled a low-class contractor. Then, the contractor might not be chosen for tendering. If within the defects liability period, the contractor will be responsible for the defects
Interviewee 3	Affects the credibility and track record of the company. The client will have a hard time trusting the company. Other than that, defects consume cost and time for maintenance
Interviewee 4	The cost for rectifications will increase and time is consumed a lot because of rectification work. The company may lose profit because the cost of maintenance and repair work will increase
Interviewee 5	Cost for maintenance increases. More money contribution, especially on reinforcement bars. It costs a lot. Delay in achieving the award of completion

Table 22.10 Answers for effects of building defects on the buildings

Interviewees	Answers
Interviewee 1	The defects such as cracks may appear and grow in the future, and there will be a need for bigger maintenance to rectify the defects
Interviewee 2	It affects the quality, finishing, and safety of the building to the occupant. The occupant may feel unsatisfied and uncomfortable with the quality of the building
Interviewee 3	Aesthetic value decreases and one defect can lead to another. The building is not safe to use and more maintenance works need to be carried out
Interviewee 4	Minor defect contributes to severe defect after few years. It also shortens the life span of the building and the building is not safe
Interviewee 5	Reduce and affect the value of the building. The company could lose profit due to defects

Table 22.11 Answers for if no action is taken against the defects

Respondents	Answers
Interviewee 1	The structure will fail and thus make the building unsafe to the occupant. The cause is casualties to the client and occupant
Interviewee 2	The defects need to do maintenance if in critical condition. It also depends on structural defects or architectural defects. If the defects are structural defects, then maintenance work needs to be done as soon as possible to prevent collapse
Interviewee 3	After many years, severe defects in the structure take place. It also affects the functionality and efficiency of buildings and equipment. Building and structure failure (minor or major) may cause a hazard to the consumer
Interviewee 4	If the defects happen to the main structure such as beams or columns, the defects can be severe and then collapse
Interviewee 5	The structure itself will collapse and we then again solve the problem by designing from the first stage again. From here the cost will increase

Table 22.12 Answers for precautions to reduce the defects

Experts	Answers
Interviewee 1	At the early stage, the consultant needs to consider safety factors, and during the construction, the contractor needs to construct according to the drawing and specification. Other than that, prevent the use of low-grade material to ensure the construction complies with design or code
Interviewee 2	Quality control and assurance must be done properly. Workmanship needs to improve with proper supervising and strictly follow the standard requirement. Monitoring and resolving issues need to be done immediately
Interviewee 3	Always ensure that the contractor follows the design as given instruction by the consultant. The contractor needs to supervise their workers regularly and the consultant or client needs to monitor the site
Interviewee 4	Regular QA/QC monitoring and inspection for the building; if any defects occurred, fix it and record it properly for further reference. Improve any method to solve defects
Interviewee 5	Do some effective QA/QC during/pre and post-construction maintenance. During construction, do the proper way of maintenance and provide quality works. Most important is intensive monitoring on-site by the client

deformation and fractures. The major effect of building defects to the building is the quality of the building. When the quality is low, the aesthetic value decreases, and the company may lose profit. The occupant may feel unsatisfied and uncomfortable living in the building. As stated by [9], customers are not satisfied with the building and can- not fully utilize it as the defects are on the building.

Based on Table 22.12, monitoring and supervising are more highlighted by the experts. Less monitoring and inspection by the supervisor also may lead to the problem [10]. Other than that, the consultant needs to make sure the usage of the material so that the contractor constructs according to the drawing and specification. The consultant needs to approve the material before purchasing it. Furthermore, the client needs regular QA/QC monitoring and inspection during and post construction.

Summary of the Comments from Experts

See Table 22.13.

22.3.6 Expert Verification

Based on the data collected through the expert interviews and verification, the building defects framework was developed as in Fig. 22.3 as a guideline for the construction industry in Malaysia regarding types of building defect, causes to building defects, the relationship between types of defect and contribution factors, effects of building defects, and action needs to be taken to minimize defects. The framework consists of four elements: types of building defects, contribution factors

Table 22.13 Highlighted comments from interviewees on the result from Tables 22.5 and 22.7

Questions	Highlighted comments from the experts
What are the consequences of the building defects to the parties involved?	• The contractor will lose profit as the cost for maintenance and rectification increase • Consume time • Affect the image and reputation of the company for tendering • If within the defects liability period, the contractor will be responsible for the defects • Delay in achieving an award of completion
What are the effects of building defects on the building?	• The major effect is the quality of the building • If quality is low, the aesthetic value decreases and the company may lose profit • The occupant may feel unsatisfied and uncomfortable with the quality of the building • Minor defect contributes to severe defect after few years. It also shortens the life span of the building and the building is not safe • Bigger maintenance to rectify the defects
What happens if no action is taken against the defects?	• Depends on structural defects or architectural defect. If the defects are structural defects, then maintenance work needs to be done as soon as possible to prevent collapse • After many years, severe defects on the structure take place • Structure collapse, then solve the problem by designing from the first stage again. Thus, the cost will increase
What are the precautions to reduce the defects?	• Less monitoring and inspection by the supervisor also may lead to the problem • Consultants need to make sure the usage of the material so that the contractor constructs according to the drawing and specification • The client needs regular QA/QC monitoring and inspection during and post-construction

to building defects, the effect of building defects, and lastly, precaution to be taken to solve or prevent the building defects problems.

The types of defects had been studied. Defects are divided into two categories: structural and non-structural defects. Cracking, foundation failure, timber decay, and corrosion of reinforced steel in the building can be categorized as structural defects, while erosion of mortar joint, peeling paint, defective plaster rendering, dampness, and roof defects are categorized as non-structural defects. Most of the building defects highly occurred due to poor workmanship and construction materials. The following show types of questions that were asked during the interview session with the experts.

Next, the defect liability period (DLP) is a common type of feature that can be normally seen in the standard form of the contracts for construction in Malaysia in the duration of the DLP to be within 12 or 24 months from the date or practical

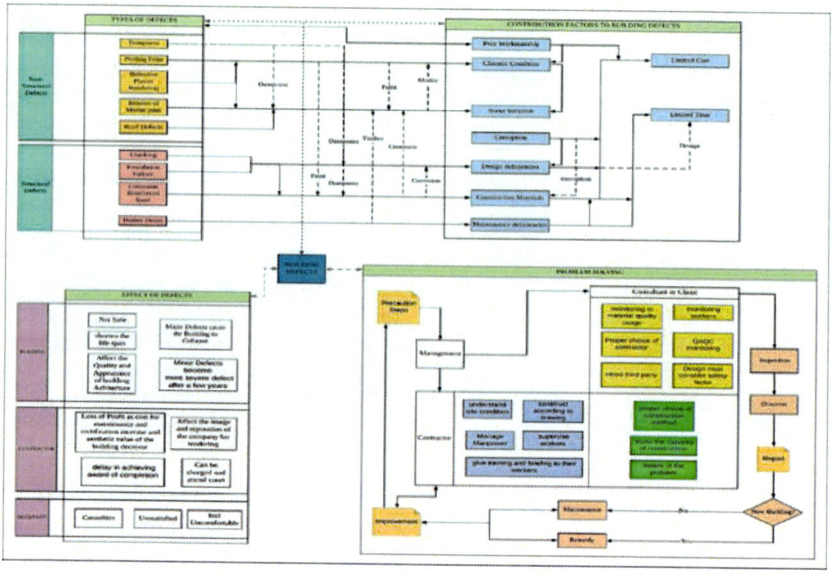

Fig. 22.3 Building defects framework

completion. According to [11], the causes of defects found mostly during the period of liability for defects are poor manufacturing, low construction material quality, and poor design. Limited cost and time will be affecting the workmanship, construction materials, design faulty, and maintenance deficiencies. Other than that, the provisions of the DLP can be found in the standard forms of the construction contracts as follows: (1) PAM 2006: Clause 15, which is practical completion and defects liability period and (2) CIDB 2000: Clause 27, which is the defects liability after completion.

According to [12], extreme climatic conditions affect the performance of the labor workers in construction. Based on [13], the moisture problem such as water intrusion that causes the non-structural defects is highlighted. The following shows the types of questions that were asked during the interview session with the experts.

22.4 Conclusion

In conclusion, it is important for parties involved such as contractor, project manager, and site supervisor to be responsible to prevent the presence of defects in building to keep their image of the company, reduce the time and cost consuming, satisfaction from the client and occupant, and ensure the safeness of the building. As supported by the experts, to prevent the growth of defects, parties involve needing to improve their management in a construction project such as proper construction methods, proper design, give proper training, strict supervision, etc. This study aims to develop an assessment framework for building defects in the Malaysian construction industry. In the nutshell, the defect assessment framework is suitable to be used as a guideline for the future construction industry to improve the awareness of building defects and management systems to minimize the defects in the construction project.

As shown in Table 22.5, cracks are the highest ranked type of building defects that occurred. Cracks occurred caused by overload, improper design, low quality of material used, and poor workmanship. The result shows in Table 22.7 that poor workmanship is the most commonly contribute to building defects. The problem involving lack of supervision, lack of experience, and lack of communication at the construction site. Therefore, the role of the contractor to supervise the activities such as concrete work and install formwork on construction site is very important.

Acknowledgements The authors would like to express their gratitude and thanks to Universiti Teknologi MARA Cawangan Pulau Pinang for funding this research.

References

1. Olanrewaju AL, Abdul-Aziz AR (2015) Building maintenance processes and practices: the case of a fast developing country. Springer, Singapore
2. Tan Wei Chuen (2008) Problem of building defects. Ministry Higher Edu 255:1–5
3. Meor HMI, Razak NNA, Enduta IR, Samah SAA, Ridzuana SSM (2016) Minimizing defects in construction project, pp 78–84
4. Wahab SNA, Hamid MY (2011) A review factors affecting building defects of structural steel construction. Case study: student accommodation in UiTM Perak. Procedia Eng 20:174–179
5. Brown S (2010) ANR program evaluation Iowa state university extension
6. Anthony LT (2013) New buildings could also have structural defects. http://www.themalaymailonline.com/malaysia/article/new-buildings-could-also-have-structural-defe
7. Clare TL (2015) Building defects have an economic, social impact. https://www.strait-stimes.com/forum/letters-in-print/building-defects-have-economic-social-impact. Accessed 28 July 2018
8. Othman AAE (2015) An international index for customer satisfaction in the construction industry. Int J Constr Manag 15(1):33–58
9. Neha B, Shruti WP (2015) Review paper on construction defects. J Mech Civil Eng 12(2 Version 3):88–91
10. Ahzahar N, Karim NA, Hassan SH, Eman J (2011) A study of contributing factors to building failures and defects in the construction industry. Procedia Eng 20:249–255. https://doi.org/10.1016/j.proeng.2011.11.162
11. Hong CH, Abdullateef QA (2016) Investigation of defects in new buildings in Malaysia. http://eprints.utar.edu.my/2310/1/CM-2016-1101687-1.pdf. Accessed 25 Aug 2018
12. Dai J, Paul MG, William FM (2009) Construction craft workers' perceptions of the factors affecting their productivity. J Construct Eng Manage 135(3):217–226
13. Othman NL, Jaafar M, Harun WMW, Ibrahim F (2015) A case study on moisture problems and building defects. Procedia Soc Behav Sci 170:27–36. https://doi.org/10.1016/j.sbspro.2015.01.011

Chapter 23
CO_2 Emission and Cost Estimates of Wastepaper Sludge Ash in Controlled Low-Strength Material Towards Sustainability

Mohd Azrizal Fauzi, Mohd Fadzil Arshad, Noorsuhada Md Nor, and Noor Idayani Noordin

Abstract Globally, the construction industry compensates for 36.9% of greenhouse gas pollution. Understanding the causes of these pollutants and their contribution to construction activity can help in climate change mitigation. CO_2 emissions and cost estimates from wastepaper sludge ash (WSA) with recycled fine aggregate (RFA) in a controlled low-strength material (CLSM) were assessed and estimated in this study (CLSM). The study on the material's entire production cycle was conducted using life cycle assessment (LCA) and Intergovernmental Panel on Climate Change (IPCC) guidelines. The LCA recognises all stages of the process, while the IPCC guideline only considers CO_2 emissions. Electricity, gasoline, and water intake are the three primary energy sources involved in this study. As a result, 21 virtual CLSM mixes were developed in this study, and the associated CO_2 emissions and costs were estimated. According to the results, using WSA in CLSM decreases CO_2 pollution and CO_2 pricing by 49.71% per 1 m^3 of CLSM. The research shows that WSA content and unconfined compressive strength (UCS) significantly affect CO_2 emissions and cost. The CO_2 emission and cost can differ significantly; similarly, the influence of WSA content on CO_2 emission and cost is vast under the same binder content. Via this study, emissions can be assessed by means of appropriate environmental management practise as well as the cost of CO_2 emissions is designed to facilitate sustainable use of the construction industry.

M. A. Fauzi (✉) · M. F. Arshad
School of Civil Engineering, College of Engineering, Universiti Teknologi MARA Shah Alam, 40450 Shah Alam, Selangor, Malaysia
e-mail: azrizal@uitm.edu.my

M. A. Fauzi · N. Md Nor
School of Civil Engineering, College of Engineering, Universiti Teknologi MARA Cawangan Pulau Pinang, 13500, Permatang Pauh, Pulau Pinang, Malaysia

N. I. Noordin
Prima Reka Konsultan, No. 39A, Jalan Ampang Baru 6A, Ampang Baru, 31350 Ipoh Perak, Malaysia

S. Abdullah et al. (eds.), *Structural Integrity Cases in Mechanical and Civil Engineering*, Structural Integrity 23, https://doi.org/10.1007/978-3-030-85646-5_23

Keywords CO_2 emission · Controlled low-strength material · IPCC · LCA · Wastepaper sludge ash · Recycled fine aggregate · Unconfined compressive strength

23.1 Introduction

The determination of CO_2 emissions is a single method for analysing the increase in greenhouse gas (GHG) atmospheric concentration correlated with special processes or products [1]. The amount of CO_2 emitted into the environment resulting from an individual's or organisation's actions is referred to as their carbon footprint. It is a metric for determining how much an individual contributes to global greenhouse gas emissions. Malaysia's sixth Prime Minister announced that the country is on target to cut CO_2 pollution by 40% in 3 years, resulting in a healthier atmosphere for future generations. He also estimated that CO_2 emissions had been lowered by 33% in 2015 relative to the country's 2005 emissions record. Malaysia is dedicated to reducing CO_2 emissions by 45% by 2030 and being CO_2 free by 2050 [2].

Natural resource-based construction materials have been scarce, and their extraction has culminated in air contamination and harmful environmental effects. Since there is such an abundant paper mill sludge produced, wastepaper sludge ash (WSA) is an economic and environmental issue for the paper and board industry. WSA is a waste by-product of the paper mill industry. Because of its low sulphate content of just 0.57% of the overall weight, WSA is a revolutionary substance used as a masonry method to promote green technologies. WSA has the potential to be used in cement manufacturing as well. The substance is well adapted for use as a cement clinker ingredient (as a Ca, Al, and Si) source [3, 4, 5].

On the other hand, burnt free lime (CaO) can render this source unsuitable for direct use in blended cement, but it can be used as a supplementary ingredient for non-structural applications, including controlled low-strength material (CLSM). CLSM, also known as flowable fill, is a poor and runny concrete mix used in construction for non-structural applications, including backfill and road foundation. In this analysis, the CLSM comprises Portland cement, water, aggregate, and WSA. CLSM has a lower strength than ordinary concrete, at less than 8.3 MPa; thus, it is a more robust backfill material and flows better than ordinary concrete [6, 7].

Two approaches to analysing CO_2 pollution are the Intergovernmental Panel on Climate Change (IPCC) and Life Cycle Assessment (LCA). LCA is based on the cradle-to-grave approach, which considers all stages of a product's life cycle, while the IPCC guideline measures CO_2 emissions [8, 9, 10, 11, 12]. The amount of CO_2 pollution was measured using the IPCC guideline, and the energy usage in CLSM output was estimated using the LCA rule.

According to the World Meteorological Organisation's (WMO) "Global Greenhouse Gas Bulletin" released on 30 October 2017, air concentrations soared at an alarmingly fast pace in 2016, reaching the highest amount in 800,000 years endangering the global environment. As a result, strategies to minimise CO_2 pollution have become the most crucial issue in the world. Following that, this study aims to evaluate CO_2 emissions and measure the amount and cost of CO_2 emissions from WSA in CLSM.

23.2 Methodology

The working method for LCA is structured along with a framework that has been formed by the four-step based on the ISO 14040:2006 guideline [13]. This framework divides the entire LCA procedure into four distinct phases, as shown in Fig. 23.1, started with goal and scope definition, inventory analysis, impact assessment, and will be interpreted as presented in the results section.

According to the cradle-to-grave approach, as shown in Fig. 23.2, the overall LCA of a WSA in CLSM product starts from the raw material extraction until production of a product consisting of four types of material which is cement, aggregate, water, and WSA.

This study focused on the LCA of the material used to produce CLSM covering from the material extraction until the production in terms of electric consumption, diesel consumption, and water consumption by each phase. Besides, to estimate the CO_2 footprint generated by the production of the WSA in CLSM at different stages, starting from the early phase of production until the end of usage.

23.2.1 Inventory Analysis

Life cycle inventory (LCI) refers to the compilation of the inventory and input data concerning the study's system and is obtained in the second phase. The LCI consists of

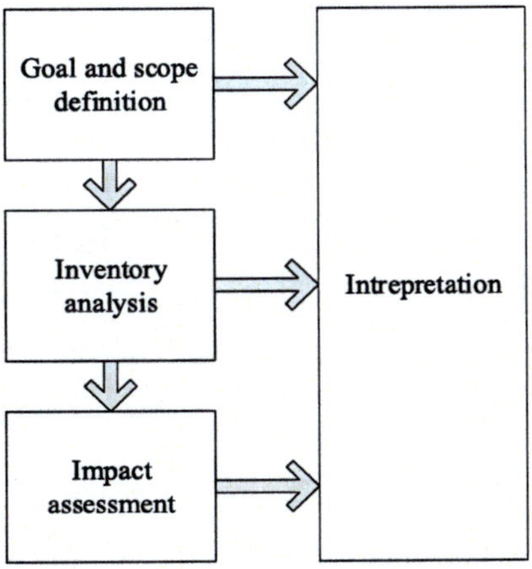

Fig. 23.1 Life Cycle Assessment (LCA)

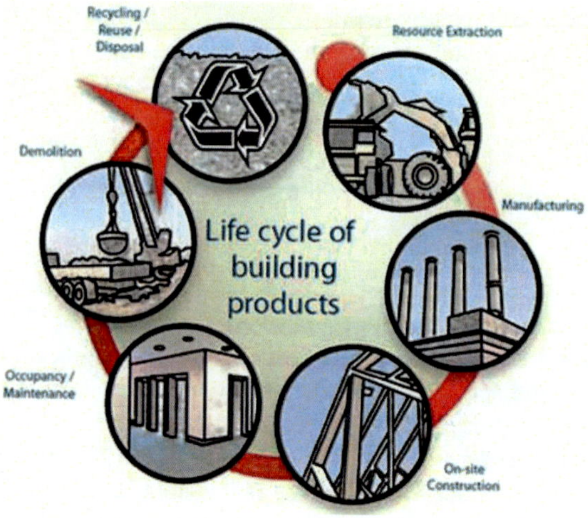

Fig. 23.2 The life cycle of building product [14]

estimates of material and energy input and air, land, and water emissions associated with product manufacturing, process activity, or service provision. In the case of WSA in CLSM, the material includes cement, recycled fine aggregate (RFA), and water. Figure 23.3 summarises each phase of the inventory data for each material.

The data in this study were gathered by a site visit to each material plant which is:

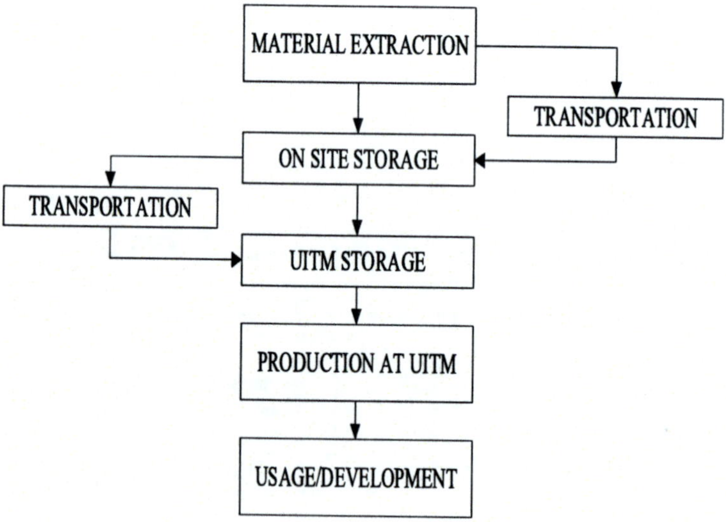

Fig. 23.3 The material life cycle process

i. Portland cement Type I: Tasek Corporation Berhad, Ipoh, Perak (Tasek Cement);
ii. Aggregate: KTK Concrete Sdn Bhd, Kulim, Kedah (RFA);
iii. Water: Sungai Dua Water Treatment Plant;
iv. WSA: Asia Honour Paper Industries (M) Sdn. Bhd., Mentakab, Pahang.

Data gathering phase by phase for each material included the electric consumption by each machine, diesel consumption, and water consumption per one tonne of material, and converted according to the mix proportion for CLSM.

23.2.2 *Impact Assessment*

The life cycle impact assessment (LCIA) evaluates the significance of potential environmental impacts based on LCI data and provides final interpretation information [10]. The LCIA of this study used the IPCC guideline emission factor in terms of four main emission factors: energy, industrial process and product use, waste, and other land use. Table 23.1 shows the emission factor for three resource types dependent on the IPCC guideline.

CO_2 emissions, the key GHGs regulated by the Kyoto Protocol, have been estimated at each phase and multiplied by the data provided in Table 23.1, which is GHG emission or removal factors per unit activity [15]. Equation (23.1) shows the mathematical formula used by this study based on Cellura et al. [8].

$$\text{Emission} = \sum (\text{resources} \times \text{emission factor}) \tag{23.1}$$

where,

Emission = emission of CO_2 ($kgCO_2$);
Resources = diesel consumption (L), electric consumption (kWh); waste consumption (m^3); Emission factor = Table 23.1.

23.2.3 *Mix Proportion*

In summary, the central composite design (CCD) approach was used to produce 20 simulated CLSM mixes as a feature of the most relevant mix design variables.

Table 23.1 IPCC guideline (2016), emission factor

Resources	Emission factor
Electricity consumption	0.618 $kgCO_2/kWh$
Diesel consumption	2.65 $kgCO_2//L$
Water Consumption	0.8 $kgCO_2/m^3$

The Design Expert v13 software aid performs the hybrid design and statistical evaluation of test results. The software-generated WSA_RFA_CLSM (WSA + RFA) mixture is a model designed and developed in this research. The three design variables were taken into consideration: (1) the addition of WSA: 50–100% replacement of cement; (2) water content: with the same addition of WSA and water-binder ratio, different mix formulas can be obtained by adjusting the contents of cement and water. The water content increased from 263 to 380 kg/m^3, resulting in a total of 20 mix selections; (3) water-binder ratio: this study designed 28-day unconfined compressive strength (UCS) with a water-binder ratio varying from 2.53 to 2.73. Ref. is the control sample for this research. While low-strength concrete is being more widely used, normal-strength concrete also has an extensive range of civil infrastructure applications. Consequently, this paper's emphasis is on assessing CO_2 emissions from CLSM applications with a 28-day UCS. Table 23.2 presents the mixture proportions produced by the model software, and then, it is used to determine the amount of CO_2 and the cost estimate of CO_2 for WSA in CLSM at 1 m^3.

Table 23.2 Mixture proportions and 28-day UCS results for WSA_RFA_CLSM

Mix ID	Cement (kg/m^3)	WSA (kg/m^3)	Water (l/m^3)	RFA (kg/m^3)	28-day UCS, kPa
Ref.	200	0	340	1310	2511
WSA-RFA1	80	80	272	1418	1351
WSA-RFA2	80	80	304	1386	613
WSA-RFA3	0	160	272	1418	1061
WSA-RFA4	0	160	304	1386	583
WSA-RFA5	100	100	340	1310	3395
WSA-RFA6	100	100	380	1270	798
WSA-RFA7	0	200	340	1310	1349
WSA-RFA8	0	200	380	1270	775
WSA-RFA9	45	135	294	1376	3785
WSA-RFA10	45	135	354	1316	143
WSA-RFA11	121	59	324	1346	1348
WSA-RFA12	0	211	324	1315	855
WSA-RFA13	37	110	263	1440	305
WSA-RFA14	53	160	385	1252	2780
WSA-RFA15	45	135	324	1346	1325
WSA-RFA16	45	135	324	1346	1330
WSA-RFA17	45	135	324	1346	1327
WSA-RFA18	45	135	324	1346	1323
WSA-RFA19	45	135	324	1346	1324
WSA-RFA20	45	135	324	1346	1328

23.2.4 CO_2 Costing

CO_2 costing is the preferred method for lowering global warming emissions by many analysts. The government decides on the amount of CO_2 tax per tonne, and it is called a CO_2 cost, which is the cost charged for the right to emit one tonne of CO_2 into the atmosphere.

Based on Teh [16], the team analyse what would happen if the government gradually implemented a CO_2 tax starting at RM145/T for 2015 and gradually increase to RM855/T in 2060.

23.3 Results and Discussions

The result was evaluated based on two types of samples—based as seen in Table 23.2, of which the first sample contains 100% of ordinary Portland Cement (Ref.); the second sample consists 100% of WSA as a binder (WSA-RFA7). The purpose of having two samples is to compare either the CO_2 emission can be reduced with the replacement of the WSA in CLSM.

23.3.1 CO_2 Emission for Material

According to Table 23.2, cement emits the most CO_2 and leads to increased emissions in CLSM mixes. Therefore, it is known that CO_2 emissions increase in line with the increase in cement content. The increase in this volume increases, indicating that the total emissions of cement and CO_2 are comparable. This also can see that the CLSM UCS increases as the cement content increases in general, as we all know. As a consequence, it is not surprising that CO_2 emissions increase in parallel with the UCS. Furthermore, CO_2 emissions for a UCS vary greatly from different mixes, with a difference of more than 100 kg CO_2-e in a 1m^3 CLSM being feasible. Considering the enormous global production of CLSM, if the CLSM can be made in a mix with the lowest CO_2 emissions, the CO_2 emissions savings might become significant.

The amount of CO_2 emitted by the three key materials listed in Table 23.3 was measured using LCA and the IPCC guidelines. Cement emits the highest CO_2 at

Table 23.3 Value of CO_2 emission for the material

Material	CO_2 emission (KgCO_2/T)
Cement	0.9804
Aggregate	0.06
WSA	0.498
Water	0.03

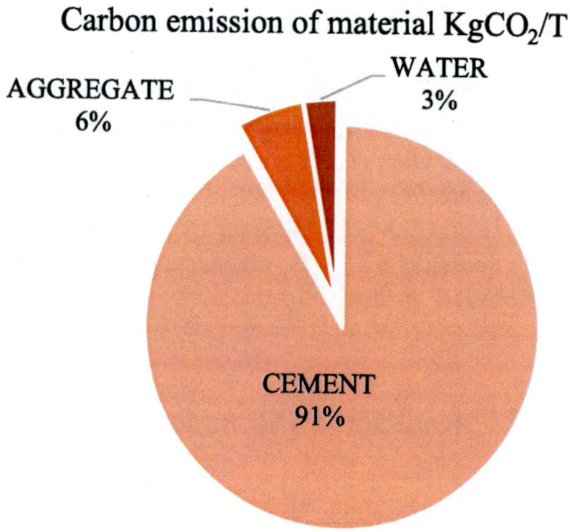

Fig. 23.4 The percentage of CO_2 emission for material

0.9804 $kgCO_2/T$, followed by aggregate 0.06 $kgCO_2/T$ and water 0.03 $kgCO_2$. Figure 23.4 shows the percentage of CO_2 emission by cement (91%), aggregate (6%), and water (3%).

The amount of CO_2 emissions per tonne for cement is substantially the same as the data from Davidovits [17] that simplifies 1 T of cement equivalent to 1 T of CO_2. The amount of CO_2 during cement manufacturing is high due to high energy. The greater the required heat is, the greater the amount of coal or fuel consumed. As a result, burning more coal or fuel increases the emission of CO_2.

Cement production emits greenhouse gas, both directly and indirectly. The heating of limestone releases CO_2 directly while fossil fuel burning heats the kiln, which then emits CO_2. Direct CO_2 emissions occur through a chemical process called calcination. Calcination occurs when limestone, made of calcium carbonate, is heated and decomposes into calcium oxide (CaO) and CO_2. This process accounts for 50% of all CO_2 production emissions. Indirect CO_2 emissions are generated through the combustion of fossil fuels to heat the kiln. Generally, kilns are heated by coal, natural gas, or oil, and the combustion of these fuels emits additional CO_2, just as they would in electricity production. That constitutes around 40% of the emissions of cement. Eventually, the energy used to power additional plant machinery and the final transport of cement is another source of indirect CO_2 emissions and accounts for 5–10% of industrial pollution.

23.3.2 CO_2 Emission of WSA in CLSM

In general, as WSA content rises when cement content decreases, cement's relative contribution to CO_2 emissions descends. Since the WSA's pozzolanic contribution to

Table 23.4 Total CO_2 emission by each sample ($1m^3$)

Material	Sample ref ($kgCO_2$)/kg	Sample WSA-RA7 ($kgCO_2$)/kg
Cement	272,156,373.20	0.00
Aggregate	1,122,678,047.00	1,122,678,047.00
Water	1,122,678,047.00	109,257.79
WSA	0.00	129,529,691.00
Total	**2,517,512,468.00**	**1,252,316,996.00**

strength occurred gradually after several days of curing, the strength of WSA-CLSM is reduced when the WSA content increases, as several researchers also reported [3, 4, 5, 18, 19, 20, 21, 22, 23]. Can we, therefore, make a related assumption, i.e., if a decrease in CO_2 emissions from WSA-CLSM can be accomplished by increasing WSA content and decreasing cement content due to a high cement emission factor?

A study of the relationship between the WSA content and emissions of CO_2 through different mixtures and strength (with the same binder content) was carried out to answer this question, such as for WSA's CO_2 emission estimate in CLSM using the LCA and IPCC guideline, the total amount of CO_2 emission for sample Ref. 1 m^3 CLSM is 2517512468 $kgCO_2$, and for sample, WSA-RA7 is 1252316996 $kgCO_2$, as mentioned in Table 23.4 were done for comparisons. Therefore, WSA can reduce CO_2 emission because the waste material has less activity and process required but can be reused as a binder. As stated results, these reuse wastes were a reduced amount of energy, and less CO_2 emissions were emitted.

The impact of WSA content on CO_2 emissions is significant, as seen in Table 23.4. CO_2 emissions in 1 m^3 CLSM can range from 100 to 200 kg/m^3, and the difference can increase as CLSM strength increases. Besides that, it can be seen that CO_2 emissions with low WSA content can be lower than those with high WSA content when appropriately designed. As a result, it is a mistake to assume that increasing WSA content in CLSM can minimise CO_2 emissions since several variables influence CLSM CO_2 emissions, and WSA content can be one of them.

23.3.3 Cost Estimation of CO_2 Emission of WSA in CLSM

As previously stated, the CLSM UCS will increase as the cement content increases. Furthermore, since the amount of cement and the cost have a positive correlation, a longitudinal fitting is used to define the relationship between cost and UCS. Besides that, the cost difference between any two adjacent UCS tends to be comparable, implying that the cost increase is proportional to the improvement of UCS.

In general, when a higher WSA content is mixed with a lower cement content, the relative contribution of cement to the cost decreases due to the lower cost of WSA compared to cement. However, if different proportions of WSA are used to replace cement, the fine aggregate (RFA) content can increase to obtain the same UCS as

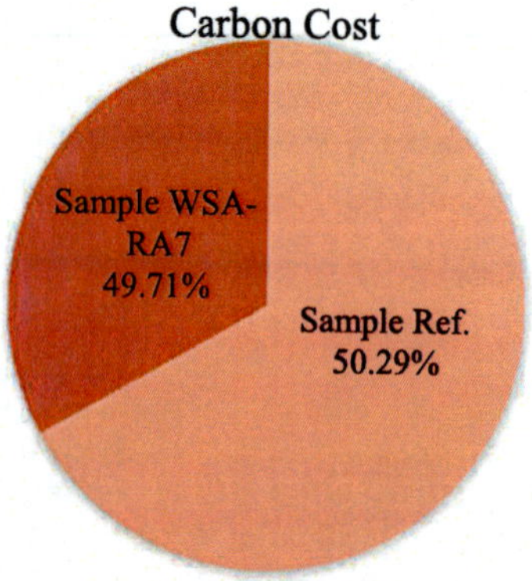

Fig. 23.5 The percentage of CO_2 costing RM/m^3/T for each sample

CLSM without WSA. As a result, an analysis of the relationship between WSA content and cost under different UCS was carried out; for example, the relationships between sample Ref. and selected mix are seen in Table 23.4. The influence of WSA content on cost is tremendous, as seen in Fig. 23.5, even with the same binder amount. Furthermore, it can see that when using a conservative design, the cost of a low WSA content material can be smaller than that of a high WSA content material.

The sample Ref. and selected mix (Sample WSA-RA7) for cost comparison were those that met the Department of Transport (DOT) and reported on controlled low-strength materials Standard by American Concrete Institute, ACI 229–13 [25] criteria. The total amount of CO_2 emitted from 1m^3 CLSM in sample Ref. is 2517512468 kgCO_2, and for sample WSA-RA7 is 1252316996 kgCO_2, as listed in Table 23.4. As the amount of cement increased, the cost increased linearly, indicating that the amount of cement and the cost are closely related to each other, similarly to the relationship between the amount of cement and the CO_2 emissions.

Table 23.5 shows the amount of CO_2 emission of WSA in CLSM in unit kg CO_2 tonne. The purpose of converting the unit to per kg is because the amount of CO_2 pricing can be determined by the tonne. Table 23.6 indicates the estimated cost of CO_2 emission (1m^3/T) is based on CO_2 tax at RM 145/tonne.

The CO_2 cost for producing 1 m^3 of CLSM with OPC and with WSA surprisingly decreases from RM 402 to RM 200 billion, with the reducing percentage 49.71% (Fig. 23.5).

Table 23.5 Total CO_2 emission by each sample (1 m^3/T)

Material	Sample ref (kgCO_2)/tonne	Sample WSA-RA7 (kgCO_2)/tonne
Cement	300,000.96	0.00
Aggregate	1,237,540.35	1,237,540.35
Water	1,237,540.35	120.43
WSA	0.00	142,782.00
Total	**2,775,081.67**	**1,380,442.79**

Table 23.6 The total cost of CO_2 emission by sample (1 m^3/T)

Material	Sample ref RM/tonne	Sample WSA-RA7 RM/tonne
Cement	43,500,139.20	0
Aggregate	179,443,350.80	179,443,350.80
Water	179,443,350.80	17,462.35
WSA	0.00	20,703,390.00
Total	402,386,842.20	200,164,204.60

This CO_2 tax can be one of the steps to control the CO_2 emission by assuming that the industries would respond to the tax by moving away from fossil fuels. The cumulative cost of climate damage over 95 years could reach RM 40.1 trillion. This amount can be reduced to RM 5.3 trillion, which could be recovered from CO_2 tax revenues amounting to RM 9.5 trillion over the same period. Similarly, the accumulated surplus of RM 4.2 trillion could be used to finance the development of environment-friendly technologies to replace fossil fuels.

23.4 Conclusion

This paper has shown that merely asserting that incorporating WSA can minimise CLSM CO_2 emissions is not robust. The amount of CO_2 emissions and cost estimates for WSA in CLSM were assessed and estimated in this study using LCA and IPCC guidelines. In conclusion, CO_2 emissions in the construction sector are mainly due to cement manufacturing. As cement is substituted with surplus materials such as WSA, CO_2 emissions are reduced as predicted, and costs are reduced suddenly from RM 402 to RM 200 billion, a savings of 49.71%. CO_2 emissions and costs vary significantly across various mixes; additionally, the difference in CO_2 emissions for any two comparable binder content seems non-uniform, while the difference in costs tends to become identical. Since the impact of WSA content on CO_2 emissions and cost is tremendous with the same binder amount, a reasonable design can lower CO_2 emissions and cost with low WSA content than with high WSA content. The Malaysian construction industry must realise that we are a part of the environment, and we are the ones who must contribute to the ecosystem's preservation by using excess resources that are entirely used as sustainability that seems to be advantageous.

Acknowledgements The authors would like to express their appreciation to the School of Civil Engineering, Universiti Teknologi MARA Cawangan Pulau Pinang, for providing financial support for this research. The authors also want to thank CIMA Sdn Bhd, PBA Sdn Bhd, and Asia Honour Paper Industries (M) Sdn Bhd for cooperating with the research data collection.

References

1. Säynäjoki A et al (2011) Carbon footprint assessment of a residential development project 2(2):116–123
2. Federal Government Administrative Centre: Malaysia Sustainable Development Goals Voluntary National Review 2017 (2017)
3. Azmi AN et al (2016) Production of controlled low strength material utilizing waste paper sludge ash and recycled aggregate concrete. In: MATEC web of conferences, vol 47, p 01011. https://doi.org/10.1051/matecconf/20164701011
4. Azrizal MF et al (2019) The properties of wastepaper sludge ash and its generic applications. J Phys: Conf Ser 1349(1):012087. https://doi.org/10.1088/1742-6596/1349/1/012087
5. Mohd Ridzuan AR et al (2011) Strength assessment of controlled low strength materials (CLSM) utilizing recycled concrete aggregate and waste paper sludge ash. In: 2011 IEEE colloquium on humanities, science and engineering. IEEE, pp 208–211. https://doi.org/10.1109/CHUSER.2011.6163718
6. Alizadeh V et al (2014) Design and application of controlled low strength materials as a structural fill. Constr Build Mater 53 425–431. https://doi.org/10.1016/j.conbuildmat.2013.12.006
7. Razak HA et al (2009) Performance appraisal of industrial waste incineration bottom ash as controlled low-strength material. J Hazard Mater 172(2–3) 862–867. https://doi.org/10.1016/j.jhazmat.2009.07.070
8. Cellura M et al (2018) Science of the total environment energy-related GHG emissions balances: IPCC versus LCA. Sci Total Environ 628–629:1328–1339. https://doi.org/10.1016/j.scitotenv.2018.02.145
9. Cho S, Chae C (2016) A study on life cycle CO_2 emissions of low-carbon building in South Korea 1–19. https://doi.org/10.3390/su8060579
10. Guinée JB, Lindeijer E (2002) Handbook on life cycle assessment: operational guide to the ISO standards. Springer Science & Business Media
11. Thakur S et al (2015) Leather pigments: towards a cleaner greener approach. progress in organic coatings 6(2):1–4+6. https://doi.org/10.1002/adfm.201403496
12. Weber S et al (2018) Life cycle assessment of a vanadium redox flow battery. Environ Sci Technol 52(18):10864–10873
13. Arvanitoyannis IS (2008) ISO 14040: life cycle assessment (LCA)–principles and guidelines. Waste management for the food industries, 97–132
14. LCA—WoodWorks. https://www.woodworks.org/sustainable-design/lca/. Accessed 1 Feb 2020
15. Bastianoni S et al (2014) The connection between 2006 IPCC GHG inventory methodology and ISO 14064–1 certification standard—a reference point for the environmental policies at sub-national scale. Environ Sci Policy 44:97–107. https://doi.org/10.1016/j.envsci.2014.07.015
16. Teh CS. Malaysian carbon tax could pave way to renewable energy future. https://phys.org/news/2016-04-malaysian-carbon-tax-pave-renewable.html. Accessed 13 Oct 2020
17. Davidovits J (2002) Environmentally driven geopolymer cement applications. Geopolymer 2002 Conf 6:1–9
18. Fauzi MA et al (2021) Statistical models to develop optimised controlled low-strength materials with wastepaper sludge ash. Constr Build Mater 286:122816. https://doi.org/10.1016/j.conbuildmat.2021.122816
19. Park J, Hong G (2020) Strength characteristics of controlled low-strength materials with waste paper sludge ash (WPSA) for prevention of sewage pipe damage. Materials 13, 19, 4238. https://doi.org/10.3390/ma13194238
20. Kaliyavaradhan SK et al (2019) Waste resources recycling in controlled low-strength material (CLSM): A critical review on plastic properties. J Environ Manage 241:383–396. https://doi.org/10.1016/j.jenvman.2019.03.017

21. Fauzi MA et al (2021) The need of statistical approach for optimising mixture design of controlled low-strength materials. Civ Eng Architect 9(5A):163–173. https://doi.org/10.13189/cea.2021.091319
22. Naoki KI et al (2017) Performance evaluation of low carbon concrete using paper sludge ash. J MMIJ 133(6):132–139. https://doi.org/10.2473/journalofmmij.133.132
23. Fauzi MA et al (2016) The effect of recycled aggregate concrete incorporating waste paper sludge ash as partial replacement of cement. In: AIP conference proceedings. pp 030007. https://doi.org/10.1063/1.496506
24. Azmi AN et al (2015) Experimental investigation in strength of sustainable concrete materials with partial replacement of waste paper sludge ash (WPSA). In: International conference on advances in civil and environmental engineering
25. ACI 229R-13 (2013) Report on controlled low-strength materials. American Concrete Institute (ACI), 38800 Country Club Drive Farmington Hills, MI 48331 USA

Chapter 24
Acoustic Emission Characterisation of Reinforced Concrete Beam at an Initiation Stage of Fatigue Damage

N. M. Nor, S. N. Mat Saliah, M. S. Tahir, and N. A. Yahya

Abstract The accumulation of fatigue loading is a significantly perilous state to the reinforced concrete structure on the formation of damage. It starts with a small crack to concrete and getting bigger as the load accumulates. Hence the monitoring of the reinforced concrete structure under fatigue loading is vital. The monitoring can be done using acoustic emission technique. This paper presents the investigation of AE activities at an initial stage of fatigue life of RC beam subjected to fatigue loading. The maximum and minimum fatigue loadings were based on the ultimate load of the average five reinforced concrete beam under static loading. Hence, the maximum fatigue loading is 127 kN, and the minimum fatigue loading is 32 kN. The acoustic emission data collected at Channel 1 as well as signal strength, intensity analysis, severity index and historical index were analysed and discussed. It was found that acoustic emission is a good approach to predict the occurrence of crack at an initial stage of fatigue damage.

Keywords Acoustic emission · Severity index · Historical index · Intensity analysis · Fatigue

24.1 Introduction

Fatigue damage of reinforced concrete (RC) structure cumulatively increases with applied load cycles. Damage is related to the irreversible process [1] in a structure with respective of cyclic accumulation that causes the significant damage led to

N. M. Nor (✉) · S. N. M. Saliah
School of Civil Engineering, College of Engineering, Universiti Teknologi MARA, Cawangan Pulau Pinang, Malaysia
e-mail: ida_nsn@uitm.edu.my

M. S. Tahir
Department of Business Management, Universiti Teknologi MARA, Cawangan Pulau Pinang, Malaysia

N. A. Yahya
School of Civil Engineering, College of Engineering, Universiti Teknologi MARA, Shah Alam, Selangor, Malaysia

S. Abdullah et al. (eds.), *Structural Integrity Cases in Mechanical and Civil Engineering*, Structural Integrity 23, https://doi.org/10.1007/978-3-030-85646-5_24

structural failure. Cumulative fatigue damage analysis is vital in the prediction of RC structures subjected to load histories [2]. Damage and failure process during load history require for the application of Non-Destructive Evaluation (NDE) such as Acoustic Emission (AE) technique that possible for real-time damage monitoring. Besides, the damage processes often generate a vast amount of AE hits or events from one damage zone [3].

The fatigue damage process divides into three stages, namely, the initiation, the steady propagation, and the coalescence of microcracks to macrocracks [4]. Md Nor et al. [5] identified the fatigue damage based on the bath-tub curve and three stages were found. The stages are burn-in, steady state, and burn-out. The burn-in is known as the infant mortality or initiation of damage. The steady state is the constant damage severity. The burn-out is wear-out, life end, or failure of a structure. Md Nor et al. [5] enhanced that the high damage severity is found to be at first point of the initiation stage of fatigue failure. Xie and Lai [6] stated that at an initiation stage, the rate of failure is high and affect to the production to be released. It is because high failure rate can be found at initiation stage on any system relates to fatigue. It gives an indication that thorough investigation needs to be carried out especially at initial stage prior to steady stage.

However, in fatigue study, most of the researchers focus on the fatigue damage process as a whole system [7–9] and lack on investigation at the initial stage. It is vital to take action the cracking process at initiation stage by accumulating the AE hits or events. AE hits or events are accumulating with crack growth [10], and it is reasonable to investigate the relationship between cumulative absolute energy and deflection. As stated by Md Nor et al. [5], the initiation stage of fatigue life of a structure is 7 and 0.8% if 0.8 and 0.7 of ultimate loads were applied to the beam.

The prediction of fatigue life of a structure can be done using AE technique. It has been extensively reported by Md Nor et al. [5] and found that AE characteristic for beam subjected to fatigue load with load ratio of 0.8 is higher than load ratio of 0.7. It is because the higher load ratio would generate the higher stress to the beam. Karimian et al. [11] inferred that the AE waveform is very reliable indicator of fatigue crack initiation. To give a better interpretation of fatigue life, the AE energy as well as signal strength is usually utilised [5, 11]. At the same time, the intensity analysis, historical index, and severity index can be exploited. From the intensity analysis, the classification of damage of the structure can be performed. Intensity analysis is a method of evaluating the structural significance of an event by a combination of historical index and severity index [12]. The intensity analysis has been applied for a structure subjected to static loading to failure for characterisation of damage for glass fibre-reinforced composites decks [8], and carbon fibre reinforced polymer reinforced concrete slabs [13]. It also has been used for damage detection of RC beam under alternating static and dynamic loads and health index evaluation of RC beam under stepwise-static loading [14].

One of the AE energy is signal strength. It is defined as "the measured area of the rectified AE signal with units proportional to volt-seconds" [15] and related to the relative energy of a specimen. The signal strength is defined in Eq. (24.1), where f_+ is the positive signal envelope function, f-is the negative signal envelope function, t_1 is time at first threshold crossing and t_2 is the time at last threshold crossing. Meanwhile, the intensity analysis is a technique for evaluating intensity fatigue damage zones by calculating historic index (HI) and severity index (Sr) [13]. The equations for HI and Sr are presented in Eqs. (24.2) and (24.3).

$$S_0 = \frac{1}{2}\int_{t_1}^{t_2} f_+(t)dt - \frac{1}{2}\int_{t_1}^{t_2} f_-(t)dt \tag{24.1}$$

$$HI = \frac{N}{N-K}\frac{\sum_{i=K+1}^{N} S_{oi}}{\sum_{i=1}^{N} S_{oi}} \tag{24.2}$$

$$S_r = \frac{1}{J}\sum_{m=1}^{J} S_{om} \tag{24.3}$$

where: N is the number of hits up to time t, S_{oi} is the signal strength of the ith event, K and J are empirical derived constant based on material [14], and S_{om} is the signal strength of the mth hit.

However, damage evaluation using acoustic emission analysis under constant fatigue loading is still lack of attention. This paper presents the investigation of AE activities at an initiation stage of fatigue life of RC beam subjected to fatigue loading. In this study, the initiation stage was based on the Md Nor et al. [5], about 7% of the fatigue life of the beam.

24.2 Experimental Programme

24.2.1 Sample Preparation

The concrete beam was prepared for concrete strength of C40. The concrete was made up of cement, water, fine aggregate and coarse aggregate with a proportion of 1: 0.43: 2.16: 2.60, respectively. Then, 1% of retarder of cement weight was added to the concrete mix to improve the workability of the fresh concrete. The maximum coarse aggregate of 20 mm was used. The beams were designed as a singly RC beam with two high yield steel bars of 16 mm to strengthen the tension part and two mild steel bars of 8 mm as hanger bars. In the stirrups, 12 diameter of mild steel with a spacing of 100 mm centre to centre was used. In the preparation of each RC beam, the reinforcement was submerged in the standard mould of 150 mm × 150 mm × 750 mm with a cover of 20 mm, then the concrete mix was cast. All beams were demoulded after +24 h and submerged in curing tank for 28 days. The compressive strength for 7 days was found to be 32.35 and 44.65 N/mm^2 at the age of 28 days. All samples were tested at the age exceeded 80 days. The reason for this was to avoid any strength increase that would influence the results.

24.2.2 Fatigue Test

Three-point fatigue tests were performed under load control with a clear span of 630 mm. A servo-hydraulic testing machine was employed for subjecting prismatic RC beams of 150 mm × 150 mm × 750 mm to the required fatigue loading. The specimen setup is represented in Fig. 24.1. The fatigue test was performed at a low frequency of 1.0 Hz, with the applied load ratio of 0.8 of the ultimate load of the beam. The ultimate load was identified from the static loading carried out on five beams. The ultimate load was 158.85 kN. Hence, the maximum fatigue loading, P_{max} was 127 kN. The minimum fatigue loading, P_{min} was taken about 20% of the ultimate load. Therefore, the P_{min} is 32 kN. The selection of the P_{min} and P_{max} was based on Md Nor et al. [5]. Then, a sine wave load cycle is used to induce the fatigue load to the beams until failed. An AE system monitored the test for all beams and the data recorded by data acquisition software.

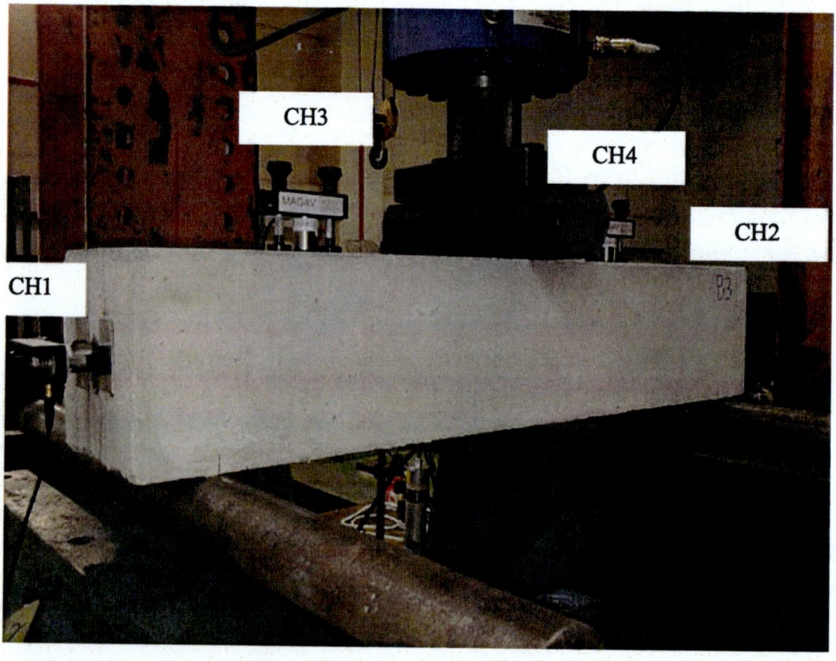

Fig. 24.1 Setup of the third point bending test for the RC beam

24.2.3 Acoustic Emission Monitoring

A total of four AE sensors (R6I—resonant 60 kHz Integral sensor) were applied on the beam, as shown in Fig. 24.1. After installation of the sensors, the calibration test was performed to check the sensitivity of both sensor and beam surfaces. Both surfaces were connected using a thin layer of high vacuum grease. The sensitivity was performed using pencil lead fracture close to the sensor on the beam surface and the results were monitored on the AE visual. The amplitude that appeared on the AE visual for each lead fracture was observed, in which the required value is above ±97 dB in accordance with [15]. In the AE hardware, the threshold level of 45 dB and wave velocity were applied throughout the test. The rearm time of 1.62 ms, duration discrimination time of 400 μs, pre-trigger samples of 200 and the sample rate of 10 MHz were set in the AE system.

In this study, only data collected on Channel 1 (CH1), as shown in Fig. 24.1, would be analysed and discussed. From the data acquisition, the signal strength corresponding to the normalised cycle at an early stage was extracted for analysis. Then, the cumulative signal strength, historical index, severity index and intensity analysis were analysed to investigate the AE activities at an earlier stage of fatigue load.

24.3 Results and Discussion on Acoustic Emission

24.3.1 Signal Strength

Figures 24.2 and 24.3 present the AE parameter of signal strength with respect to the normalized cycle. The signal strength is used to observe the damage activities in the RC beams under fatigue load at an earlier stage. The normalised cycle is the present cycle divided by the fatigue life, N/Nf [8]. From the fatigue test, the number of cycles at failure, Nf of RC beam subjected to a load ratio of 0.8 is 848461 cycles on Channel 1 (CH1). In this analysis, the AE hits, or events in the first 3 s of loading are not considered for analysis. It is due to the signal collected at the beginning of fatigue loading, which may originate from complex sources such as the contact of supports and the specimen [7]. Otherwise, Gostautas et al. [10] were discarded the first 100 hits in the analysis. It is because within the first 100 hits usually are associated with cracking of excessive matrix material under initial loading. In Fig. 24.2, high AE signal strength is produced at the early stage of cracking due to hasty of crack propagation of concrete. The highest signal strength at an early stage is 3.6×10^9 nVs, as presented in Figs. 24.2 and 24.3. Therefore, the cumulative signal strength becomes higher in the early stage of the cycle, as shown in Fig. 24.4. A similar finding has been found by El Batanouny et al. [16] that the presence of high signal strength is closely related to the formation of an initial crack in the specimen as it can be seen through the data collected by the fixed sensor on specimen's surface.

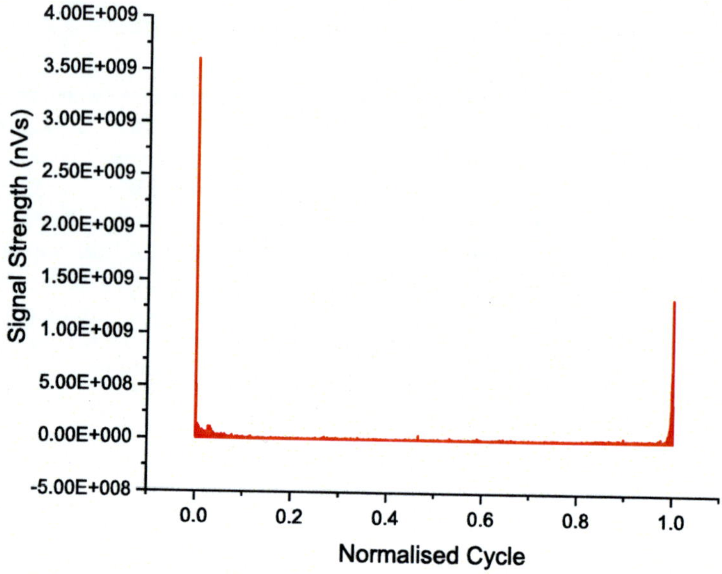

Fig. 24.2 Relationship between signal strength and normalised cycle

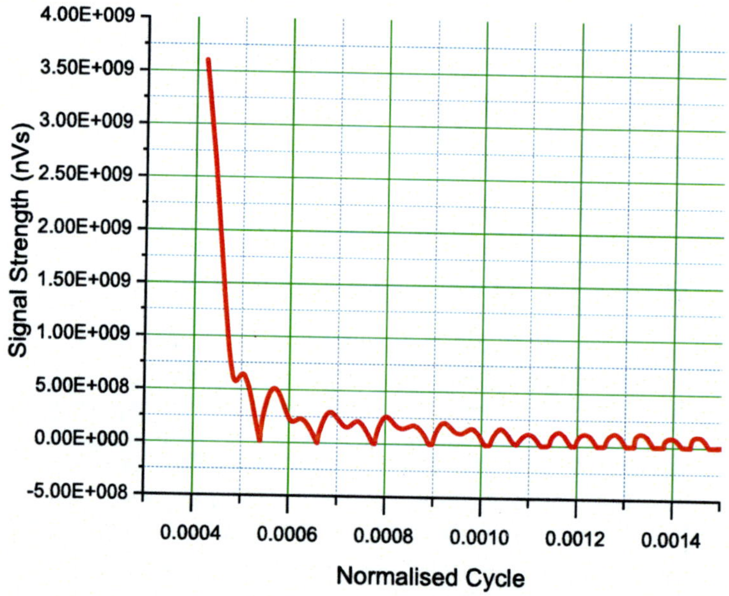

Fig. 24.3 Relationship between signal strength and normalised cycle at an early stage

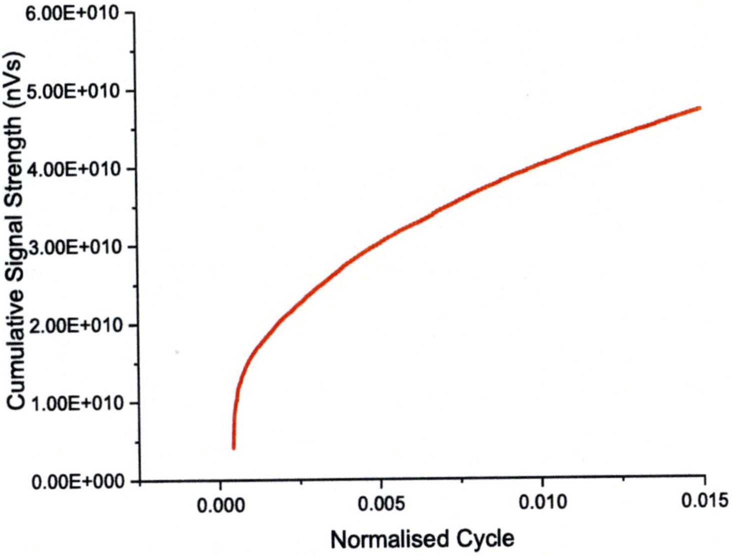

Fig. 24.4 Cumulative signal strength at an early stage

This trend of the curve of cumulative signal strength—normalised cycles—is quite similar to the curve represented by Roberts and Talebzadeh [8] on absolute cumulative energy—normalised cycles.

24.3.2 *Intensity Analysis*

The presence of a crack in the beam specimen at an early stage of load cycle is pronounced by the relationship between the severity index and historical index, which is known as AE intensity analysis. The intensity analysis provides a quantitative assessment of damage classifications [10]. Five damage zones, namely, Zone A (insignificant emission), Zone B (minor), Zone C (intermediate), Zone D (follow-up) and Zone E (major) were adopted to classify the damage. Figure 24.5 shows the intensity chart of RC beam subjected to cyclic loading at an early stage. In the chart, the plots divide into three parts as depicted in Table 24.1: P1, P2 and P3. P1 is the AE hits at an early stage of 1000 cycles. P2 is the AE hits data collected from cycles 1001–2000. Meanwhile, P3 is the AE hits from cycles 2001–3000. The three parts identified from the investigation were carried out by Md Nor et al. [5] on the early stage of fatigue failure.

From the chart, most of the plots for data collected at P1 indicate the beam in damage Zone E. It indicates that the beam is suffering to significant major damage. Gostautas et al. [10] stated that, as the structure suffering to major damage, the

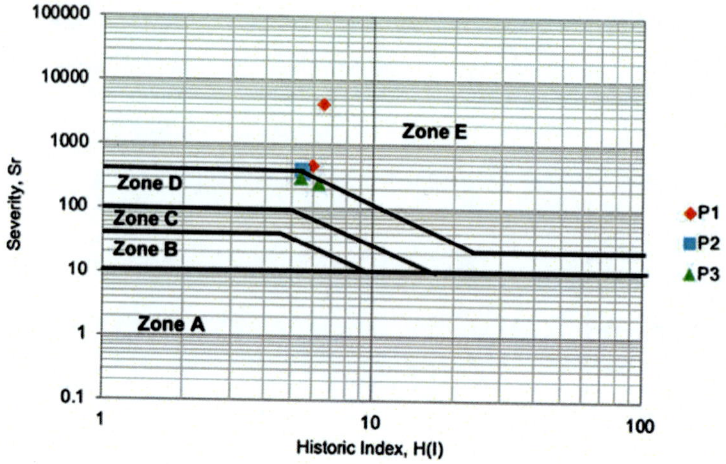

Fig. 24.5 Intensity analysis of the RC beam under fatigue

Table 24.1 The cycles for each part

Parts	Cycles
Part 1, P1	0–1000
Part 2, P2	1001–2000
Part 3, P3	2001–3000

structure is required an immediate shut-down and follow-up inspection. The major defect (Zone E) is produced by the first 1000 events of AE signals, and it is generated by the first few loading cycles under high load ratio. It is due to the occurrence of crack initiation in the concrete at an early stage of the fatigue load. As the load cycles increase, from load cycles 1001–2000, the plots are in the middle of Zones E and D. Meanwhile, as the load continuously cycles, the plots turn to Zone D. Zone D means to follow up where the significant defect is requiring to carry out further inspection. The inspection is carried out by analysing the additional data collected from the testing. Gostautas et al. [10] recommended a non-destructive examination for further data analysis. Usually, the inspection is carried out periodically to inspect the condition of the structure.

The plots in the intensity chart are reflected in the high signal strength at an early stage, as presented in Fig. 24.2. It was also pronounced by the robust increase in the accumulated signal strength, which becomes a hint on the occurrence of the second stage and the third stage of fatigue failure. At the same time, due to rapid crack initiation, it produces high AE activities. According to Roberts and Talebzadeh [8], an early stage of fatigue damage has occurred in the first 10% of Nf before the curve stabilises. Meanwhile, Md Nor et al. [5] found that 7% of Nf was the endpoint of the early stage of fatigue damage under the same load ratio of 0.8 of the ultimate loads.

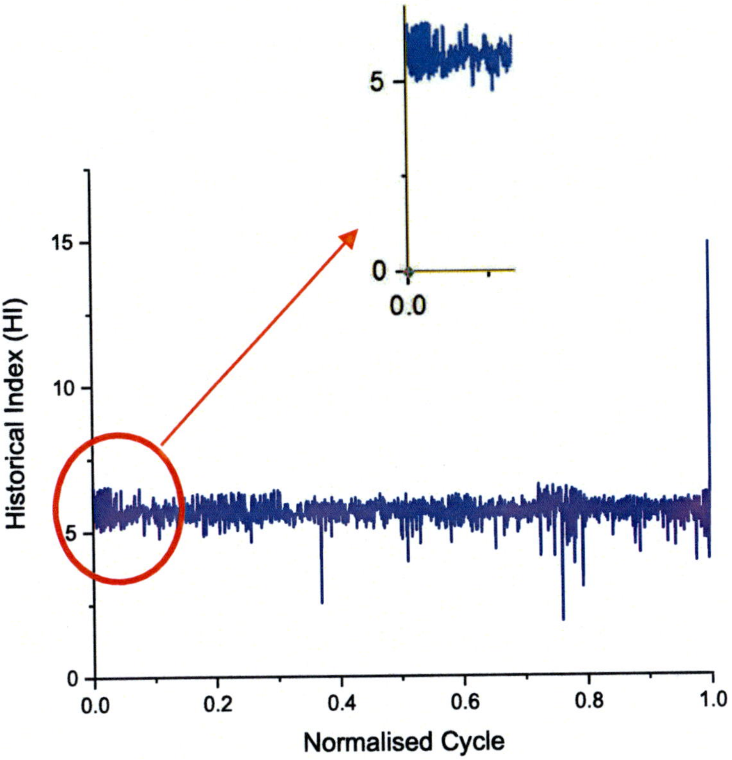

Fig. 24.6 Relationship between historical index and normalised cycle

24.3.3 Intensity Analysis

Figures 24.6 and 24.7 show the historical index and severity index with respect to normalised cycle for beam subjected to fatigue loading, which are representing the overall performance of the beam condition. From the historical index, as shown in Fig. 24.6, high historical index can be observed at initiation stage as shown in small graph in Fig. 24.6. This high historical index is due to the presence of crack at the concrete beam. However, as the loading continuously cycles at the same range of 32 and 127 kN, several peaks of historical index were noticed until the beam failed. It is because as generally known that the historic index has been found to be a sensitive parameter to quantify changes in the slope of the cumulative signal strength by weighing the average signal strength of the last 20% or 200 hits. From Fig. 24.7, high severity index is found at the beginning of load cycle. The severity reduces as the cycle increases. A significant high in severity at the beginning of fatigue loading indicates the presence of damage in the structure [17, 18].

At an early stage of fatigue damage, the high historical index is found in Fig. 24.8 with a value of 6.15. The similar pattern can be found through the severity index

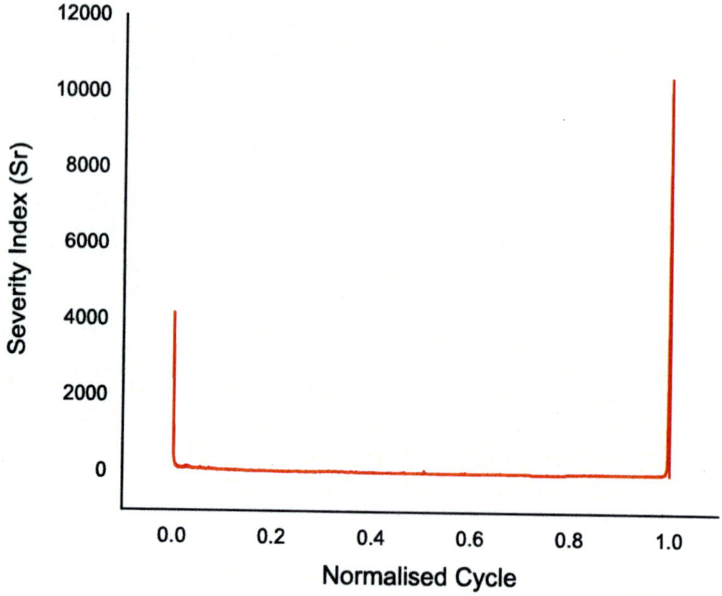

Fig. 24.7 Relationship between severity index and normalised cycle for the whole testing

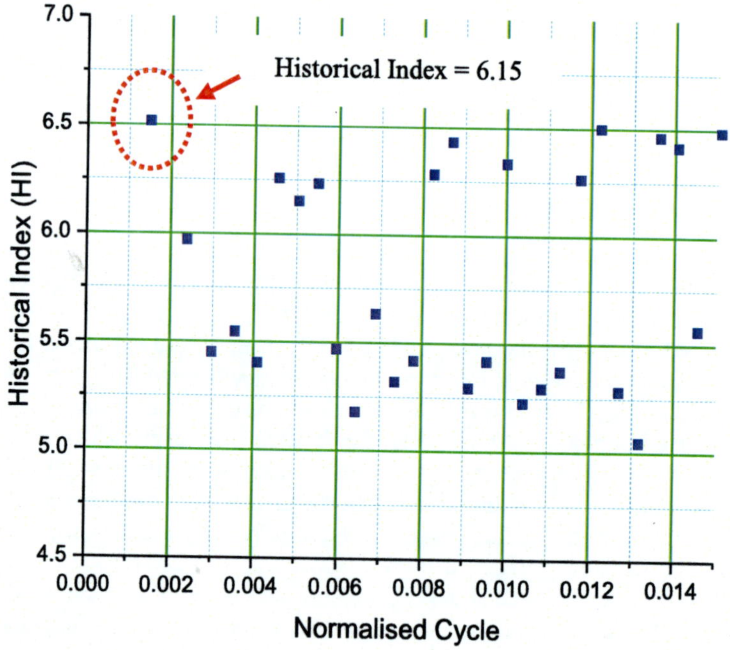

Fig. 24.8 High historical index at an early stage

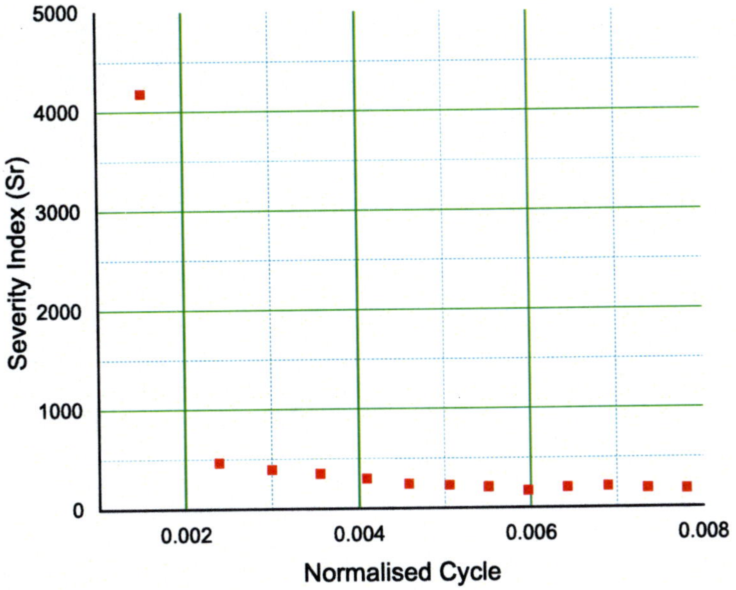

Fig. 24.9 Severity index at an early stage

as presented in Fig. 24.9 with the value of more than 4200. According to Xie and Lai [6], at the beginning of fatigue failure, the initial failure rate is too high for the product to be released directly after production. It gives an indication to any system that relates to fatigue that would have a high failure rate or damage at the initial load cycle. When turning back to this study, it gives a clear picture of the development of an initial crack in the concrete, which would reflect on the production of high historical index and severity index.

24.4 Conclusions

The acoustic emission activities at an initiation stage of fatigue damage for reinforced concrete beam were investigated. It is found that high signal strength was noticed at an earlier fatigue damage of the beam with the value of 3.6×10^9 nVs. This high value is closely related to the formation of initial crack in the concrete. A rapid increase in the cumulative signal strength at this stage was presented. The results were verified by the intensity analysis with the plots at first 1000 load cycles were on Zone E. The Zone E means that major damage and rigorous inspection of the beam needs to be carried out. As the load cycles accumulated, the plots were reduced to Zone D, which requires follow-up inspection of the beam.

Similar pattern was found on the severity index and historical index with respect to normalised cycle, where high value was found at first load cycle. All data from the acoustic emission activities can be used as a hint to the formation of the next stage in the fatigue damage process. From the investigation of acoustic emission activities at an initiation stage of fatigue damage, a prediction of the beam performance can be identified earlier.

References

1. Khan S, Wilde F, Beckmann F, Mosler J (2012) Low cycle fatigue damage mechanism of the lightweight alloy Al2024. Int J Fatigue 38:92–99
2. Fatemi A, Yang L (1997) Cumulative fatigue damage and life prediction theories: a survey of the state of the art for homogeneous materials. Int J Fatigue 20(1):9–34
3. Crosse CU, Finck F (2006) Quantitative evaluation of fracture processes in concrete using signal-based acoustic emission techniques. Cement Concr Compos 28:330–336
4. Wang C, Zhang Y, Ma A (2011) Investigation into the fatigue damage process of rubberised concrete and plain concrete by AE analysis. J Mater Civ Eng 23(7):953–960
5. Md Nor N, Ibrahim A, Muhamad Bunnori N, Mohd Saman H, Mat Saliah SN, Shahidan S (2014) Diagnostic of fatigue damage severity on reinforced concrete beam using acoustic emission technique. Eng Fail Anal 41:1–9
6. Xie M, Lai CD (1995) Reliability analysis using an additive Weibull model with bath-tub shaped failure rate function. Reliab Eng Syst Saf 52:87–93
7. Li W, Sun W, Jiang J (2011) Damage of concrete experiencing flexural fatigue load and closed freeze/thaw cycles simultaneously. Constr Build Mater 25:2604–2610
8. Roberts TM, Talebzadeh M (2003) Acoustic emission monitoring of fatigue crack propagation. J Constr Steel Res 59:695–712
9. Yu J, Ziehl P, Zárate B, Caicedo J (2011) Prediction of fatigue crack growth in steel bridge components using acoustic emission. J Constr Steel Res 67:1254–1260
10. Gostautas RS, Ramirez G, Peterman RJ, Meggers D (2005) Acoustic emission monitoring and analysis of glass fiber-reinforced composites bridge decks. J Bridg Eng 10(6):713–721
11. Karimian SF, Modarres M, Bruck HA (2020) A new method for detecting fatigue crack initiation in aluminum alloy using acoustic emission waveform information entropy. Eng Fract Mech 223
12. Degala S, Rizzo P, Ramanathan K, Harries KA (2009) Acoustic emission monitoring of CFRP reinforced concrete slabs. Constr Build Mater 23:2016–2026
13. Nair A, Cai CS (2009) Damage detection of concrete structures using acoustic emission. Key Eng Mater 400–402:101–106
14. Shahidan S, Muhamad Bunnori N, Md Nor N, Basri SR (2012) Health index evaluation on acoustic emission signal for concrete structure by intensity analysis method. Adv Mater Res 403–408:3729–3733
15. ASTM E. 1316 (2006) Standard terminology for nondestructive examinations. ASTM International
16. ElBatanouny MK, Larosche A, Mazzoleni P, Ziehl PH, Matta F, Zappa E (2014) Identification of cracking mechanisms in scaled frp reinforced concrete beams using acoustic emission. Exp Mech 54:69–82
17. Gostautas RS (2007) Identification of failure prediction criteria using acoustic emission monitoring and analysis of GFRP bridge deck panels, University of Kansas
18. Soffian Noor MS, Md Nor N, Abd Rahman N, Abdullah S, Tahir MS (2021) Evaluation of severely damaged reinforced concrete beam repaired with epoxy injection using acoustic emission technique. Theor Appl Fract Mech 112